Organic Synthesis with Enzymes in Non-Aqueous Media

Edited by
Giacomo Carrea and
Sergio Riva

Related Titles

Gotor, V., Alfonso, I.,
García-Urdiales, E. (eds.)

Asymmetric Organic Synthesis with Enzymes

approx. 400 pages
2007
Hardcover
ISBN: 978-3-527-31825-4

Aehle, W. (ed.)

Enzymes in Industry

Production and Applications

approx. 520 pages with approx. 120 figures
2007
Hardcover
ISBN: 978-3-527-31689-2

Bornscheuer, U. T., Kazlauskas, R. J.

Hydrolases in Organic Synthesis

Regio- and Stereoselective Biotransformations

368 pages with 255 figures and 23 tables
2006
Hardcover
ISBN: 978-3-527-31029-6

Breslow, R. (ed.)

Artificial Enzymes

193 pages with 81 figures and 8 tables
2005
Hardcover
ISBN: 978-3-527-31165-1

Bommarius, A. S.,
Riebel-Bommarius, B. R.

Biocatalysis

Fundamentals and Applications

634 pages with 251 figures and 78 tables
2004
Hardcover
ISBN: 978-3-527-30344-1

Organic Synthesis with Enzymes in Non-Aqueous Media

Edited by
Giacomo Carrea and Sergio Riva

WILEY-
VCH

WILEY-VCH Verlag GmbH & Co. KGaA

The Editors

Dr. Giacomo Carrea
Istituto di Chimica del
Riconoscimento Molecolare C.N.R.
Via Mario Bianco 9
20131 Milano
Italy

Dr. Sergio Riva
Istituto di Chimica del
Riconoscimento Molecolare C.N.R.
Via Mario Bianco 9
20131 Milano
Italy

All books published by Wiley-VCH are carefully produced. Nevertheless, authors, editors, and publisher do not warrant the information contained in these books, including this book, to be free of errors. Readers are advised to keep in mind that statements, data, illustrations, procedural details or other items may inadvertently be inaccurate.

Library of Congress Card No.:
applied for

British Library Cataloguing-in-Publication Data
A catalogue record for this book is available from the British Library.

Bibliographic information published by the Deutsche Nationalbibliothek
Die Deutsche Nationalbibliothek lists this publication in the Deutsche Nationalbibliografie; detailed bibliographic data are available in the Internet at <http://dnb.d-nb.de>.

© 2008 WILEY-VCH Verlag GmbH & Co. KGaA, Weinheim

All rights reserved (including those of translation into other languages). No part of this book may be reproduced in any form – by photoprinting, microfilm, or any other means – nor transmitted or translated into a machine language without written permission from the publishers. Registered names, trademarks, etc. used in this book, even when not specifically marked as such, are not to be considered unprotected by law.

Typesetting SNP Best-set Typesetter Ltd., Hong Kong
Printing Strauss GmbH, Mörlenbach
Binding Litges & Dopf GmbH, Heppenheim

Printed in the Federal Republic of Germany
Printed on acid-free paper

ISBN: 978-3-527-31846-9

Contents

Preface *XIII*
List of Contributors *XV*

Part One Biocatalysis in Neat Organic Solvents – Fundamentals

1 **Fundamentals of Biocatalysis in Neat Organic Solvents** *3*
Patrick Adlercreutz
1.1 Introduction *3*
1.2 Effects of Water on Biocatalytic Reactions *3*
1.2.1 Quantification of Water in Low-Water Systems *3*
1.2.2 Water Activity Control *4*
1.2.2.1 Water Activity Control Using Saturated Salt Solutions *4*
1.2.2.2 Water Activity Control Using Sensors *5*
1.2.2.3 Water Activity Control Using Pairs of Salt Hydrates *5*
1.2.3 Distribution of Water *5*
1.2.3.1 Hysteresis Effects *8*
1.2.4 Water Effects on Activity *8*
1.2.5 Water Effects on Selectivity *10*
1.2.6 Water Effects on Stability *12*
1.3 Solvent Effects *13*
1.3.1 Solvent Effects on Enzyme Activity *13*
1.3.2 Solvent Effects on Stability *16*
1.4 Effects on Equilibria *18*
1.4.1 Water Effects on Equilibria *19*
1.4.2 Solvent Effects on Equilibria *20*
1.5 Effects of pH in Organic Solvents *21*
1.6 Concluding Remarks *22*

2 **Effects of Organic Solvents on Enzyme Selectivity** *25*
Jaap A. Jongejan
2.1 Introduction *25*
2.2 Enzyme Enantioselectivity *26*
2.3 Effects of Organic Solvents on the E-value *27*

Organic Synthesis with Enzymes in Non-Aqueous Media. Edited by Giacomo Carrea and Sergio Riva
Copyright © 2008 WILEY-VCH Verlag GmbH & Co. KGaA, Weinheim
ISBN: 978-3-527-31846-9

2.4	Possible Causes of the Complexity of Solvent Effects on E	30
2.5	The Accuracy of E-value Determinations	31
2.6	Kinetic and Thermodynamic Analysis of the Specificity Constants	33
2.7	Solvents Effects on Non-Hydrolytic Enzymes	36
2.8	Major Achievements	38
2.9	Concluding Remarks	41
3	**Activating Enzymes for Use in Organic Solvents**	**47**
	Anne L. Serdakowski, Jonathan S. Dordick	
3.1	Introduction	47
3.2	Water–A Unique and Necessary Solvent for Enzymatic Catalysis	49
3.2.1	Challenges for Enzymatic Catalysis in Water	49
3.2.2	Enzymes do Function Without Water as a Bulk Solvent–Lessons from Extreme Halophiles	49
3.2.3	Behavior of Enzymes in the Absence of Bulk Water	50
3.2.4	Removing Water from Enzymes–the Effect of Lyophilization on Enzyme Structure and Function	51
3.3	Enzyme Activation in Nonaqueous Media	53
3.3.1	Addition of Water and Water Mimics	53
3.3.2	Immobilization	55
3.3.3	Solid-State Buffers	58
3.3.4	Lyophilization in the Presence of Excipients	59
3.3.4.1	Polymers	60
3.3.4.2	Crown Ethers and Cyclodextrins	60
3.4	Salt-Activated Enzymes	61
3.4.1	Salt Activation is not due to a Relaxation of Diffusional Limitations	63
3.4.2	Mechanism of Salt Activation	63
3.4.3	The Structural and Molecular Dynamics of Salt Activation	66
3.5	Conclusions	67

Part Two Biocatalysis in Neat Organic Solvents – Synthetic Applications

4	**Exploiting Enantioselectivity of Hydrolases in Organic Solvents**	**75**
	Hans-Erik Högberg	
4.1	Introduction	75
4.1.1	Enantioselectivity	75
4.1.2	Desymmetrization	78
4.1.3	The Reversibility Problem	79
4.1.4	Determining and Optimizing Enantioselectivity	81
4.2	Enantioselective Reactions in Organic Solvents	82
4.2.1	Reactions of Alcohols	82
4.2.1.1	Primary Alcohols	82
4.2.1.2	Secondary Alcohols	88
4.2.1.3	Tertiary Alcohols	100
4.2.1.4	Resolution of Dihydroxybiaryls	100

4.2.2	Reaction of Amines	*101*
4.2.2.1	Alkylamines	*101*
4.2.2.2	Arylalkylamines	*101*
4.2.2.3	Amino Acid Derivatives	*104*
4.2.3	Reaction of Acid Derivatives	*104*
4.2.3.1	Acids	*104*
4.2.3.2	Other Acid Derivatives	*107*
4.3	Summary and Outlook	*108*
5	**Chemoenzymatic Deracemization Processes**	*113*
	Belén Martín-Matute and Jan-E. Bäckvall	
5.1	Introduction	*113*
5.1.1	Asymmetric Synthesis Methods	*113*
5.1.2	Dynamic Kinetic Resolution (DKR)	*115*
5.1.3	Cyclic Deracemizations	*116*
5.2	Dynamic Kinetic Resolution	*117*
5.2.1	Base-Catalyzed Racemization	*117*
5.2.1.1	DKR of Thioesters	*118*
5.2.1.2	DKR of Activated Esters	*118*
5.2.1.3	DKR of Oxazolones	*118*
5.2.1.4	DKR of Hydantoins	*119*
5.2.1.5	DKR of Acyloins	*119*
5.2.2	Acid-Catalyzed Racemization	*119*
5.2.3	Racemization through Continuous Reversible Formation–Cleavage of the Substrate	*120*
5.2.3.1	DKR of Cyanohydrins	*121*
5.2.3.2	DKR of Hemithioacetals	*122*
5.2.4	Racemization Catalyzed by Aldehydes	*122*
5.2.5	Enzyme-Catalyzed Racemization	*124*
5.2.6	Racemization through S_N2 Displacement	*125*
5.2.7	Radical Racemization	*125*
5.2.8	Metal-Catalyzed Racemization	*126*
5.2.8.1	DKR of Allylic Acetates and Allylic Alcohols	*127*
5.2.8.2	DKR of *sec*-Alcohols	*128*
5.2.8.3	DKR of Amines	*132*
5.2.9	Other Racemization Methods	*133*
5.2.9.1	DKR of 5-Hydroxy-2-(5*H*)-Furanones	*133*
5.2.9.2	DKR of Hemiaminals	*133*
5.2.9.3	DKR of 8-Amino-5,6,7,8-Tetrahydroquinoline	*134*
5.3	Cyclic Deracemizations	*135*
5.3.1	Chemoenzymatic Cyclic Deracemizations	*136*
5.3.1.1	Cyclic Deracemizations of Amino Acids	*136*
5.3.1.2	Chemoenzymatic Cyclic Deracemizations of Hydroxy Acids	*137*
5.3.1.3	Chemoenzymatic Cyclic Deracemizations of Amines	*138*
5.3.2	Biocatalyzed Deracemizations	*139*
5.4	Concluding Remarks	*140*

6	**Exploiting Enzyme Chemoselectivity and Regioselectivity** 145
	Sergio Riva
6.1	Introduction 145
6.2	Chemoselectivity of Hydrolases 146
6.3	Regioselectivity of Hydrolases 148
6.3.1	Regioselective Acylation of Polyols 148
6.3.2	Regioselective Acylation of Carbohydrates 150
6.3.3	Regioselective Acylation of Natural Glycosides 153
6.3.3.1	Terpene Glycosides 154
6.3.3.2	Flavonoid Glycosides 155
6.3.3.3	Nucleosides 156
6.3.3.4	Other Glycosides 158
6.3.4	Regioselective Alcoholyses 158
6.3.5	Rationales 160
6.4	Closing remarks 162

7	**Industrial-Scale Applications of Enzymes in Non-Aqueous Solvents** 169
	David Pollard and Birgit Kosjek
7.1	Introduction 169
7.2	Ester Synthesis by Esterification 170
7.3	Resolution of Racemic Alcohols 170
7.3.1	Synthesis of a Chiral *R*-(+) Hydromethyl Glutaryl Coenzyme A Reductase Inhibitor (Anticholesterol Drug) 171
7.3.2	The Synthesis of *S*-N(*tert*-Butoxycarbonyl)-3-Hydroxymethylpiperidine (Tryptase Inhibitor) 172
7.3.3	The Synthesis of (*S*)-[1-(Acetoxyl)-4-(3-Phenyl)Butyl]Phosphonic Acid Diethyl Ester (Anticholesterol Drug) 172
7.3.4	Synthesis of Desoxyspergualin (Immunosuppressant and Antitumor Drug) 173
7.3.5	Dynamic Kinetic Resolution of Racemic Alcohols 174
7.4	Kinetic Resolution of Racemic Amines 174
7.4.1	Resolution of Rac-2-Butylamine 175
7.4.2	Synthesis of a Selective, Non-Peptide, Non-Sulfhydryl Farnesyl Protein Transfer Inhibitor (Antitumor Agent) 177
7.5	Resolution of Amino Alcohols and Methyl Ethers 178
7.6	Resolution of an Ester 178
7.6.1	Ester Resolution for the Synthesis of Emtricitabine (Antiviral Drug) 178
7.7	Desymmetrization by Transesterification 179
7.7.1	Synthesis of an Antifungal Azole Derivative 179
7.8	Regioselective Acylation 180
7.8.1	Regioselective Acylation of Drug Intermediate for an Antileukaemic Agent 180
7.8.2	Regioselective Acylation of Ribavirin: Antiviral Agent 180
7.9	Asymmetric Ring Opening of Racemic Azlactone 181

7.9.1	Synthesis of (S)-Benzyl-L-*tert*-Leucine Butyl Ester	181
7.9.2	Synthesis of (S)-γ-Fluoroleucine Ethyl Ester	182
7.10	Cyanohydrin Formation	183
7.10.1	Synthesis of (R)-2-Chloromandelic Acid: Intermediate for Clopidogrel	183
7.10.2	Synthesis of Chiral Phenoxybenzaldehyde Cyanohydrin: Intermediate for the Synthesis of Pyrethroid Insecticides	184
7.10.3	Hydrocyanation of 3-Pyridinecarboxyaldehyde	185
7.11	Outlook	186

Part Three Biocatalysis in Biphasic and New Reaction Media

8	**Biocatalysis in Biphasic Systems: General**	**191**
	Pedro Fernandes and Joaquim M. S. Cabral	
8.1	Introduction	191
8.2	Organic-Aqueous Biphasic Systems: General Considerations	191
8.3	Classification of the Systems: Macro- and Microheterogeneous Systems	193
8.4	Mechanisms of Enzyme Inactivation	194
8.5	Approaches to Protection of the Enzyme	195
8.6	Solvent Selection	196
8.7	Operational Parameters	198
8.7.1	Interfacial Area and Volume Phase Ratio	198
8.7.2	Mass Transfer Coefficients and Process Modeling	200
8.8	Reactor Types	205
8.9	Downstream Processing	206
8.10	Recent Applications	207
8.11	Conclusions	207
	Acknowledgments	207

9	**Biocatalysis in Biphasic Systems: Oxynitrilases**	**211**
	Manuela Avi and Herfried Griengl	
9.1	Introduction	211
9.2	Biphasic Systems	212
9.2.1	Liquid/Liquid Biphasic Systems	212
9.2.1.1	Buffer/Organic Solvent	212
9.2.1.2	Buffer/Ionic Liquids	215
9.2.2	Liquid/Solid Biphasic Systems	216
9.2.2.1	Crude Enzyme Preparations	216
9.2.2.2	Inert Carriers	217
9.2.2.3	Cross-Linking of Oxynitrilases	219
9.2.3	Other Biphasic Systems	220
9.2.3.1	Encapsulation in Sol-Gel Matrices	222
9.2.3.2	Comparison of CLEAs and sol-gel HNLs	223
9.3	Conclusion	223

10	**Ionic Liquids as Media for Enzymatic Transformations**	*227*
	Roger A. Sheldon and Fred van Rantwijk	
10.1	Introduction	*227*
10.2	Solvent Properties of Ionic Liquids	*228*
10.3	Enzymes in Ionic Liquids	*230*
10.4	Enzymes in Nearly Anhydrous Ionic Liquids	*231*
10.5	Stability of Enzymes in Nearly Anhydrous Ionic Liquids	*235*
10.6	Whole-Cell Biotransformations in Ionic Liquids	*236*
10.7	Biotransformations in Ionic Liquid Media	*237*
10.7.1	Lipases and Proteases	*237*
10.7.2	Dynamic Kinetic Resolution of Chiral Alcohols	*243*
10.7.3	Glycosidases	*244*
10.7.4	Oxidoreductases	*245*
10.8	Downstream Processing	*247*
10.9	Conclusions	*248*
11	**Solid/Gas Biocatalysis**	*255*
	Isabelle Goubet, Marianne Graber, Sylvain Lamare, Thierry Maugard and Marie-Dominique Legoy	
11.1	Introduction	*255*
11.2	Operating Solid/Gas Systems	*258*
11.2.1	Examples of the Use of Solid/Gas Systems	*263*
11.2.1.1	Example 1: Synthesis of Chiral Compounds Using Enzymes in Solid/Gas Reactors	*263*
11.2.1.2	Example 2: Conversion of Gaseous Halogenated Compounds by Dehydrated Bacteria	*267*
11.3	Influence of Operational Parameters on the Dehalogenase Activity of Whole Dehydrated Cells	*269*
11.3.1	Effect of Temperature	*269*
11.3.2	Effect of Water Activity	*269*
11.3.3	Effect of Buffer pH Before Dehydration	*271*
11.3.4	Range of Substrates	*271*
11.3.4.1	Example 3: Scaling up the System; Application to Industrial Production	*272*
11.4	Conclusion	*274*
12	**Biocatalysis with Undissolved Solid Substrates and Products**	*279*
	Alessandra Basso, Sara Cantone, Cynthia Ebert, Peter J. Halling and Lucia Gardossi	
12.1	Introduction	*279*
12.2	The Reaction System: Classification	*280*
12.3	Theory	*280*
12.4	Factors Affecting the Reactions in the Presence of Undissolved Substrates/Products	*284*
12.5	Precipitation-Driven Synthesis of Peptides	*287*

12.6	Precipitation-Driven Synthesis of Esters and Surfactants	*292*
12.7	The Synthesis of β-Lactam Antibiotics in the Presence of Undissolved Substrates	*294*
12.8	Conclusion	*298*

Index *303*

Preface

Once upon a time ... there was water. Long after enzymes began to be used for preparative-scale transformations of organic compounds, this peculiar solvent was still considered to be not only the most suitable but the "only" medium in which the natural biocatalysts could exert their activity. Then both scientific curiosity and practical problems (the need to overcome the poor solubility of most of the substrates in aqueous solutions) prompted different scientists to add small amounts of water-miscible organic solvents to the aqueous medium. And it worked: in most cases the enzymes were still fully alive.

The next step was the addition of water-immiscible organic solvents to give biphasic systems in which the biocatalysts were still solubilized in water and the organic phase was acting both as a substrate reservoir and a product extractor. Again the outcome was surprisingly positive, in some cases allowing inhibition of substrate(s) and/or product(s) to be overcome. This happened in the seventies.

At this point it became logical to raise the question "how much water do enzymes really need?". In trying to answer to this question it was found that some enzymes – particularly lipases and proteases – were still active even in neat organic solvents. This happened in the early eighties, and was the beginning of so-called "non-aqueous enzymology".

In 1996 Klibanov (the "father" of non-aqueous enzymology) and Koskinen published a book discussing progress in this exciting area of research and highlighting the successes, the questions that were still open, and the foreseeable developments [1].

When we received the invitation from Wiley to act as the Editors for a new book, we thought that now was the time, about ten years after the publication of the earlier book, for a general reflection on what had been achieved in the meantime.

As well as biocatalysis in neat organic solvents and biphasic systems (fundamentals and synthetic applications), the present volume covers new and promising aspects of non-aqueous enzymology that have emerged in recent years, including biocatalysis with undissolved solid substrates or vaporized compounds, the use of ionic liquids as solvents, and the preparative-scale exploitation of oxynitrilases and "dynamic kinetic resolutions". For the sake of completeness and comparison,

Organic Synthesis with Enzymes in Non-Aqueous Media. Edited by Giacomo Carrea and Sergio Riva
Copyright © 2008 WILEY-VCH Verlag GmbH & Co. KGaA, Weinheim
ISBN: 978-3-527-31846-9

some of the chapters devoted to synthetic applications necessarily also include a few examples of biotransformations performed in aqueous systems.

It has been a privilege to compile this volume. We have first of all to acknowledge the invited contributors for the timeliness of their contributions: despite the tightness of the schedule, they have done their best to assemble all the published information relevant to their assigned topics and to discuss all aspects exhaustively. Thanks are also due to Dr. Elke Maase, Dr. Steffen Pauly, and Dr. Andreas Sendtko of Wiley-VCH Publishers for their invitation and for their painstaking technical editorial support.

Finally, we would like to thank our readers in advance, with the hope that they might find this volume useful and stimulating for their research and/or teaching activities.

Milano, July 2007

Giacomo Carrea
Sergio Riva

[1] "Enzymatic reactions in organic media", A. M. P. Koskinen and A. M. Klibanov, Eds., Blackie Academic & Professional, Glasgow (UK), 1996.

List of Contributors

Patrick Adlercreutz
Lund University
Department of Biotechnology
P.O. Box 124
SE-22100 Lund
Sweden

Manuela Avi
Institute of Organic Chemistry
Graz University of Technology
Stremayrgasse 16
8010 Graz
Austria

Alessandra Basso
Università degli Studi
Dipartimento di Scienze
Farmaceutiche
Piazzale Europa, 1
34127 Trieste
Italy

Jan-E. Bäckvall
Stockholm University
Department of Organic Chemistry
Arrhenius Laboratory
10691 Stockholm
Sweden

Joaquim M. S. Cabral
IBB-Institute for Biotechnology and
Bioengineering
Centre for Biological and Chemical
Engineering
Instituto Superior Técnico
Avenida Rovisco Pais
1049-001 Lisboa
Portugal

Sara Cantone
Università degli Studi
Dipartimento di Scienze Farmaceutiche
Piazzale Europa, 1
34127 Trieste
Italy

Jonathan S. Dordick
Rensselaer Polytechnic Institute
Department of Chemical and Biological
Engineering
110 8th Street
Troy
NY 12180
USA

Cynthia Ebert
Università degli Studi
Dipartimento di Scienze Farmaceutiche
Piazzale Europa, 1
34127 Trieste
Italy

Organic Synthesis with Enzymes in Non-Aqueous Media. Edited by Giacomo Carrea and Sergio Riva
Copyright © 2008 WILEY-VCH Verlag GmbH & Co. KGaA, Weinheim
ISBN: 978-3-527-31846-9

Pedro Fernandes
IBB-Institute for Biotechnology and
Bioengineering
Centre for Biological and Chemical
Engineering
Instituto Superior Técnico
Avenida Rovisco Pais
1049-001 Lisboa
Portugal

Lucia Gardossi
Università degli Studi
Dipartimento di Scienze
Farmaceutiche
Piazzale Europa, 1
34127 Trieste
Italy

Isabelle Goubet
Laboratoire de Biotechnologies et de
Chimie Bio organique
FRE CNRS 2766
Université de La Rochelle
Bâtiment M. Curie
Avenue M. Crépeau
17042 LA ROCHELLE cedex 1
FRANCE

Marianne Graber
Laboratoire de Biotechnologies et de
Chimie Bio organique
FRE CNRS 2766
Université de La Rochelle
Bâtiment M. Curie
Avenue M. Crépeau
17042 LA ROCHELLE cedex 1
FRANCE

Herfried Griengl
Research Centre Applied Biocatalysis
Petersgasse 14
8010 Graz
Austria

Peter J. Halling
University of Strathclyde WestCHEM
Department of Pure and Applied
Chemistry
Thomas Graham Building
285 Cathedral Street
G1 1XL Glasgow
UK

Hans-Erik Högberg
Mid Sweden University
Department of Natural Science
S-851 70 Sundsvall
Sweden

Jaap A. Jongejan
Delft University of Technology
Biotechnology Department
Julianalaan 67
2628 BC Delft
The Netherlands

Birgit Kosjek
Merck Research Laboratories
Merck & Co. Inc. 126 Lincoln Avenue
Rahway
NJ 07065
USA

Sylvain Lamare
Laboratoire de Biotechnologies et de
Chimie Bio organique
FRE CNRS 2766
Université de La Rochelle
Bâtiment M. Curie
Avenue M. Crépeau
17042 LA ROCHELLE cedex 1
FRANCE

Marie-Dominique Legoy
Laboratoire de Biotechnologies et de
Chimie Bio organique
FRE CNRS 2766
Université de La Rochelle
Bâtiment M. Curie
Avenue M. Crépeau
17042 LA ROCHELLE cedex 1
France

Belén Martín-Matute
Stockholm University
Department of Organic Chemistry
Arrhenius Laboratory
10691 Stockholm
Sweden

Thierry Maugard
Laboratoire de Biotechnologies et de
Chimie Bio organique
FRE CNRS 2766
Université de La Rochelle
Bâtiment M. Curie
Avenue M. Crépeau
17042 LA ROCHELLE cedex 1
FRANCE

David Pollard
Merck Research Laboratories
Merck & Co. Inc. 126 Lincoln Avenue
Rahway
NJ 07065
USA

Fred van Rantwijk
Laboratory of Biocatalysis and Organic
Chemistry
Delft University of Technology
Julianalaan 136
2628 BL Delft
The Netherlands

Sergio Riva
Istituto di Chimica del Riconoscimento
Molecolare
C.N.R.
Via Mario Bianco 9
20131 Milano
Italy

Anne L. Serdakowski
Abbott Laboratories
Process Sciences
100 Research Drive
Worcester
MA 01605
USA

Roger A. Sheldon
Laboratory of Biocatalysis and Organic
Chemistry
Delft University of Technology
Julianalaan 136
2628 BL Delft
The Netherlands

Part One Biocatalysis in Neat Organic Solvents – Fundamentals

1
Fundamentals of Biocatalysis in Neat Organic Solvents
Patrick Adlercreutz

1.1
Introduction

It is well established that enzymes can express catalytic activity in predominantly organic media [1–8]. In some cases the reported catalytic activities in organic media are several orders of magnitude lower than those in aqueous solution. However, by careful selection of the type of enzyme preparation to use and the reaction conditions, it is often possible to achieve catalytic activities in the same order of magnitude as in water. Types of enzyme formulations for catalysis in organic solvents are treated in a separate chapter, while this chapter is largely concerned with reaction conditions for enzymatic catalysis in organic media. The most important points to consider are the solvent, the water content, and enzyme ionization effects (corresponding to pH effects in aqueous media).

1.2
Effects of Water on Biocatalytic Reactions

This book concerns the use of enzymes in non-aqueous media, which means that the major part of the medium surrounding the enzyme is non-aqueous. However, it is extremely difficult to remove water completely from enzymes. Even after extensive drying, a few water molecules remain tightly bound to the enzyme [6], and in most cases enzyme activity can be increased considerably by supplying an optimal amount of water to the system. It is thus very important to control the amount of water in the reaction mixtures in a proper way.

1.2.1
Quantification of Water in Low-Water Systems

The most straightforward way to quantify the amount of water in a reaction mixture is to use the water concentration (in $mol\,L^{-1}$ or % by weight or volume). However, the properties of the enzyme (catalytic activity, etc.) are much more

Organic Synthesis with Enzymes in Non-Aqueous Media. Edited by Giacomo Carrea and Sergio Riva
Copyright © 2008 WILEY-VCH Verlag GmbH & Co. KGaA, Weinheim
ISBN: 978-3-527-31846-9

influenced by the amount of water bound to the enzyme than by the total water concentration [9]. Unfortunately, it is difficult to measure the amount of water bound to the enzyme directly. On the other hand, it has been found that the amount of water bound to the enzyme is largely influenced by the thermodynamic water activity (relative humidity) [10], which can be measured by sensors and controlled in reaction mixtures using methods described below. The water activity has been widely accepted as the best way to quantify water in low-water systems for enzymatic synthesis [1, 5]. At equilibrium, the water activity is the same in all phases. This means that it can be measured wherever practical, which often means in the gas phase.

1.2.2
Water Activity Control

In order to study the effects of water activity on an enzymatic reaction, there is a need for practical methods to adjust the water activity in the reaction mixture. Likewise, it is highly desirable to keep this parameter close to the optimal value during large-scale conversions. A range of water activity control methods have been developed [11], and which one to choose depends on the scale of reaction, the quantities of water to be removed or added, and availability of equipment.

1.2.2.1 Water Activity Control Using Saturated Salt Solutions
A water activity control method that does not require special equipment and is useful on laboratory scale involves equilibration with saturated salt solutions via the gas phase [11]. Since the solubility of a salt in water has a fixed value (at a fixed temperature), the saturated solution has a fixed water activity (and water concentration). Small containers with enzyme preparation or substrate solution can be put into larger containers that are partially filled with saturated salt solutions. No salt solution should enter the small containers, but water in the form of vapor should be allowed to move between the saturated salt solutions and the material in the small containers. After the pre-equilibration period, both the enzyme preparation and the substrate solution will have the same water activity as the saturated salt solution. After mixing, the rate at this water activity can be measured. The saturated salt solutions are used as "buffers" of the water activity. There should be some solid salt present which can dissolve if water comes into the system (for example, as a product formed in an enzymatic condensation reaction). On the other hand, if water is consumed in the system (for example in a hydrolysis reaction) some salt will crystallize from the solution. As long as there is both a saturated solution and solid salt, the water activity will be kept constant provided that the mass transfer is fast enough in the system. By using different salts, a range of water activities can be obtained (Table 1.1). If the reaction is slow, the equilibration through the gas phase as described above can be used to maintain the water activity, but if large amounts of water must be removed or added to the reaction at fixed water activity, a more efficient system is needed. One way to achieve this is to pump the saturated salt solution through silicone tubing immersed in the

Table 1.1 Saturated salt solutions suitable for water activity control. Values are given for 25 °C [77].

Salt	Water activity
LiCl	0.113
$MgCl_2$	0.225
K-acetate	0.328
K_2CO_3	0.432
$Mg(NO_3)_2$	0.529
$SrCl_2$	0.708
KCl	0.843
KNO_3	0.936
K_2SO_4	0.973

reactor [12]. The surface area of the silicone tubing can be chosen to match the water transport capacity required.

1.2.2.2 Water Activity Control Using Sensors

When it is important to control the water activity in a reactor, a water activity sensor is quite useful. The sensor should ideally measure the water activity in the liquid reaction medium. However, the sensors available are designed for gas phase measurements, and, provided there is effective enough equilibration between the liquid and gaseous phases, they can be used to control the water activity in the reactor. If the measured water activity is above the set point, drying is initiated, for example, by passing dry air through the reactor. On the other hand, if the water activity is too low, water can be added, either as liquid water or as humid air. Automatically controlled systems of this kind have been successfully used to monitor and control enzymatic reactions in organic media [13, 14].

1.2.2.3 Water Activity Control Using Pairs of Salt Hydrates

An alternative method for water activity control is based on the fact that salt hydrates containing different numbers of water molecules are interconverted at fixed water activities [15]. The first salt hydrate used was $Na_2CO_3 \cdot 10H_2O$. This is converted to $Na_2CO_3 \cdot 7H_2O$ at a water activity of 0.74 at 24 °C. The salt hydrates act as a buffer of the water activity. As long as both salt hydrates are present, the water activity remains at 0.74. If another water activity is desired, another pair of salt hydrates should be chosen. The salt hydrates can be added directly to the organic reaction mixture. One should be careful that the salt hydrates do not interfere with the enzyme or the enzymatic reaction.

1.2.3
Distribution of Water

During biocatalysis in organic media, the small amounts of water present are associated with various components of the system: dissolved in the solvent, bound to the enzyme and bound to support materials and other additives that might be present. Water is exchanged between these different components, and at equilibrium the water activity is the same throughout the reaction mixture and the gas phase above it. Data on the amount of water associated to the various components as a function of water activity therefore helps in understanding the behavior of enzymes in such environments. The amount of water bound to the enzyme is described by the water adsorption isotherm. The adsorption isotherms of different enzymes are often relatively similar: a typical one is shown in Figure 1.1.

Support materials used for enzymes in organic media bind various amounts of water. An example of a support material that binds quite low amounts of water is Celite (Figure 1.2). It should be noted that additives, such as buffer salts, present in the enzyme preparation can bind substantial amounts of water (Figure 1.2).

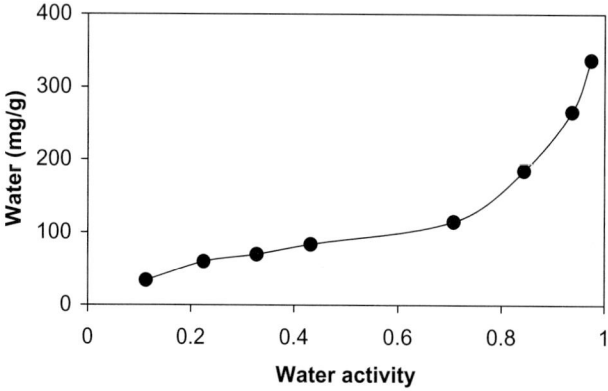

Figure 1.1 Water adsorption isotherm of α-chymotrypsin at 25 °C. Reprinted from Ref. [18].

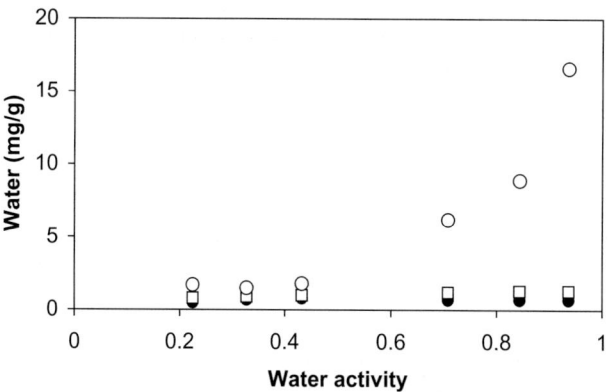

Figure 1.2 Water adsorption isotherms at 25 °C for Celite and preparations obtained by mixing Celite with different solutions (1.0 mL g^{-1} Celite) and drying. Pure Celite (●); Celite and α-chymotrypsin in water (4.0 mg mL^{-1}) (□); Celite and α-chymotrysin (4.0 mg mL^{-1}) in 50 mM sodium phosphate buffer, pH 7.8 (○). Reprinted from Ref. [18].

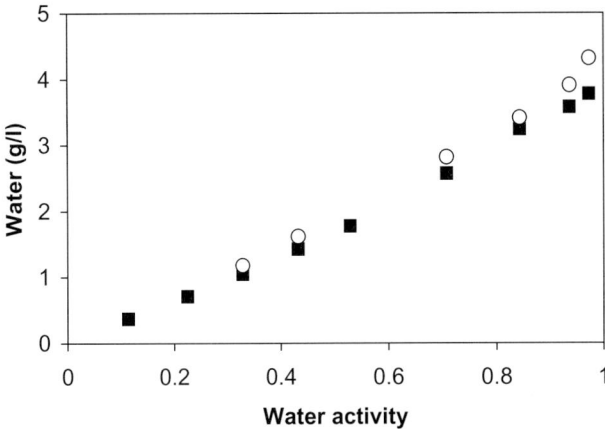

Figure 1.3 Solubility of water at different water activities in diisopropyl ether (■) and a substrate solution containing 10 mM Ac-Phe-OEt and 100 mM 1-butanol in diisopropyl ether (○). Reprinted from Ref. [18].

When working at fixed water concentration, the water-absorbing capacity of the support, called aquaphilicity [16], gives an indication of how well the support can compete with the enzyme for the water in the system. A high aquaphilicity means that the support absorbs a lot of water, leaving little for the enzyme, and this results in a low enzyme activity in most cases. When working at fixed water activity, more equal activities of enzymes on different supports are observed [17] although some differences still appear [18].

The solubility of water in a water-immiscible solvent at water activity 1 can be determined by equilibrating the solvent with pure water, followed by water analysis, for example, by Karl Fischer titration. At water activities lower than 1, lower amounts of water dissolve in the solvents, as shown in Figure 1.3. It should be noted that the solubility of water in the solvent changes when solvent composition is changed, for example, by dissolving substrates (Figure 1.3). In a hydrophobic solvent the increase in solubility of water can be substantial when substrates are dissolved in it.

The amounts of water associated with various components in a typical reaction mixture are shown in Table 1.2. Most of the water is dissolved in the reaction medium, and the amount of water bound to the enzyme is obviously just a minor fraction of the total amount of water. If the solvent was changed to one able to dissolve considerably more water and the same total amount of water was present in the system, the amount of water bound to the enzyme would decrease considerably and thereby its catalytic activity as well. Changing solvent at fixed water activity would just increase the concentration of water in the solvent and not the amount bound to the enzyme. Comparing enzyme activity at fixed enzyme hydration (fixed water activity) is thus the proper way of studying solvent effects on enzymatic reactions.

Table 1.2 Amounts of water associated with various components of a reaction mixture containing Celite-immobilized enzyme in diisopropyl ether at $a_w = 0.7$. Data from [18].

Component	Amount of water (mg)
1 mL diisopropyl ether with dissolved substrates	2.8
0.4 mg enzyme (α-chymotrypsin)	0.04
100 mg Celite	0.1
Buffer salts on the Celite	0.5
Total reaction mixture	3.44

1.2.3.1 Hysteresis Effects

Sometimes it can be difficult to know if the system has come to a true equilibrium concerning water distribution. It has been noted that water adsorption isotherms sometimes show hysteresis effects, which means that the water content, for example, that bound to the enzyme, depends not only on the water activity, but also on the hydration history [6]. More water is thus bound if a specified water activity is approached from a higher value (dehydration direction) than if the enzyme is hydrated from a drier state. The hysteresis effects might be due to slow conformational changes in the enzyme.

1.2.4
Water Effects on Activity

The catalytic activity of an enzyme in an organic medium often varies by several orders of magnitude depending on the degree of enzyme hydration [9, 19]. Control of enzyme hydration or water activity is thus a key issue when optimizing enzymatic conversions in organic solvents.

All enzymes to be used in organic media have at a previous stage been in an aqueous phase. They are then transferred to the organic medium, and this transfer process involves removal of water. This can be achieved by lyophilization or just drying of the enzyme solution, possibly in the presence of a support material or other additives. Another possibility is to dilute the aqueous enzyme solution with a water-miscible organic solvent which dissolves the water and causes the enzyme to precipitate. In one version, the enzyme solution also contains a crystal-forming solute such as an inorganic salt or an amino acid [20]. In this case, crystals are formed and the enzyme covers the crystals.

Lyophilization is a very common method to prepare enzymes for use in organic media, but the procedure often results in preparations having low catalytic activity. The method can cause inactivation of the enzyme both in the freezing step and in the drying step [7]. FT-IR spectroscopy has been used to study secondary structure in enzyme preparations, and lyophilization has been shown to decrease the α-helix content and increase the β-sheet content compared to native enzyme and enzyme in aqueous solution [21]. The conformational changes during lyophilization can be prevented by lyoprotectants, such as substrate analogs or polyethylene glycol [22, 23]. Many other additives, including inorganic salts and crown ethers, have been used to improve enzyme activity after lyophilization, but methods to make enzyme preparations for use in organic media are treated in more detail in another chapter in this book and will not be discussed further here.

Rehydration of the enzyme preparation before use in organic media usually increases the catalytic activity considerably. Water is often called a molecular lubricant of enzymes in organic media [19]. Correlations between internal protein flexibility and degree of hydration have been shown using time-resolved fluorescence anisotropy studies [24] and electron paramagnetic resonance (EPR) spectroscopy [25]. A major effect of the increase in flexibility is probably that enzymes can reverse the conformational changes causing inactivation during lyophilization or other drying procedures for the preparation of the enzyme for use in organic media. In addition, increased flexibility can be beneficial for movements required in the catalytic process itself.

Relatively few detailed studies of enzyme kinetics in organic media have been carried out. Preferably, full kinetics should be studied, allowing the determination of K_m and k_{cat} values, but it is much more common to see just reports on the catalytic activity at fixed substrate concentrations as a function of water activity. That such studies can be misleading was shown in an investigation of lipase-catalyzed esterification [26]. When the reaction rate in the esterification reaction was plotted versus the water activity at three different substrate concentrations, maxima were obtained at three different water activities (Figure 1.4). Such maxima should not be used to claim that the optimal water activity of the enzyme was found. Detailed kinetic studies showed that both the k_{cat} and the K_m values (for the alcohol substrate) varied with the water activity. The K_m value of the alcohol increased with increasing water

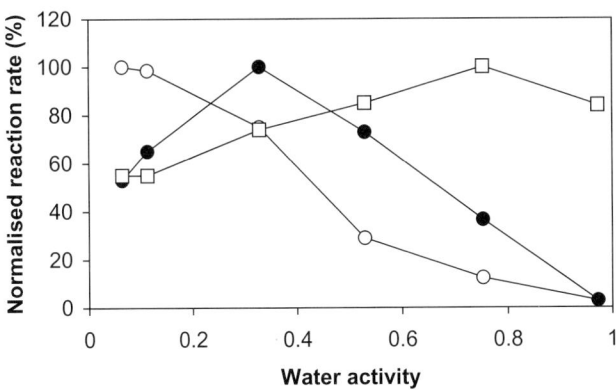

Figure 1.4 Normalized reaction rate as function of water activity for the esterification of dodecanol with decanoic acid catalyzed by *Rhizopus arrhizus* lipase at three different concentrations: 20 mM (○), 200 mM (●) and 800 mM (□) of each substrate. Data obtained from Ref. [26].

Table 1.3 Water activity at which the enzymes express 10% of their maximal activity.

Enzyme type	Water activity
Glycosidases	0.5–0.8
Lipases	0.0–0.2
Oxidoreductases	0.1–0.7

activity, mainly because water competed with the alcohol in the deacylation of the acyl enzyme. Variation of both substrate concentrations made it possible to estimate true k_{cat} values [26]. The maximal k_{cat} values were obtained at much higher water activities than the maxima observed at low substrate concentrations.

The response of enzymes to an increase in water activity varies considerably. There are examples of lipases that express considerable activity at water activities as low as 0.0001 [27], while many enzymes require considerably higher water activity. When comparing water activity dependence of enzymes it can be useful to compare water activities at which each enzyme expresses 10% of its maximal activity. Table 1.3 shows the results of a literature survey on this topic. It is clear that glycosidases require considerably higher water activity than lipases and that there is a large variation within the group oxidoreductases. In the majority of the experimental studies made, it has been found that enzymatic activity increases with increasing water activity. Maximal activity is often found quite close to a water activity of 1. As indicated above, this is the case also for most lipases, provided that measurements are made at high enough substrate concentrations.

It is not known which molecular features of lipases keep them active at low water activity. It is worth pointing out that when they are used to catalyze various reactions, such as hydrolysis, reversed hydrolysis, and transesterification reactions, the water activity dependence is similar in the different reactions [26, 28]. The same is true for phospholipase A_2 [29]. This shows that the effect of water on the enzyme is more important than the effect of water as a reactant when determining the reaction rate at different water activities.

1.2.5
Water Effects on Selectivity

When water molecules interact with an enzyme, it is natural that conformational changes can occur, which in turn can cause changes in the selectivity of the enzyme. Since enantioselectivity of enzymes is of major importance for many applications, it is a common task to investigate how to choose reaction conditions providing the maximal enantioselectivity. As might be expected, because water can interact with enzymes in many ways, it is difficult to generalize the effects. In some studies of lipase-catalyzed esterification reactions, no effects of water activity on enantioselectivity were observed [30]. In a similar study, no effects were observed in most cases, while the enantioselectivity of one lipase-catalyzed reaction decreased with increasing water activity [31]. The enantioselectivity in reductions catalyzed by alcohol dehydrogenase from *Thermoanaerobium brockii* increased with increasing water activity in hexane [32]. The formation rate for both enantiomers of 2-pentanol increased with increasing water activity, but the formation of the S enantiomer was enhanced more, resulting in higher enantiomeric purity of the product. When even more water was added, forming a two-phase system with hexane, the enantioselectivity decreased [32]. A more detailed study showed that the formation of the S enantiomer was enthalpically favored in the whole range of water activities, while formation of the R enantiomer was entropically favored. In the competition between the two pathways leading to the R and S enantiomers, the enthalpy effect dominated, which resulted in the formation of an excess of the S enantiomer [33].

Another type of important selectivity is that between hydrolysis and transferase reactions (transesterification, transglycosylation, etc.) catalyzed by hydrolases. In this case, water can act both as a reactant and as a substance that modifies the properties of the enzyme. Effects of water as a reactant can be expected to be governed by the concentration or activity of water, as with other substrates. The effects of water as an enzyme modifier are considerably more difficult to predict.

The most straightforward way to quantify the competition between the transferase reaction and hydrolysis is to measure the initial ratio of these two reactions. Intuitively, one would assume the transferase/hydrolysis ratio to decrease with increasing water activity because of the effect of water as a reactant. This is often the case when lipases are used as catalysts [34–36]. However, in reactions catalyzed by glycosidases and proteases the transferase/hydrolysis ratio can either increase or decrease with increasing water activity [37, 38].

The competition between transferase and hydrolysis reactions can be described in terms of nucleophile (acceptor) selectivities of the enzymes, and selectivity constants can be defined. These constants are meant to quantify the intrinsic selectivity of the enzymes. Selectivity constants in combination with the concentrations (or thermodynamic activities) of the competing nucleophiles give the transferase/hydrolysis ratio. The selectivity constants are defined as follows [38, 39]:

$$\frac{r_s}{r_h} = S_c \frac{[\text{nucleophile}]}{[\text{water}]} \qquad (1)$$

$$\frac{r_s}{r_h} = S \frac{a_{\text{nucleophile}}}{a_w} \qquad (2)$$

where r_s and r_h are the rates of the transferase and hydrolysis reactions and S_c and S are the selectivity constants based on concentrations and activities, respectively.

The selectivity constants of glycosidases generally increase with increasing water activity [38], while those of lipases often decrease with increasing water activity [34]. The combination of high catalytic activity at low water activity and the good selectivity for transferase reactions at low water activity makes lipases very efficient transesterification catalysts [34]. When glycosidases are used for transglycosylation, it is surprisingly beneficial to use a water activity close to 1.0, which provides both high reaction rate and high selectivity for transglycosylation [38].

1.2.6
Water Effects on Stability

Water is often considered as the best solvent for enzymatic reactions. However, water is able to react in many ways with the enzyme, thereby causing its inactivation. Examples of such reactions include hydrolysis of peptide bonds in the enzyme, deamidation of amino acid side chains, and destruction of cystine residues [40]. Furthermore, the increased flexibility of the enzyme caused by water, which results in high catalytic activity, facilitates these inactivation reactions. It was thus realized that elimination of the major part of the water from the surroundings of the enzyme can cause pronounced stabilization compared to the situation in aqueous solution. It was shown that normal enzymes in organic media can express considerable activity even at 100 °C [41]. Several later studies have confirmed that enzyme stability generally decreases with increasing water activity [36, 42]. It should be pointed out that it is the absence of water that is important: other water-poor enzyme preparations, such as dry enzyme powders surrounded by gas, also exhibit high thermostability. When the *operational stability* of various protease preparations was studied, no conclusive effect of the water activity was observed when comparing water activities between 0.22 and 0.76 [43].

Since the catalytic activity of enzymes increases with increasing temperature, the possibility to use enzymes at high temperatures in low-water media might indicate that very high catalytic activities could be obtained. However, in one study of chymotrypsin-catalyzed reactions it was found that the reduction in catalytic activity due to the decrease in water activity was larger than the increase caused by the higher reaction temperature (Figure 1.5) [42]. Of course, there might be

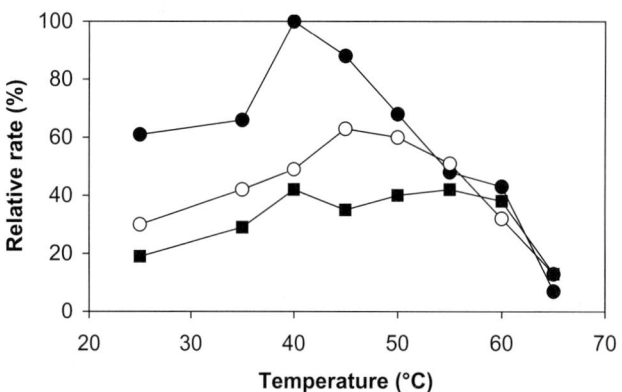

Figure 1.5 Relative activity of α-chymotrypsin in 5-methyl-2-hexanone at different temperatures and water contents: 100% water saturation (●), 75% water saturation (○), and 50% water saturation (■). Reprinted from Ref. [42] with kind permission of Springer Science and Business Media.

other cases where indeed higher overall catalytic activities can be obtained by a reduction in water activity combined with an increase in temperature.

1.3
Solvent Effects

Ever since it was discovered that enzymes can be catalytically active in neat organic solvents, the question of how to select the correct solvent for a specified enzymatic conversion has been of crucial importance. The solvent can influence an enzymatic reaction both by direct interaction with the enzyme and by influencing the solvation of the substrates and products in the reaction medium. An example of direct interaction between solvent and enzyme is when the solvent acts as an inhibitor of the enzyme. In other cases the solvent causes conformational changes in the enzyme, thereby changing its catalytic properties. The solvent can also influence the amount of water bound to the enzyme, but this effect can largely be avoided by the use of fixed water activity as described above. Direct interaction between solvent and enzyme can influence enzyme stability as well as activity.

A very common and important effect of the solvent on enzymatic reactions is that of affecting the solvation of the substrates and products of the reaction catalyzed. The solvation of the substrate influences its free energy and thereby its reactivity. Solvents which are able to dissolve a substrate very efficiently lower the free energy, and the rate of the catalyzed reaction is thereby decreased. The solvent also influences the equilibrium position of reactions, and here the solvation of both substrates and products must be considered.

Since there are so many solvents to choose from, it is natural that the search for guidelines for solvent selection has been intense. Researchers have tried to correlate enzyme activity, stability, and selectivity with different solvent descriptors, such as log P, dielectric constant, dipole moment, Hildebrand solubility parameters, and many others. When this approach is successful, the search for the optimal solvent can be limited to those having suitable values of the selected solvent descriptor(s). A list of solvent descriptors of a range of commonly used solvents is given in Table 1.4.

Below, solvent effects on activity and stability of enzymes will be discussed, while solvent effects on enzyme selectivity is a large topic which is treated in a separate chapter. Solvent effects on equilibria are treated in Section 1.4.

1.3.1
Solvent Effects on Enzyme Activity

It was early discovered that enzyme activity in organic solvents depends very much on the nature of the solvent. It was realized that the polarity or hydrophobicity of the solvent had a large influence, with non-polar hydrophobic solvents often providing higher reaction rates than more polar, hydrophilic solvents. When the kinetics of enzymatic reactions is studied, it is often found that K_m values in organic solvents are much higher than those in water for the corresponding

Table 1.4 Solvent descriptors of organic solvents commonly used for biocatalysis. Sw/o (solubility of water in solvent, wt%) So/w (solubility of solvent in water, wt%) and ε (dielectric constant) values from [78], log P (P = partition coefficient between octanol and water), ET (empirical polarity parameter by Reichardt-Dimroth) and HS (Hildebrand solubility parameter, $J^{1/2}cm^{-3/2}$) from [79].

Solvent	log P	ε	ET	HS	Sw/o	So/w
DMF	−1.01	36.71	0.404	20.3	100	
Methanol	−0.77	32.66	0.762	29.7	100	100
Ethanol	−0.31	24.55	0.654	26.1	100	100
1,4-Dioxane	−0.27	2.21	0.164	20.7	100	100
Acetone	−0.24	20.56	0.355	20.5	100	100
2-Butanone	0.29	18.51	0.327	19	10	24
Pyridine	0.65	12.91	0.302	21.7	100	100
Ethyl acetate	0.73	6.02	0.228	18.6	2.94	8.08
1-Butanol	0.88	17.51	0.506	23.7	20.5	7.45
Diethyl ether	0.89	4.2	0.117	15.1	1.47	6.04
Diisopropyl ether	1.52	3.88	0.102	14.4	0.57	1.2
Butyl acetate	1.7	5.01		17.4	1.2	0.68
Benzene	2.13	2.27	0.111	18.7	0.0635	0.179
1,1,1-Trichlorethane	2.49	7.25	0.17	17.4	0.034	0.132
Toluene	2.73	2.38	0.099	18.2	0.0334	0.0515
Hexane	3.98	1.88	0.009	14.9	0.0111	0.00123
Heptane	4.57	1.92	0.012	15.2	0.0091	0.000357

reactions. These effects are due to the effective solvation of the substrate in the organic solvent [44], reducing its free energy, so that the free energy of activation of the enzymatic reaction increases, resulting in a lower reaction rate. The reaction rate can as usual be increased by increasing the substrate concentration. However, if the substrate concentration is fixed at a moderate level, one should avoid solvents able to dissolve much higher substrate concentrations or else there is a severe risk that the reaction rate will be low due to the increase in apparent K_m value.

One way of measuring the solvation of a substance in a range of solvents is to study the partitioning of the substance between those solvents and a standard solvent which is immiscible with the others. The partitioning of two different substrates between water and a range of water-immiscible solvents was thus studied and was correlated with the relative rate of enzymatic conversion of these two substrates [45]. In this case, solvent effects on enzyme *specificity* were studied, but the same principles apply to rates of enzymatic conversions of single substrates. It is useful to contemplate the situation when an enzyme is acting in a two-phase system consisting of water and the organic solvent under study. The partitioning of the substrate between the phases will influence the substrate concentration in the aqueous phase, which in turn determines the reaction rate. A solvent that very efficiently extracts the substrate from the aqueous phase will thus lower the aqueous substrate concentration and thereby lower the reaction rate (unless the aqueous substrate concentration is much higher than the K_m value). The same kind of reasoning can be used even though the aqueous phase is small or even non-existent. What really matters is the partitioning of the substrate between the bulk solvent and the active site of the enzyme, and this is not influenced by the introduction of an aqueous phase in between.

Partitioning between water and water-immiscible organic solvents is thus a straightforward way to get quantitative data for predicting solvent effects on the conversion of a certain substrate. However, the method is not applicable to water-miscible solvents. An alternative way to quantify solvation is to carry out theoretical calculation of interactions between the various components in the reaction mixture. The most frequently used method to do this is the UNIFAC group contribution model which can be used to calculate activity coefficients of all components in mixtures [46]. In this model, a molecule is treated as the sum of its different building blocks, such as methyl groups, carbonyl groups, etc. To calculate the difference in solvation between the two substrates N-Ac-L-Phe-OEt and N-Ac-L-Ser-OEt, activity coefficients of methanol and toluene in a range of organic solvents were calculated. Methanol and toluene represented the amino acid side chains of the substrate molecules, which otherwise were identical. There was a good correlation between the ratios of reactions rates and the ratios of the partition coefficients calculated from activity coefficients in the different solvents [47] (Figure 1.6). The UNIFAC model is under constant development to increase the range of compounds that can be handled and to improve the accuracy [48]. Some version of it is included in several calculation software packages.

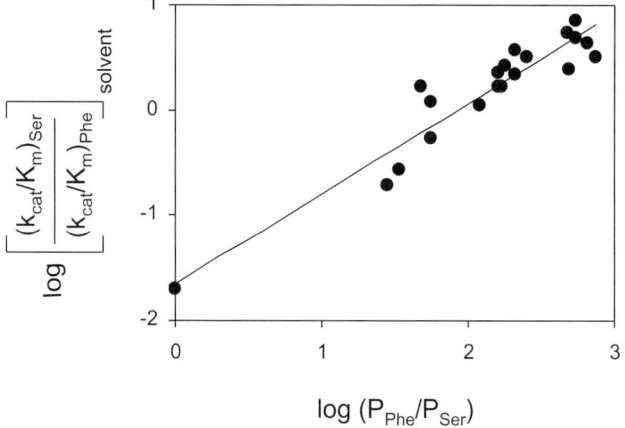

Figure 1.6 Dependence of the substrate specificity of subtilisin Carlsberg on the calculated ratio of the solvent-to-water partition coefficients of N-Ac-Phe-OEt and N-Ac-Ser-OEt in various solvents. Reprinted with permission from [47]. Copyright (1993) American Chemical Society.

When substrate activities are used instead of substrate concentrations in studies of enzyme kinetics in organic media, solvent effects due to substrate solvation disappear. Remaining solvent effects should be due to direct interactions between the enzyme and the solvent. In a study of lipase-catalyzed esterification reactions, it was found that K_m values based on activities were indeed more similar than those based on concentrations in different solvents, but still some differences remained [49].

As discussed in Section 1.2.3, it is crucial that the effects of solvents are studied at fixed water activity, or else indirect effects due to competition for water between enzyme and solvent will cause strong effects and mask the true solvent effects. In general, when correcting for substrate solvation, hydrophobic solvents seem to give higher rates than other solvents [5].

1.3.2
Solvent Effects on Stability

Enzyme stability in organic solvents depends on the direct interactions between enzyme and solvent. In addition to water, solvents like DMSO, formamide, and ethyleneglycol are able to dissolve proteins [50, 51]. A thorough study on the solubility of lysozyme revealed several other polar solvents able to dissolve more than $10\,g\,L^{-1}$ of the enzyme, including glycerol, 2,2,2-trifluoroethanol, methanol, and phenol [51]. Recent studies have shown that enzymes can also be dissolved in ionic liquids that are good H-bond acceptors, such as those containing acetate, lactate, or nitrate as anions [52]. In the majority of all these cases, dissolution results in inactivation of the enzyme due to disruption of its tertiary and sometimes also secondary structure. However, there are interesting exceptions. Lysozyme can fold correctly in glycerol [53]. Furthermore, morphine dehydrogenase was found to be catalytically active when dissolved in the ionic liquid 1-(3-hydroxypropyl)-3-methyl-imidazolium glycolate [54]. In organic solvents other than those mentioned, enzymes are practically insoluble. However, these undissolved enzyme preparations are often catalytically active and very useful for synthesis. The fact that the enzyme is present in a solid phase makes it easy to separate the reaction product from the enzyme, which simplifies both product purification and enzyme re-use.

Organic solvents can interact with enzymes in several ways. There can be specific interactions between isolated solvent molecules and enzyme molecules. This kind of interaction also occurs in water containing dissolved solvent molecules. When a separate organic solvent phase is present, interfacial inactivation can also occur. This is sometimes called phase toxicity [55] to distinguish it from the molecular toxicity of isolated solvent molecules. Interfacial inactivation can be studied in detail by bubbling solvent through an aqueous enzyme solution under controlled conditions [56].

In most experimental studies of the influence of organic solvents on enzyme stability, the remaining catalytic activity after exposure to different solvents has been measured. In such a study it was found that the remaining activity of a

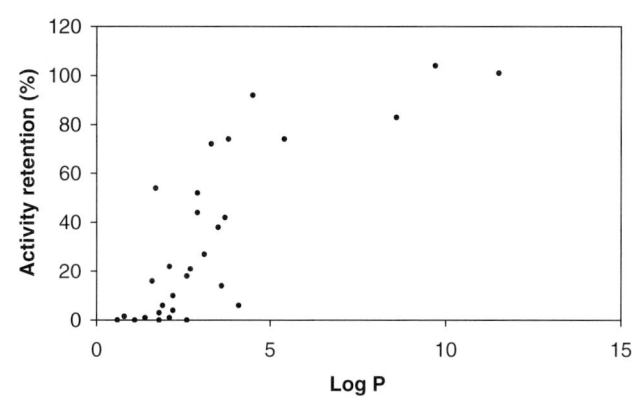

Figure 1.7 Activity retention of immobilized cells catalyzing an epoxidation reaction after exposure to organic solvents with different log P values. Data from Brink and Tramper [76] plotted according to Laane et al. [57]. Reprinted with permission of Wiley-Liss, a subsidiary of John Wiley and Sons, Inc.

whole-cell biocatalyst catalyzing epoxidation reactions correlated well with the log P value of the solvent (Figure 1.7) [57]. Solvents having log P values above 4 caused negligible inactivation, while those with log P values below 2 were highly inactivating. In the intermediate range of log P values a large scatter in the results was observed, making generalizations difficult. Similar results have been obtained in several subsequent studies, and log P is the most frequently used solvent parameter in this kind of investigations. Among studies giving contradictory results, one investigation of two glycosyl hydrolases and one phosphatase can be mentioned [58]. In this study, solvents with high log P values, like n-alkanes between n-octane (log P = 4.5) and n-hexadecane (log P = 8.8), were highly inactivating, while good stability was observed with butyl acetate having a log P value of 1.7. It is not surprising that a single solvent descriptor, such as log P, cannot give good predictions for all combinations of solvents and enzymes. In an attempt to create a more accurate model for predicting the inactivation of enzymes by solvents, principal component analysis was applied and 14 solvent descriptors were taken into account, resulting in good correlation between predicted and observed remaining activity of free and Celite-immobilized horse liver alcohol dehydrogenase [59]. A slightly worse model was obtained using only three solvent descriptors (log P, dielectric constant, and melting point), but a correlation coefficient of 0.957 between predicted and observed remaining activity was still obtained [59].

Large differences in sensitivity toward interfacial inactivation were observed between α-chymotrypsin and *Candida rugosa* lipase [56]. The lipase was most rapidly inactivated by 1-butanol and tolerated the hydrophobic hydrocarbons quite well, while the opposite was true for α-chymotrypsin. A detailed study of interfacial inactivation by 12 different solvents, all having log P values around 4, revealed large differences between the solvents and also between the different enzymes studied [60]. The previously mentioned tendency of hydrocarbons to cause interfacial inactivation of α-chymotrypsin was confirmed. For both α- and β-chymotrypsin the interfacial inactivation increased with increasing interfacial tension of the solvents [60]. Proteins which are rigid and less prone to structural changes are less sensitive to interfacial inactivation, ribonuclease and papain being typical examples [60].

In conclusion, most protein-dissolving solvents cause enzyme inactivation, and, among the other solvents, those having high log P values are less inactivating than others in many cases. However, the chemical nature of the solvents is certainly also of importance, and there are clearly large individual variations between different enzymes.

1.4
Effects on Equilibria

When choosing reaction conditions, such as water activity and solvent, for an enzymatic reaction, possible effects on the equilibrium position of the reaction should be considered. When the aim is to produce an equilibrium mixture as the final product, the position of this equilibrium is of course of vital importance. It is, however, also important that the biocatalyst expresses sufficient catalytic activity under the conditions used, so that equilibrium is reached within a reasonable time. In practice, it often happens that a compromise must be made between high reaction rate and high equilibrium conversion.

The thermodynamic equilibrium constant of reaction $A + B \leftrightarrows C + D$ is defined as follows:

$$K = \frac{a_C \cdot a_D}{a_A \cdot a_B} \quad (3)$$

The value of K does not depend on the medium composition, and in principle this is the only equilibrium constant needed. However, since activities of substrates and products are often hard to get at, concentration-based equilibrium constants are often used instead. Concentrations can, for example, be expressed as molar ratios (x_A, etc.). For each substrate or product, mole ratio and activity can be interconverted using activity coefficients (γ_A, etc), where $a_A = \gamma_A \cdot x_A$.

The different equilibrium constants are thus related as follows:

$$K = K_\gamma \cdot K_x = \frac{\gamma_C \cdot \gamma_D}{\gamma_A \cdot \gamma_B} \cdot \frac{x_C \cdot x_D}{x_A \cdot x_B} \quad (4)$$

where K_x is the molar ratio-based equilibrium constant and $K\gamma$ is the combination of activity coefficients shown in the formula. The effects of the medium on the concentration-based equilibrium constant are described by the activity coefficients.

For reactions of the type acid + alcohol ⇋ ester + water, catalyzed by hydrolases, an equilibrium constant involving both concentrations and the water activity has been suggested [61]:

$$K_0 = \frac{[\text{ester}] \cdot a_w}{[\text{acid}] \cdot [\text{alcohol}]} \tag{5}$$

The reason for using an equilibrium constant like this is that the water activity is easily measured and/or controlled in the reaction mixtures and is often fixed to provide good conditions for the enzymatic reaction. Concentrations are more practical to use than activities for the other reactants.

1.4.1
Water Effects on Equilibria

One of the most common reasons for using low-water media for enzymatic reactions is that one wants to use a hydrolase for catalyzing reactions other than hydrolysis. In low-water media these enzymes can be efficiently used to catalyze reversed hydrolytic reactions and various types of tranferase reactions.

It is clear that the water activity is of crucial importance for the equilibrium yield in a reversed hydrolysis reaction. As expected, the equilibrium yield increases with decreasing water activity. This has been shown, for example, for the condensation of glucose and octanol [62], esterification of lysophospholipids with fatty acids [29, 63], and in normal lipase-catalyzed esterification reactions [64, 65]. The same situation is observed in ionic liquids [66].

In addition to its direct influence via the water activity in the system, the amount of water also influences the activity coefficients of the other components in the mixture, and therefore equilibrium constants like K_0 can vary with the water activity in the system (Table 1.5) [29, 63, 64]. This can be seen as a solvent effect on the equilibrium constant. The tendency in esterification reactions is that K_0 increases with decreasing water activity, which means that it is favorable to use low water activity because of both the direct effect of water activity on the equilibrium and the influence of water on K_0.

Table 1.5 Equilibrium yield and equilibrium constants (K_0) in the synthesis of phosphatidylcholine from lysophosphatidylcholine and oleic acid (800 mM) at different water activities. Data from Ref. [29].

Water activity	Equilibrium yield (%)	K_0 (M^{-1})
0.22	37.2	0.163
0.33	29.9	0.176
0.43	20.8	0.141
0.53	13.4	0.102

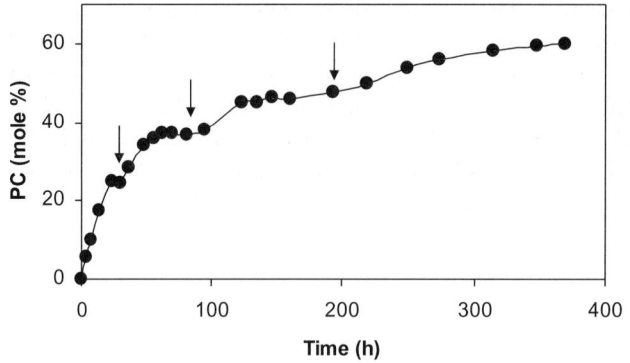

Figure 1.8 Time course of the synthesis of phosphatidylcholine from lysophosphatidylcholine and oleic acid catalyzed by phospholipase A_2. The water activity was initially 0.43 and was decreased stepwise to 0.33, 0.22 and finally to 0.11 (at times indicated by arrows). Reprinted with permission from Elsevier [29].

Reversed hydrolysis reactions should ideally be carried out at a water activity close to 0. One reason for not obeying this general rule is that many enzymes require a water activity considerably above 0 to express reasonable catalytic activity (see Section 1.2.4). Therefore compromises are often made between high activity and high yield. In order to optimize the synthesis reaction, it can be worthwhile to start the reaction at a relatively high water activity to achieve a high reaction rate and then stepwise or continuously decrease the water activity to obtain a high final yield (Figure 1.8) [29].

It is worth pointing out that the equilibrium positions in many esterification reactions in hydrophobic solvents are quite favourable, so that high yields can be obtained even if the water activity is close to 1. In these cases, it is the effective solvation of the ester product in the medium that is a main driving force the reaction. This is further discussed below.

1.4.2
Solvent Effects on Equilibria

Solvent effects on enzymatic reactions have been most thoroughly studied for esterification reactions. It has been observed that those reactions are favorably carried out in relatively hydrophobic solvents, while the equilibrium position is less favorable for esterification in more hydrophilic solvents. Correlations between equilibrium constants and solvent parameters have been evaluated. It was shown that the solubility of water in the solvent ($S_{w/o}$) gave better correlation with esterification equilibrium constants than log P and other simple solvent descriptors [61].

Solvent effects on esterification equilibria have also been described using UNIFAC calculations of activity coefficients. This method was claimed to give somewhat better correlation than the one with solubility of water ($S_{w/o}$) [67]. A drawback with the UNIFAC method is that it does not consider the location of the various groups in the molecules and therefore it cannot account for the influence from neighboring groups. An alternative method to determine activity coefficients is the COSMO-RS method. This is based on the approximated continuum description of a solvent [68]. It was shown to give somewhat better correlation with experimental results for esterification reactions than the UNIFAC method [68].

1.5
Effects of pH in Organic Solvents

When enzymatic reactions are carried out in aqueous solutions, it is common practice to incorporate buffering species that keep pH constant. The reason is that the pH of the solution influences the ionization of important functional groups in the enzyme and thereby has a direct impact on the rate of the enzymatic reaction. Adding or removing a proton can turn the enzyme on or off or at least change the activity drastically. In a similar way, the ionization of these groups is important when enzymes are used in organic solvents. The buffers that are used in aqueous media are insoluble in most organic solvents, so there is a need for other methods to control the "pH" in such reaction media. The questions concerning "pH" control in organic media have been investigated by Halling and coworkers, and useful guidelines have been presented [69]. When ions cannot move freely in the medium, counter ions will also have an influence on the protonation of ionizable groups. Two separate equilibria are of importance:

1. Exchange of hydrogen ions and cations, such as sodium ions, with acidic groups of the enzyme
2. Transfer of both hydrogen ions and anions, such as chloride, onto basic groups of the enzyme.

In many cases, an enzyme in an organic solvent keeps the ionization state from the aqueous solution in which it was present before removal of water and transfer to organic medium. This is sometimes called the "pH memory" of enzymes in organic media [41]. If the reactions catalyzed do not involve the formation or consumption of acidic or basic substances, it can be sufficient to use a suitable pH value in the aqueous solution used to prepare the enzyme. In most of the procedures used, the enzyme will also keep some counter ions and other ions around it, and some of those will act as pH buffers to some extent. In cases when more pH buffering capacity is needed, it is worthwhile to consider the use of buffers suitable for organic media. Some buffers are soluble in organic media, triphenylacetic acid and its sodium salt being a typical example of a pair controlling the exchange of hydrogen ions and cations [70, 71]. A corresponding pair controlling exchange of hydrogen ions and anions is tri-*iso*-octylamine and its hydrochloride [70]. These buffers are soluble mainly in relatively polar organic solvents, like pentanone and acetonitrile. Dendritic polybenzyl ether derivatives have been developed as alternatives suitable for more hydrophobic media [72].

A drawback of using organo soluble buffers for pH control is that in order to obtain the reaction product in a pure form after the enzymatic reaction, the buffer substances must be removed, which complicates the procedure. The use of solid-state buffers for organic media has thus been proposed, lysine and its hydrochloride being a typical example [73]. In addition, a wide range of "biological buffers" such as PIPES, MOPS, TES, HEPES, HEPPSO, TAPS, and AMPSO have been used in combination with their sodium or potassium salts [74]. Transfer of ions between the solid-state buffer and the enzyme can be slow in hydrophobic solvents, resulting in lag phases of up to 30 min [69].

1.6
Concluding Remarks

In many cases it is easy to detect some catalytic activity of an enzyme in an organic solvent. However, in order to get practically useful reaction rates there is normally a need to design the enzyme preparation and the reaction conditions with much care and thought. Hopefully, this chapter and its references can help in this task. A more extensive collection of practical advice on this topic can be found in [75].

References

1 P. J. Halling, *Enzym. Microbiol. Technol.* 1994, **16**, 178–206.
2 Y. L. Khmelnitsky, J. O. Rich, *Curr. Opin. Chem. Biol.* 1999, **3**, 47–53.
3 G. Carrea, S. Riva, *Angew. Chem., Int. Ed.* 2000, **39**, 2226–2254.
4 A. M. Klibanov, *Nature* 2001, **409**, 241–246.
5 P. Halling, *Enzym. Catal. Org. Synth.* (2nd Ed.) 2002, **1**, 259–285.
6 P. J. Halling, *Philos. Trans. R. Soc. London, Ser. B: Biol. Sci.* 2004, **359**, 1287–1297.
7 M. N. Gupta, I. Roy, *Eur. J. Biochem.* 2004, **271**, 2575–2583.
8 E. P. Hudson, R. K. Eppler, D. S. Clark, *Curr. Opin. Biotechnol.* 2005, **16**, 637–643.
9 A. Zaks, A. Klibanov, *J. Biol. Chem.* 1988, **263**, 8017–8021.
10 P. J. Halling, *Biochim. Biophys. Acta* 1990, **1040**, 225–228.
11 G. Bell, P. J. Halling, M. Lindsay, B. D. Moore, D. A. Robb, R. Ulijn, R. Valivety, in *Enzymes in nonaqueous solvents* (Eds.: E. N. Vulfson, P. J. Halling, H. L. Holland), Humana Press, Totowa, New Jersey, 2001, pp. 105–126.
12 E. Wehtje, I. Evensson, P. Adlercreutz, B. Mattiasson, *Biotechnol. Tech.* 1993, **7**, 873–878.
13 K. Won, S. B. Lee, *Biotechnol. Prog.* 2001, **17**, 258–264.
14 A. E. V. Petersson, P. Adlercreutz, B. Mattiasson, *Biotechnol. Bioeng.* 2007, **97**, 235–241.
15 E. Zacharis, I. C. Omar, J. Partridge, D. A. Robb, P. J. Halling, *Biotechnol. Bioeng.* 1997, **55**, 367–374.
16 M. Reslow, P. Adlercreutz, B. Mattiasson, *Eur. J. Biochem.* 1988, **172**, 573–578.
17 D. K. Oladepo, P. J. Halling, V. F. Larsen, *Biocatalysis* 1994, **8**, 283–287.

18 P. Adlercreutz, *Eur. J. Biochem.* 1991, **199**, 609–614.
19 J. L. Schmitke, C. R. Wescott, A. M. Klibanov, *J. Am. Chem. Soc.* 1996, **118**, 3360–3365.
20 M. Kreiner, M. C. Parker, B. D. Moore, *Chem. Commun.* 2001, 1096–1097.
21 K. Griebenow, A. M. Klibanov, *Acad. Sci. USA* 1995, **92**, 10969–10976.
22 K. Dabulis, A. M. Klibanov, *Biotechnol. Bioeng.* 1993, **41**, 566–571.
23 F. Secundo, G. Carrea, *J. Mol. Catal. B: Enzym.* 2002, **19–20**, 93–102.
24 J. Broos, A. J. W. G. Visser, J. F. J. Engbersen, W. Verboom, A. van Hoek, D. N. Reinhoudt, *J. Am. Chem. Soc.* 1995, **117**, 12657–12663.
25 D. S. Clark, *Phil. Trans. R. Soc. London, Ser. B: Biol. Sci.* 2004, **359**, 1299–1307.
26 E. Wehtje, P. Adlercreutz, *Biotechnol. Bioeng.* 1997, **55**, 798–806.
27 R. H. Valivety, P. J. Halling, A. R. Macrae, *FEBS Lett.* 1992, **301**, 258–260.
28 E. Wehtje, P. Adlercreutz, *Biotechnol. Lett.* 1997, **19**, 537–540.
29 D. Egger, E. Wehtje, P. Adlercreutz, *Biochim. Biophys. Acta* 1997, **1343**, 76–84.
30 E. Wehtje, D. Costes, P. Adlercreutz, *J. Mol. Catal. B: Enzym.* 1997, **3**, 221–230.
31 M. Persson, D. Costes, E. Wehtje, P. Adlercreutz, *Enzym. Microbiol. Technol.* 2002, **30**, 916–923.
32 Å. Jönsson, W. van Breukelen, E. Wehtje, P. Adlercreutz, B. Mattiasson, *J. Mol. Catal. B: Enzym.* 1998, **5**, 273–276.
33 Å. Jönsson, E. Wehtje, P. Adlercreutz, B. Mattiasson, *Biochim. Biophys. Acta* 1999, **1430**, 313–322.
34 L. Ma, M. Persson, P. Adlercreutz, *Enzym. Microbiol. Technol.* 2002, **31**, 1024–1029.
35 P. Vidinha, N. Harper, N. M. Micaelo, N. M. T. Lourenco, M. D. R. Gomes da Silva, J. M. S. Cabral, C. A. M. Afonso, C. M. Soares, S. Barreiros, *Biotechnol. Bioeng.* 2004, **85**, 442–449.
36 D. Pirozzi, G. Greco, *Enzym. Microbiol. Technol.* 2004, **34**, 94–100.
37 P. Clapes, G. Valencia, P. Adlercreutz, *Enzym. Microb. Technol.* 1992, **14**, 575–580.
38 T. Hansson, M. Andersson, E. Wehtje, P. Adlercreutz, *Enzym. Microbiol. Technol.* 2001, **29**, 527–534.
39 F. van Rantwijk, M. Woudenberg-van Oosterom, R. A. Sheldon, *J. Mol. Cat. B: Enzym.* 1999, **6**, 511–532.
40 T. J. Ahern, A. M. Klibanov, *Science* 1985, **228**, 1280–1284.
41 A. M. Klibanov, *Chemtech.* 1986, **16**, 354–359.
42 E. Wehtje, H. de Wit, P. Adlercreutz, *Biotechnol. Tech.* 1996, **10**, 947–952.
43 J. F. A. Fernandes, P. J. Halling, *Biotechnol. Prog.* 2002, **18**, 1455–1457.
44 K. Ryu, J. S. Dordick, *Biochemistry* 1992, **31**, 2588–2598.
45 C. R. Wescott, A. M. Klibanov, *J. Am. Chem. Soc.* 1993, **115**, 1629–1631.
46 A. Fredenslund, R. L. Jones, J. M. Prausnitz, *AIChE Journal* 1975, **21**, 1086–1099.
47 C. R. Wescott, A. M. Klibanov, *J. Am. Chem. Soc.* 1993, **115**, 10362–10363.
48 J. Gmehling, J. Li, M. Schiller, *Ind. Eng. Chem. Res.* 1993, **32**, 178–193.
49 J. B. A. van Tol, R. M. Stevens, W. J. Veldhuizen, J. A. Jongejan, J. A. Duine, *Biotechnol. Bioeng.* 1995, **47**, 71–81.
50 S. J. Singer, C. B. Anfinsen, M. L. Anson, Jr., K. Bailey, J. T. Edsall, Editors, *Advan. Protein Chem.* 1962, **17**, 1–68.
51 J. T. Chin, S. L. Wheeler, A. M. Klibanov, *Biotechnol. Bioeng.* 1994, **44**, 140–145.
52 F. van Rantwijk, F. Secundo, R. A. Sheldon, *Green Chem.* 2006, **8**, 282–286.
53 R. V. Rariy, A. M. Klibanov, *Proc. Natl. Acad. Sci. USA* 1997, **94**, 13520–13523.
54 A. J. Walker, N. C. Bruce, *Tetrahedron* 2004, **60**, 561–568.
55 R. Bar, *Trends Biotechnol.* 1986, **4**, 167.
56 A. S. Ghatorae, M. J. Guerra, G. Bell, P. J. Halling, *Biotechnol. Bioeng.* 1994, **44**, 1355–1361.
57 C. Laane, S. Boeren, K. Vos, C. Veeger, *Biotechnol. Bioeng.* 1987, **30**, 81–87.
58 M. Cantarella, L. Cantarella, F. Alfani, *Enzym. Microbial Technol.* 1991, **13**, 547–553.
59 M. Andersson, H. Holmberg, P. Adlercreutz, *Biocatal. Biotransform.* 1998, **16**, 259–273.

60 A. C. Ross, G. Bell, P. J. Halling, *J. Molec. Catal. B: Enzym.* 2000, **8**, 183–192.

61 R. Valivety, G. Johnston, C. Suckling, P. Halling, *Biotechnol. Bioeng.* 1991, **38**, 1137–1143.

62 G. Ljunger, P. Adlercreutz, B. Mattiasson, *Enzym. Microbiol. Technol.* 1994, **16**, 751–755.

63 D. Adlercreutz, H. Budde, E. Wehtje, *Biotechnol. Bioeng.* 2002, **78**, 403–411.

64 I. Svensson, E. Wehtje, P. Adlercreutz, B. Mattiasson, *Biotechnol. Bioeng.* 1994, **44**, 549–556.

65 E. Wehtje, D. Costes, P. Adlercreutz, *J. Am. Oil Chem. Soc.* 1999, **76**, 1489–1493.

66 D. Barahona, P. H. Pfromm, M. E. Rezac, *Biotechnol. Bioeng.* 2006, **93**, 318–324.

67 E. C. Voutsas, H. Stamatis, F. N. Kolisis, D. Tassios, *Biocatal. Biotransform.* 2002, **20**, 101–109.

68 M. Fermeglia, P. Braiuca, L. Gardossi, S. Pricl, P. J. Halling, *Biotechnol. Prog.* 2006, **22**, 1146–1152.

69 J. Partridge, N. Harper, B. D. Moore, P. J. Halling, *Methods Biotechnol.* 2001, **15**, 227–234.

70 A. D. Blackwood, L. J. Curran, B. D. Moore, P. J. Halling, *Biochim. Biophys. Acta* 1994, **1206**, 161–165.

71 K. Xu, A. M. Klibanov, *J. Am. Chem. Soc.* 1996, **118**, 9815–9819.

72 M. Dolman, P. J. Halling, B. D. Moore, *Biotechnol. Bioeng.* 1997, **55**, 278–282.

73 E. Zacharis, B. D. Moore, P. J. Halling, *J. Am. Chem. Soc.* 1997, **119**, 12396–12397.

74 J. Partridge, P. J. Halling, B. D. Moore, *J. Chem. Soc., Perkin Trans. 2* 2000, **2000**, 465–471.

75 E. N. Vulfson, P. J. Halling, H. L. Holland, Editors, *Enzymes in Nonaqueous Solvents: Methods and Protocols. [In: Methods Biotechnol., 2001; 15]*, 2001.

76 L. E. S. Brink, J. Tramper, *Biotechnol. Bioeng.* 1985, **27**, 1258–1269.

77 L. Greenspan, *J. Res. Natl. Bureau Standards A. Phys. Chem.* 1977, **81A**, 89–96.

78 J. A. Riddick, W. B. Bunger, T. K. Sakano, *Organic solvents. Physical properties and methods of purification*, John Wiley & Sons, Inc., 1986.

79 R. Carlson, *Design and optimization in organic synthesis*, Vol. 8, Elsevier Science Publisher B.V., Amsterdam, 1992.

2
Effects of Organic Solvents on Enzyme Selectivity

Jaap A. Jongejan

2.1
Introduction

As Nature's solution to the intricate catalytic demands of the living cell, enzymes have reached high levels of sophistication [1–3]. Although enzymes did not evolve in order to accommodate the catalytic needs of the organic chemist, their efficiency and in particular their (enantio)selectivity have made them challenging options for synthetic applications [4, 5]. While enzymes function to perfection in aqueous solutions, their (generally diminished) performance in alternative media was long considered to be a curiosity of biophysical interest only [6]. With early reports dating back to the beginning of the last century (see [7, 8] for a historical overview), this view has since been seriously challenged. The actual breakthrough occurred in the 1980s with landmark papers by Klibanov and coworkers [9–11].

It has now been firmly established that replacing the bulk of the water with an organic solvent (or neat substrate) is a serious option for both laboratory-scale and industrial biocatalytic applications [12–14]. Non-aqueous or low-water environments provide many potential advantages for biocatalytic applications, including higher substrate solubility, reversal of hydrolytic reactions, and modified enzyme properties (see [15, 16] for reviews). In addition to the catalytic rate and the stability of the biocatalyst, the selectivity (including chemo-, regio-, and enantioselectivity) can be affected to a greater or lesser extent. Early examples of the effects of organic solvents on enzyme enantioselectivity appeared in the 1990s [17–23], and many more have been reported since*.

* Literature searches for the occurrence of combinations of the key words "enantioselectivity", "resolution", "enzyme", and "organic media" score several hundreds of hits. Most of the relevant papers have been published in the last 15 years.

Organic Synthesis with Enzymes in Non-Aqueous Media. Edited by Giacomo Carrea and Sergio Riva
Copyright © 2008 WILEY-VCH Verlag GmbH & Co. KGaA, Weinheim
ISBN: 978-3-527-31846-9

2.2
Enzyme Enantioselectivity

Selectivity is an intrinsic property of enzymatic catalysis. [3] Following the nomenclature proposed by Cleland [24, 25], the pseudo second-order rate constant for the reaction of a substrate with an enzyme, k_{cat}/K_M, is known as the *specificity constant*, k_{sp}. [26] To express the relative rates of competing enzymatic reactions, involving any type of substrates, the ratio of the specificity constants appears to be the parameter of choice [3]. Since the authoritative proposition by Sih and coworkers [27], the ratio of specificity constants for the catalytic conversion of *enantiomeric substrates*, R and S, is commonly known as the *enantiomeric ratio* or *E*-value (Equation 1):

$$E = \frac{k_{sp}^R}{k_{sp}^S} = \frac{k_{cat}^R/K_M^R}{k_{cat}^S/K_M^S} = \frac{V_{max}^R/K_M^R}{V_{max}^S/K_M^S} \tag{1}$$

For enzymes that obey the rate equation $r = k_{sp} \cdot c_S \cdot c_E$, with c_S the substrate concentration and c_E the concentration of free, uncomplexed, enzyme, this leads directly to Equation 2 for a catalytic reaction involving both enantiomers:

$$\frac{r^R}{r^S} = E \cdot \frac{c_S^R}{c_S^S} \tag{2}$$

The notation E^R or E^{RS} to express the preferential conversion of R over S, has not gained widespread popularity. Any ambiguity arising from the use of the unsuperscripted notation is addressed in the accompanying text. With $0 < E < \infty$ and $E^{RS} = 1/E^{SR}$, it is common practice to express the enantiomeric ratio as $E \geq 1$, with separate indication of the preferred enantiomer.

The enantiomeric ratio is an intrinsic feature of enzyme-enantiomer couples. The actual realization of this property in a resolution reaction affects the *enantiomeric excess value*, ee_S (for the substrate) and ee_P (for the product) (Equation 3):

$$ee_{S,P} = \left| \frac{c_{S,P}^R - c_{S,P}^S}{c_{S,P}^R + c_{S,P}^S} \right| \tag{3}$$

The *ee*-values are also commonly expressed with textual reference to the predominant enantiomer. An alternative format uses $ee_{S,P}$ (%) = $ee_{S,P} \cdot 100\%$

Following the definition (Equation 1), *E*-values can be obtained by measurement of the k_{sp}-values for the enzymatic reaction of the enantiomers. Considering practical limitations, i.e. the availability of chirally pure enantiomers, *E*-values are more commonly determined by evaluating the *ee*-values as a function of the extent of conversion in batch reactions (see [28, 29] and [30] for overviews). It must be emphasized that the relationship between *E*, the intrinsic property, and *ee*, the realization of this property, shows a (complex) dependence on the concentrations (more precisely, the mass-action equivalents or thermodynamic activities) of the

2.3 Effects of Organic Solvents on the E-value

reacting species, on the kinetics of the catalyst (enzyme), and on the process mode (see [27] for some basic cases). In the case of the catalytic conversion of a prochiral substrate into the reactant enantiomers, this relation takes the relatively simple form (Equation 4):

$$E = (1 + ee_P)/(1 - ee_P). \tag{4}$$

When the catalyst (enzyme) acts on a racemic mixture, Equation 4 also applies in the limit of low conversion (with ξ the extent of conversion) (Equation 4a):

$$E = \lim_{\xi \to 0} \{(1 + ee_P)/(1 - ee_P)\}, \tag{4a}$$

However, in more general cases, the extent of conversion needs to be taken into account explicitly (Equation 5 and Equation 6):

$$E = \frac{\ln\{1 - \xi(1 + ee_P)\}}{\ln\{1 - \xi(1 - ee_P)\}} \tag{5}$$

$$E = \frac{\ln\{(1 - \xi)(1 - ee_S)\}}{\ln\{(1 - \xi)(1 + ee_S)\}} \tag{6}$$

Although Equation 4, Equation 5, Equation 6, and Equation 7 may be used to establish the E-value of a catalyst in many cases of interest, they certainly do not cover the full range of possible situations. This holds in particular for the commonly employed hydrolases that show bi bi ping pong kinetics. For enzymes that follow this kinetic scheme, the ratio of reaction rates in the case of a so-called A-P resolution, $A^{R,S} + B \Leftrightarrow P^{R,S} + Q$ (see [31]) takes the following form (Equation 7):

$$\frac{r_A^R}{r_A^S} = E \frac{(K_{eq}c_B + \alpha_{SQ}c_P^S)c_A^R - (c_Q + \alpha_{SQ}c_A^S)c_P^R}{(K_{eq}c_B + \alpha_{RQ}c_P^R)c_A^S - (c_Q + \alpha_{RQ}c_A^R)c_P^S} \tag{7}$$

with $c_A^R, c_A^S, c_P^R, c_P^S$ the concentrations of the chiral reactant, A, and its chiral product, P, respectively, c_B, c_Q the concentrations of the (non-chiral) second substrate and second product, respectively, and α_{RQ}, α_{SQ} the ratio of the specificity constants for the forward ($A^R \to P^R$, viz. $A^S \to P^S$) and the backward reaction ($Q \to B$). Accordingly, the validity of the model choice must be carefully checked in each case. For a comprehensive discussion see [30].

2.3
Effects of Organic Solvents on the E-value

From the point of view of organic synthetic applications, stereoselectivity is the most valuable property of enzymes [32]. However, whereas enzymes commonly show high (prochiral- and enantio-) selectivity when processing their natural substrates, they may show insufficient selectivity in non-natural but practically important reactions. With $ee \geq 0.995$ as a generally required quality criterion for enantiopurity, the catalyst applied in batch mode should have $E \geq 400$ for prochiral substrates and an even larger value to produce a reasonable yield of a single enantiomer product from a racemic substrate mixture. With few exceptions, such high enantioselectivities have not been observed for enzymes catalyzing the conversion of non-natural substrates. However, when the process is designed to enrich a racemic substrate mixture by converting the unwanted enantiomer into a separable product, lower E-values suffice. Typically, $E \geq 20$ will be acceptable in a batchwise resolution of a racemic substrate. For enzymes that do catalyze the reaction of a non-natural low-molecular-weight substrate but have not been optimized with respect to chiral recognition, one tends to find $1 \leq E < 5$. The challenge will then be to increase the enantioselectivity by at least one order of magnitude to bring it within the required range.

A number of options have been exploited for optimization strategies. Including, in no particular order, "engineering" of (1) the enzyme [33, 34], (2) the substrate [35–39], (3) the ionic composition or ionic strength [40, 41], (4) the pH [42–44], (5) the temperature [45–49], (6) the pressure [50, 51], (7) the extent of conversion [27], (8) the method (dynamic resolution, combi processes including non-enzymatic racemization) [52], (9) the enzyme preparation [53–55], i.e. CLECs [56] CLEAs [57] or immobilization, and (10) perhaps even gravity [58]. Finally, of all such strategies, "engineering" the medium (including additives, organic solvents, ionic liquids, and neat substrates), certainly is among the options that are easily realized in practice.

Early reports on the effects of the choice of solvent on enzymatic enantioselectivity showed that substantial changes may be observed. For the transesterification reaction of sec-phenethyl alcohol with vinyl butyrate catalyzed by subtilisin Carlsberg, a 20-fold increase in the E-value was reported when the medium was changed from acetonitrile to dioxane [59]. Similar changes were recorded for the prochiral selectivity of *Pseudomonas* sp. lipase in the hydrolysis of 2-substituted 1,3-propanediol diesters [60]. Not only were (one) order-of-magnitude changes observed: in some cases even the preference of the enzyme could be inverted [61–63]. Following the proposed explanations for these effects, notably the differential desolvation of the diastereomeric transition states, it appeared to be only a matter of time to establish a set of rules for "rational medium engineering".

Since then, a number of propositions to modify (if possible, to enhance) the enantioselectivity of enzymes (mostly hydrolases) by "medium engineering" have been put forward [20, 51, 60, 64–69], providing a challenging mix of scientifically and practically interesting questions.

Many attempts have been made to correlate the enantioselective properties of enzymes with structure and/or process conditions [61, 70–73]. Attempts to correlate the effect of a particular medium on the enantioselectivity of an enzyme-catalyzed reaction with physico-chemical descriptors of the solvent have also been reported by a number of groups [22, 59, 64, 74]. Correlations with solvent size [75], dielectric constant [59], polarizability, electron pair acceptance index [76], log$P$ [17], flexibility (by EPR of spin-labeled enzyme [77]), have been investigated. In most cases, however, these attempts have been largely unsuccessful.

Despite this failure to consistently correlate changes in enantioselectivity with individual solvent characteristics, a multi-parameter correlation method has been proposed to provide a valid means of predicting the effects of solvents on enantioselectivity in non-aqueous media ([78]). Using the Linear Solvation Energy Relationship (LSER), which incorporates dispersion, polarity, acidity, basicity, and Hildebrand solubility parameters (R, π^*, α, β, and δ_H^2), calculated enantioselectivity values showed a reasonably strong positive correlation with literature values ($r >$ 0.9). Although, as the saying goes, the number of parameters used would be enough to "fit an elephant", the authors claim that the LSER method may be used generally to predict the effect of solvent on enzyme enantioselectivity. However, even though it is widely applicable, this approach hardly provides detailed information on the mechanisms of solvent interaction.

In view of the predictive properties of the octanol-water partition coefficient, $\log P$, in the description of enzymatic activity [79], this parameter has received much attention. However, as argued in an early but still authoritative review compiled by Carrea and coworkers [80], its usefulness in the correlation of solvent effects on enantioselectivity appears to be limited. The problem is exemplified by the entries in Figure 2.1. Various enzyme-catalyzed enantioselective reactions in a series of organic media and ionic liquids have been collected and plotted with respect to the $\log P$ value of the medium. Clearly, correlations, if any(!), are far from obvious. Although it might appear that these data merely disqualify $\log P$ as a useful solvent descriptor for enantioselectivity in organic media, the problem is more general. Despite occasional successes in individual cases, it would seem that the use of any of the

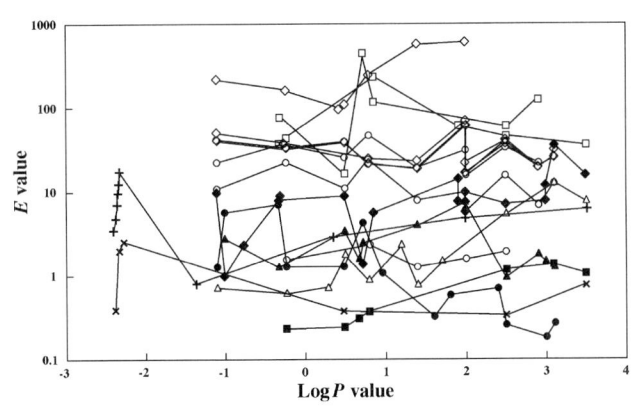

Figure 2.1 Enantioselectivity of enzyme-catalyzed reactions in organic media plotted as a function of the $\log P$-value of the solvent. Data obtained from the literature 1. (open circles) [81], 2. (open diamonds) [17], 3. (closed triangles) [19], 4. (closed circles) [20], 5. (open squares) [82], 6. (closed squares) [23], 7. (open triangles) [63], 8. (closed diamonds) [83], 9. (plusses) [84], 10. (crosses) [85]. Notice the reversal of enantiopreference for the data sets that cross the line $E = 1$. Details of solvents, enzymes, substrates, and conditions are reported in the original papers.

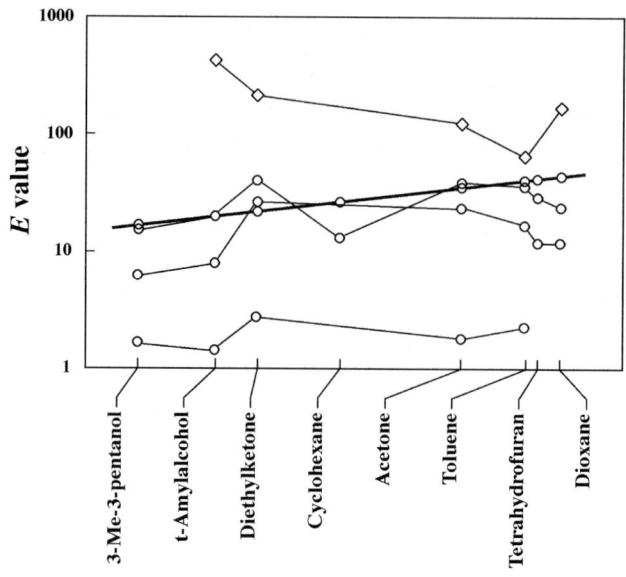

Figure 2.2 Enantioselectivity of enzyme-catalyzed reactions in a set of organic media. The order and relative separation of the solvents plotted on the abscissa was obtained by assuming a strict correlation between the particular solvent and the E-value for the porcine pancreatic lipase-catalyzed transesterification of sulcatol (6-methyl-5-hepten-2-ol) with butyric acid trifluoroethyl ester in this solvent [81]. Data from the literature 1. (open circles) [81], 2. (open diamonds) [17].

established QSAR-related solvent descriptors, either singly or in combination, does not provide a portable format to correlate the observed enantioselectivities.

The situation is summarized from a slightly different perspective in Figure 2.2. Here, data obtained using the same set of organic solvents have been plotted. The porcine pancreatic lipase-catalyzed transesterification of sulcatol (6-methyl-5-hepten-2-ol) with butyric acid trifluoroethyl ester [81] has been chosen (arbitrarily) to calibrate the solvents. It is clear that, irrespective of the solvent descriptors that one may aim to investigate, different systems are seen to respond differently to the same (change of) medium.

2.4
Possible Causes of the Complexity of Solvent Effects on E

A change of solvent may have an effect on the substrate, the enzyme, or the enzyme-substrate couple. In isotropic solvents the thermodynamic activity coefficients of the enantiomers will be equal. To a first approximation, one may assume that the activity coefficients of all the enzyme species relevant for enantioselection will be equal as well. This assumption is certainly less audacious than it may seem. For many enzymes, the substrate, once bound, is almost completely protected from the surrounding medium, making the solute-solvent interactions virtually equal for the free enzyme, for the Michaelis complexes of that enzyme, and for the transition state species. In addition, and probably more importantly, in many organic media the enzyme is *not* physically dissolved. Either as a lyophilized powder, or as an immobilized preparation, including CLECs and CLEAs, it may be considered to form a separate solid phase, with activity coefficients equal to 1 (by definition). On the basis of this assumption, the minimal Michaelis-Menten equation takes the following form (Equation 8):

$$r = \frac{k_{sp} \cdot \gamma_S c_S}{1 + \gamma_S c_S / K_M^{int}} \cdot \gamma_E c_{E_0} \qquad (8)$$

with the activity coefficients, γ_S for the substrate, γ_E for the enzyme, and K_M^{int} the intrinsic Michaelis constant [86]. Compared with the value for K_M that is determined as the substrate *concentration* at half-maximal velocity, one finds (Equation 9):

$$K_M = K_M^{int} / \gamma_S \qquad (9)$$

Since the substrates for which organic solvents are attractive media normally have lower activity coefficients in organic solvents than in water, the K_M-values determined from plots of rate vs. substrate concentration will generally be higher in organic media. The validity of this approach has been ascertained for several systems [81, 86].

When this reasoning is applied to enantioselective enzymatic reactions, it follows that the ratio of specificity constants should not be affected by a change of medium that leads to different (but of course identical for the two enantiomers) values for the substrate activity coefficients. Indeed, solvent effects were *not* observed for, e.g., the chymotrypsin-catalyzed hydrolysis of 3-phenyl lactate ethyl ester [87], pig pancreatic lipase-catalyzed transesterification of glycidol [88], and the CaLB-catalyzed hydrolysis of 3-phenyl-substituted glutaric acid anhydrides [89], to mention a few examples. The implication that "solvent-induced changes of enzyme enantioselectivity are an exception rather than a rule" [87] has been challenged by Klibanov and coworkers [90] on the obvious grounds that solvent-induced changes are the experimental reality. However, this does not necessarily discredit the model assumption of equal activity coefficients for all species of which the enzyme is a part. Instead, it lends credibility to the view that the focus should perhaps be on the overall (structural and electronic) impact of organic media on the enzyme rather than on the relatively straightforward events of substrate and active-site solvation-desolvation effects.

2.5
The Accuracy of *E*-value Determinations

The investigation by Wolff and coworkers into the solvent effects of the enantioselectivity of chymotrypsin-catalyzed hydrolyses [87] raises another question. The research aimed at corroborating the results reported by Jones and Mehes [91] on the more than 100-fold reduction of the enantioselectivity of chymotrypsin upon addition of 5–30% methanol to the aqueous medium. Despite the fact that, for both practical and semantic reasons, marginally different substrates were considered by Wolff et al. [87], their results cast serious doubts on the accuracy of the former data. In fact, there are numerous reasons why E-value determinations reported in the literature may be severely biased [29, 30]. Considering the analytical protocol that is based on progression curve analysis of the ee-value as a function of the extent of conversion [27], one may distinguish at least four levels of importance: (1) acquiring accurate values for the ee-value and for the extent of conversion, (2) correct choice of the relevant kinetic model, (3) proper definition of the system, including chemical and chiral purity of the substrate and various physical parameters, i.e. water activity, pH, temperature, pressure, and (4) enzyme-related issues, including purity of the enzyme, physical state, and history. The first of these issues reflects the analytical problems of obtaining accurate data for extremes of the ee-value. It is now commonly appreciated that the limitations of the quantitative measurement of the lower one of the two enantiomer concentrations sets an upper limit to the accurate determination of E-values of the order of several hundreds. For the present argument, however, such high values are of lesser concern. What remains is the statistical noise that will not, in general, exceed the differences observed, as depicted in Figure 2.1. The second issue, on the other hand, may be of greater importance. Although it can be argued that partial neglect of the kinetic intricacies (compare, e.g., Equations 6 and 7 with the more elaborate model, Equation 8) would constitute a systematic error that might leave the observed differences intact, it cannot be discounted [92]. The main reason for concern is related to the fact that enzymes in organic media normally show dramatically lower catalytic rates [93]. On average, data points for the media that have a relatively low effect on the rate are obtained at higher extents of conversion, whereas data for slow-reacting systems rely on low extents of conversion with a concomitantly high inaccuracy. In addition, a shift of the position of the equilibrium may lead to an erroneous interpretation of the data [94]. The third issue may again be of limited concern provided that proper care has been taken to chemically define the substrate composition [95]. Except maybe for the water activity, these parameters can be managed relatively straightforwardly, while the effect of changes of the water activity on the enantioselectivity may be limited [96, 97]. However, see also [98, 99].

Finally, the purity-related performance of the enzyme preparation would appear to be a constant throughout a series of measurements. This, however, may not be the case when several enzyme species are present and different activities are expressed as a function of the medium composition [100]. Although it would appear that this source of error is abolished by the present-day availability of highly pure enzyme preparations (but see also [101, 102]), the intrinsic properties of i.e. lipases may lead to different E-values as a result of interfacial activation [103] and the conformation of the lid structure of lipases [56].

In summary, it may well be that a substantial number of cases where effects of solvent on the enantioselectivity have been reported are erroneous rather than erratic. Still, many instances of unbiased results do confirm the effect. The question then remains as to why the effects of organic solvents on enantioselectivity are apparently unpredictable using the tools that have been successful in other areas where solvent and medium effects occur?

2.6 Kinetic and Thermodynamic Analysis of the Specificity Constants

It can be readily shown that the specificity constant $k_{sp} = k_{cat}/K_M$ can be taken to act as a (pseudo) second-order rate constant in the rate equation for an enzymatic reaction that follows minimal Michaelis-Menten kinetics:

$$S + E \underset{k_{-1}}{\overset{k_1}{\Leftrightarrow}} "ES" \overset{k_2}{\to} P + E \tag{10}$$

$$r = \frac{k_{cat} \cdot c_S}{K_M + c_S} \cdot c_{E_0} \tag{11}$$

$$r = k_{sp} \cdot c_S \cdot c_E \tag{12}$$

It should be emphasized that Equation 11 and Equation 12 represent identical rates as a result of the purely neutral substitutions $k_{sp} = k_{cat}/K_M$ and $c_E = c_{E_0}/(1 + c_S/K_M)$.

Based on the format of Equation 12, it is rather tempting to treat the specificity constant as a true second-order rate constant for which one may assume an Eyring-type relationship (with the corresponding Gibbs free energy difference) to hold under the conditions assumed for the derivation [104, 105] (Equation 13):

$$k_{sp} = \kappa \frac{k_B T}{h} \exp\left(-\frac{\Delta G^{\#}_{sp}}{RT}\right) \tag{13}$$

with $\Delta G^{\#}_{sp}$ representing the Gibbs free energy difference between the reacting species, substrate and free enzyme, and a (virtual) transition state free energy for the reaction, and all other symbols with their regular meaning. Since we also have Equation 14

$$k_{sp} = \frac{k_1 k_2}{k_{-1} + k_2} = \frac{\exp(-\beta \Delta G^{\#}_I) \exp\{-\beta(\Delta G^{\#}_{II} - \Delta G_M)\}}{\exp\{-\beta(\Delta G^{\#}_I - \Delta G_D)\} + \exp\{-\beta(\Delta G^{\#}_{II} - \Delta G_D)\}} \tag{14}$$

where ΔG_M denotes the stability of the Michaelis complex, $\beta = 1/RT$, and the pre-exponentials have been assumed to be identical, the relationship between $\Delta G^{\#}_{sp}$, $\Delta G^{\#}_I$ and $\Delta G^{\#}_{II}$ takes the following form (Equation 15):

$$\Delta G^{\#}_{sp} = \ln\{\exp(\Delta G^{\#}_I) + \exp(\Delta G^{\#}_{II})\} \tag{15}$$

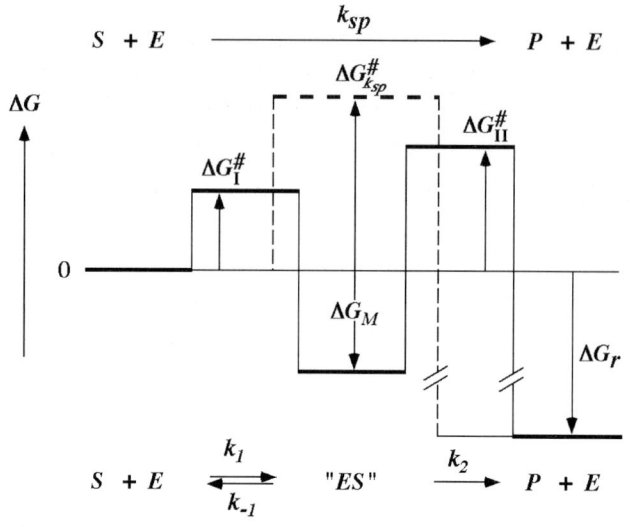

Figure 2.3 Comparison of the Michaelis-Menten model for a minimal kinetic scheme (bottom equation) with the pseudo second-order format (top equation). Relationship between the kinetic barriers for the formation of the Michaelis complex and the chemical transformation S → P, and the Gibbs free energy of the (virtual) barrier for the pseudo second-order reaction S + E → P + E.

A kinetic profile that illustrates this relationship is shown in Figure 2.3.

Before examining the results for a more realistic kinetic scheme, it is helpful to consider the magnitudes of the Gibbs free energies involved (at 300 K). If one assumes k_1 to represent the diffusional approach, one has $\Delta G_I^\# \cong 30\,kJ \cdot mol^{-1}$. For a reasonably fast enzyme we may take k_2 to be of order $1000\,s^{-1}$, leading to $(\Delta G_{II}^\# - \Delta G_M) \cong 55\,kJ \cdot mol^{-1}$. With the equilibrium constant for the formation of the Michaelis complex arbitrarily set at $K_{aff} = 10^4\,M^{-1}$, one has $\Delta G_M = -25\,kJ \cdot mol^{-1}$, leaving $\Delta G_{II}^\# \cong 30\,kJ \cdot mol^{-1}$. In this example, both barriers are seen to contribute equally to the specificity rate constant! Notice that it is not the stability of the Michaelis complex that plays a role, but its rate of formation.

By what appears to be a convenient coincidence, it turns out that the barriers that contribute to k_{sp} for a more realistic kinetic scheme, notably the bi bi ping pong scheme adopted by the majority of hydrolases that are currently employed in biocatalytic resolutions reactions, are equally simple to identify. Figure 2.4 shows the barriers that contribute. By straightforward manipulation of the kinetic equations one obtains Equation 16:

$$\Delta G_{sp}^\# = \ln\{\exp(\Delta G_I^\#) + \exp(\Delta G_{II}^\#) + \exp(\Delta G_{III}^\#)\} \tag{16}$$

for the formation of the acyl-enzyme complex (and a similar expression for the deacylation stage) (Figure 2.4).

2.6 Kinetic and Thermodynamic Analysis of the Specificity Constants | 35

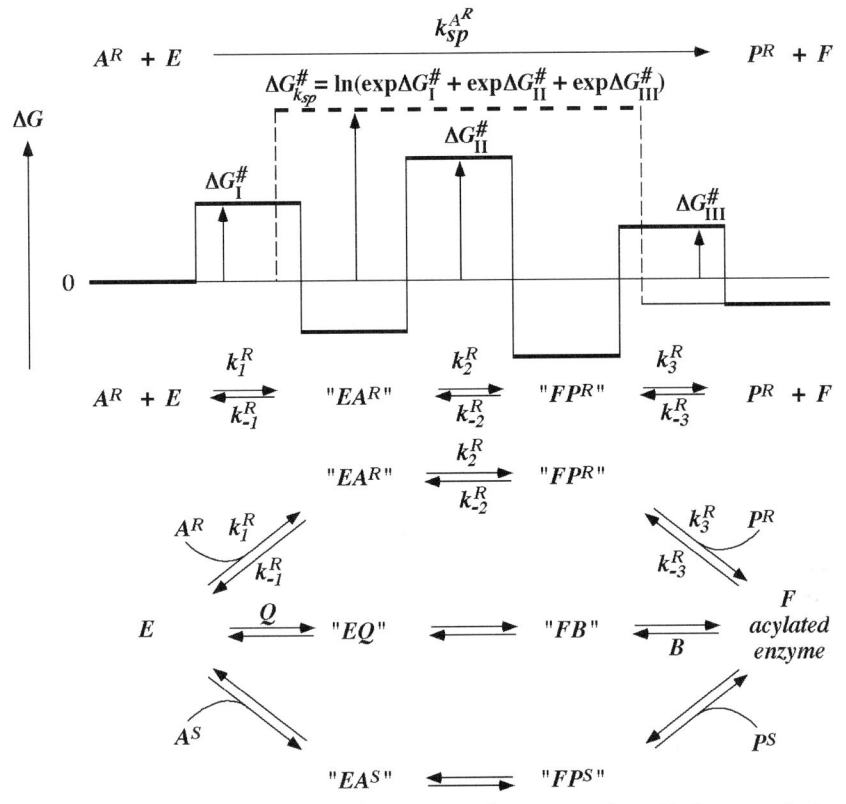

Figure 2.4 Schematic profile for the acylation stage involving the R-enantiomer of an enantioselective reaction catalyzed by an enzyme that follows bi bi ping pong kinetics. The reaction is an example of the so-called A-P resolution type, where chirality occurs in the first substrate to enter and the product to leave the cycle. Similar profiles can be drawn for the reaction of the S-enantiomer and the deacylation reaction, $F + B \rightarrow Q + E$.

Since the enantiomeric ratio is defined as the ratio of the specificity constants for the two enantiomers, one has (Equation 17):

$$\Delta G^{\#}_{sp,R} - \Delta G^{\#}_{sp,S} = \Delta_{R-S}\Delta G^{\#}_{sp} = -RT \ln E \tag{17}$$

At this point, it is instructive to notice that the numerical example given above for the minimal Michaelis-Menten scheme will probably be very relevant to the situation for the majority of the enzymes considered so far. In consequence, it appears that the effects of organic solvents on the enantioselectivity are not restricted to their relative effects on the ground state system and the transition state. Instead, substantial contributions from the diffusional process parameters have to be taken into account as well. Since these contributions are probably better described by the Einstein-Smoluchovski relation,

$$k_{diff} = \frac{8RT}{6\pi\eta rN} \tag{18}$$

replacing them by the corresponding free energy barrier will only be numerically consistent. Whereas solvent effects on stabilities may be addressed by using linear free energy relationships, this will probably not hold for the diffusional barrier. In addition, k_1 may contain contributions from substrate-induced changes of the enzyme, especially in the case of lipases where a flexible lid structure is involved [56].

Considering the thermodynamic differences between the various levels relevant to enantioselection (Figure 2.3 and Figure 2.4), several scenarios can be explored. On the assumption that the diffusional barriers will be largely unselective (= of equal magnitude for both enantiomers), raising these barriers would be expected to lower the E-value. This may be caused, e.g., by mass transport limitations, particularly in lyophilized powders at low substrate concentrations [106]. As the physical interpretation of the low catalytic rates of enzymes in organic media is still rather unclear, other factors may also contribute to the increase of these barriers. If, on the other hand, the lower rates are the result of an increase of $\Delta G_{II}^{\#}$, higher values of E are to be expected. Clearly, if the activation energy of the chemical step (the formation of the transition state from the Michaelis complex, i.e. the sum of $\Delta G_{II}^{\#}$ and $-\Delta G_M$), is unaffected, the generally lower stability of the Michaelis complex in organic media will tend to rise $\Delta G_{II}^{\#}$, leading again to higher E-values. It must be anticipated that all of these factors will be operative to various extents when a series of organic solvents is explored.

In summary, the failure of solvation-desolvation theory to explain the observed changes of enzyme enantioselectivity may well be caused by the fact that several barriers of physically different character contribute to the E-value. For additional considerations the reader is referred to [107] and [30].

2.7
Solvents Effects on Non-Hydrolytic Enzymes

Most of the data reviewed in the preceding paragraphs have been obtained for enzymes with hydrolytic activities. In aqueous media the equilibrium position of the reactions catalyzed by these enzymes normally favors the hydrolyzed products. Replacing the bulk of the water by an organic solvent makes it possible to steer the reaction in the opposite direction. For other types of enzymes, notably redox enzymes, replacing the water or adding cosolvents will hardly affect the equilibrium position of the reaction. Still, the increase in solubility of substrates and products may be advantageous. So far, only a limited number of non-hydrolytic enzymes have been studied under such conditions [108, 109]. An interesting example has been reported by Cowan and coworkers [110]. Using a thermostable secADH [111] from a thermophilic anaerobic bacterium, they demonstrated that low concentrations of water-miscible organic solvents have a significant effect on enzyme enantioselectivity. In the oxidation of chiral butan-2-ols, increasing concentrations (10 to 40% v/v) of acetonitrile were observed to induce a 2.5-fold increase in selectivity for the R-enantiomer [112]. A similar effect was found in the reverse reaction, the reduction of butan-2-one.

De Gonzalo and coworkers reported the use of organic cosolvents to oxidize organic sulfides using Baeyer–Villiger monooxygenases, BVMOs, from different sources: phenylacetone monooxygenase (PAMO) from *Thermobifida fusca*, 4-hydroxyacetophenone monooxygenase (HAPMO) from *Pseudomonas fluorescens* ACB and ethionamide monooxygenase (EtaA) from *Mycobacterium tuberculosi* [113].

They showed that despite the fact that lower activities were generally observed, significant improvements of enantioselectivity in the oxidation of thioanisole by PAMO and EtaA could be induced by the addition of short-chain alcohols such as methanol and ethanol. Remarkably, methanol was able to cause a reversal of PAMO enantiopreference in the case of several substrates. Reversal of enantiopreference was also observed with EtaA when using *t*-BuOMe. The authors hypothesize that in these enzymes solvents exert their influence on enantioselectivity by binding in or near the enzyme active site and, depending on their structure, interfere with the association of the substrate.

A similar implication can be deduced from the report that the lipase-catalyzed transesterification of sulcatol with vinyl acetate in the presence of organic solvents was inhibited more by solvents with cyclic and branched-chain structures than by those with linear structures, again suggesting specific binding of solvent molecules in the active site [114]. In this respect, both the molecular shape and the overall (or regional) hydrophobicity of the solvent molecule might be expected to influence the affinity of the solvent for a hydrophobic pocket in the enzyme active site or elsewhere [115]. These suggestions are in agreement with the observation that solvents can bind specifically in enzyme active sites, as has been demonstrated for subtilisin and chymotrypsin [116].

A third example has been described by Yun and coworkers [117]. They synthesized cis-(1S,2R)-1-amino-2-indanol, an important chiral auxiliary and a chiral resolving agent for, e.g., HIV protease inhibitors, using the ω-transaminase, ω-TA, from *Vibrio fluvialis* JS17 expressed in *E. coli*, by enantioselective transamination of the substrate (R)-2-hydroxy indanone. In this case, a substantial rise in stereoselection occurred when relatively large amounts (up to 5% by weight) of γ-cyclodextrin were added, and the reaction temperature was brought down to 4 °C. Although host-guest molecules like crown ethers and cyclodextrins are not organic solvents in a strict sense, their effect on enantioselectivity may parallel some of the effects of organic (co)solvents [68, 118].

Based on a suggestion by Odell and Earlam [119] that crown ethers and cryptands can cause proteins to dissolve in methanol, Broos and coworkers [120] investigated the effects of crown ethers on the enzymatic activity of α-chymotrypsin in the transesterification reaction of N-acetyl-L-phenylalanine ethyl ester with n-propanol in organic solvents. They observed a 30-fold rate acceleration when 18-crown-6 was used in octane. At that time, it was proposed that the water- and cation-complexing ability of crown ethers was responsible for the rate enhancement. Meanwhile, effects of crown ethers and cyclodextrins on the enantioselectivity of enzymes have been observed in several cases. Mine and coworkers used *Pseudomonas cepacia* lipase lyophilized in the presence of (partially methylated) cyclodextrins in transesterification reactions of sulcatol and solketal with enol esters in ether solutions. Substantial effects on the E-value were noted [121].

A highly interesting investigation by Griebanow and coworkers [122, 123] describes the use of methyl-β-cyclodextrin (MβCD) as a macrocyclic additive to simultaneously enhance the activity and enantioselectivity of dehydrated subtilisin Carlsberg suspended in neat organic solvents. MβCD not only raised the activity for the transesterification between *sec*-phenethyl alcohol and vinyl butyrate (>150-fold) but also significantly improved the enantioselectivity of subtilisin in co-lyophilizates when compared to the powder lyophilized from buffer alone. In addition, marked solvent effects were noted. To investigate the possible relationship between enzyme structure and these kinetic data, the secondary structure of subtilisin was investigated by Fourier transform infrared (FTIR) spectroscopy under all relevant conditions.

It was found that MβCD is able to restore dehydration-induced structural perturbations to a certain extent. Results were interpreted to show that enantioselectivity and structural intactness in the various solvents investigated were clearly related. Increased enzyme activity, in contrast, is mainly caused by increased structural flexibility of subtilisin in the solvents by MβCD. Similar results have been reported for subtilisin Carlsberg [124].

Considering these results and the contributions by Broos [125], Watanabe [126], and Ueji [55, 77], it may be concluded that the relation between enzyme flexibility and enantioselective performance in organic solvents is now firmly established. Molecular dynamic simulations on the flexibility of subtilisin and the mobility of bound water molecules in carbon tetrachloride corroborate the idea that organic solvents reduce molecular flexibility via interactions at specific binding sites [127]. Whether predictive tools can be developed on the basis of this knowledge remains to be seen.

2.8
Major Achievements

From the results reviewed above, one might get the impression that the choice of an organic solvent that optimizes the enantioselectivity of the enzyme in a given resolution reaction is a matter of tedious trial and error, with little guidance from established rules or insights. In practice, however, one has to consider several mitigating circumstances. In many cases of interest, the choice will be limited to a relatively small number of solvents that are either industrially approved or readily available in the laboratory. Since most practical resolutions start from a racemic mixture obtained by chemical synthesis, batch-mode enrichment requiring relatively modest E-values will be an attractive method. In that case, solubility and easy separation of substrates and products will probably be more important factors in the choice of medium. For industrial applications, the choice of solvent will be a trade-off between the enantioselectivity and the catalytic rate.

To illustrate the potential of the use of organic solvents to increase (among other process objectives) the enantioselective performance of the enzyme, three cases of interest will be discussed in some detail.

An early success story, which has been pivotal for the advancement of enantioselective biocatalysis among organic synthetic chemists, touches on the work of K. Barry Sharpless [128] on the catalytic asymmetric epoxidation of alkenes using a titanium tartrate catalyst. Starting with (prochiral) alkenes, a series of interesting (chiral) epoxides could be readily obtained with remarkably high *ee*-values (>98% in most cases) [129]. The epoxidation of allyl alcohol, however, formed an annoying exception: despite intensive research, the *ee*-values of the resulting alcohol, glycidol, could not be raised above 90% [130]. As glycidol is an attractive building block for a number of pharmaceutical compounds (e.g., propranolol) this *ee*-value is not acceptable. A few years earlier, a short note from the laboratory of G.M. Whitesides had gained widespread attention [131]. Using materials and methods that appeared to be taken straight from a student's practical: racemic glycidyl butyrate and porcine pancreatic lipase (pancreatin, PPL), enantiomerically enriched (*ee* > 98%) glycidyl butyrate could be obtained in good yield by enzyme-catalyzed enantioselective hydrolysis (Scheme 2.1).

Stimulated by industrial and academic interest, the properties of PPL in the enantioselective hydrolysis of glycidyl esters were investigated more closely [88, 132–134]. It was found that the system harbored many hidden intricacies, most disturbingly reflected by the occurrence of a plateau-like *ee*-value at 98% when industrially relevant conditions were used. Whereas the reason for this behavior could be traced to the subtleties of the kinetics of the lipase [103], other factors were evident, notably the presence of different hydrolytic activities in the commercial PPL preparation [103]. The observed *E*-values for the main activity were rather low, $E = 8$ on dissolved substrate and $E = 17$ at the aqueous-ester interface, whereas a minor fraction (33 kDa) showed $E = 17$ throughout. Curiously, the use of crude PPL as source for proteins with carboxyl esterase activity, other than the triacylglycerol hydrolase (EC 3.1.1.3), had already been recognized by Ghisalba and coworkers [135]. They used partially purified fractions with high esterase and

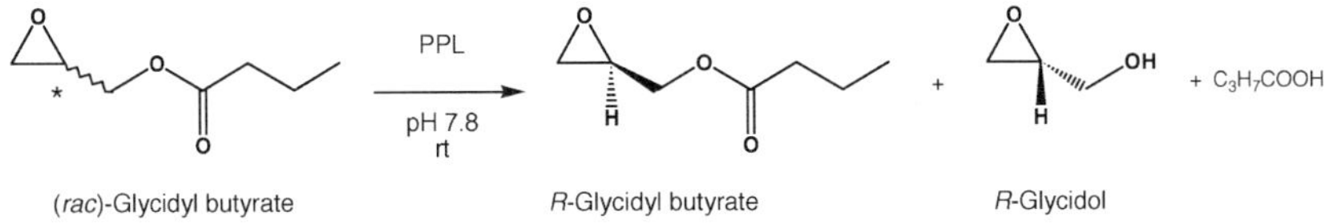

Scheme 2.1 Kinetic resolution of glycidyl butyrate catalyzed by porcine pancreatic lipase, PPL. The preferential conversion of (S)-glycidyl butyrate into (R)-glycidol and butyric acid results from the stereochemical nomenclature rules under the Cahn, Ingold, Prelog convention, the configuration around the chiral centre is not affected.

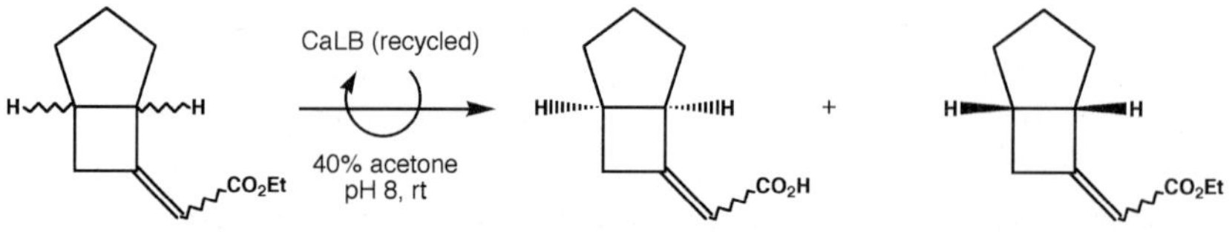

racemic and *cis/trans*, **1** | *cis/trans*-(1*R*,5*R*), **2** >98.5% ee, 40-45% yield | *cis/trans*-(1S,5S), **3**

Scheme 2.2 Enantioselective kinetic resolution of *cis/trans*-(1*R*,5*R*)-bicyclo[3.2.0]hept-6-ylidene-acetate ethyl ester, 1, catalyzed by immobilized lipase B from *Candida antarctica* (Novozyme 435). The reaction is run in a pH-stat at pH 8.0 and monitored by reversed-phase HPLC checking the extent of conversion and the enantiomer purity of the product *cis/trans*-(1*R*,5*R*)-bicyclo[3.2.0]hept-6-ylidene-acetic acid, **2** (adapted from [136]).

acylase activity to obtain 2-substituted 1,3-propanediol monoacetates of good chiral purity.

Recently, a report appeared of what is probably a conclusive optimization of the system [102]. After isolating a (25 kDa) "lipase-like enzyme" from crude PPL, applying intricate immobilization methods, and carefully checking experimental conditions, the authors report *E*-values in the kinetic resolution of glycidyl butyrate that are still no higher than 10. However, illustrating the importance of organic solvents on enzyme enantioselectivity, when dioxane (10% v/v) is added as a cosolvent, $E > 100$ is found!

Despite the fact that solvent effects on enzyme enantioselectivity appear to resist our efforts to rationalize their outcome using commonly accepted solvent descriptors, the effects are certainly there. An impressive example is provided in a report on the successful resolution of *cis/trans*-(1*R*,5*R*)-bicyclo[3.2.0]hept-6-ylidene-acetate ethyl esters, intermediates in the synthesis of GABA (γ-aminobutyric acid) analogs, by the Pfizer Biotransformations and Global R&D groups (Scheme 2.2) [136]. From a screening protocol, CaLB was identified as a reactive catalyst for the hydrolysis of the racemic mixture of *E*/*Z*-ester enantiomers with approximately equal activity for the *cis*- and *trans*-isomers and a rather modest ($E = 2.7$) preference for the *E*/*Z*-(1*R*,5*R*)-enantiomers. Application of "medium engineering" resulted in a phenomenal increase in the enantioselectivity (addition of 40% acetone, $E > 200$), while the *cis*- and *trans*-isomers were still converted at an almost equal rate.

The dramatic solvent effect observed in this case prompted the authors to investigate possible conformational changes of the protein backbone as the cosolvent concentration was raised. They prepared ^{15}N-labeled CaLB and used NMR spectroscopy to monitor the changes. As a many-parameter NMR probe for studying intermolecular interactions of the protein they used 2D TROSY spectroscopy to obtain a fingerprint of the protein amide region that is highly sensitive to changes in the protein environment. Structural and functional changes within a protein created by a chemical or physical event, such as solvation or binding to another molecule, can be probed by monitoring the chemical shift changes in a TROSY spectrum. The 2D, ^{1}H/^{15}N TROSY spectra for CaLB in the absence and presence of acetonitrile showed that many of the CaLB amide resonances were shifted upon the addition of acetonitrile. As a possible interpretation of these data, it is proposed that CaLB adopts a very different conformation in the presence of acetonitrile as compared to the normal H_2O/buffer state.

Not only does this example show that solvent effects should be taken seriously, it also provides a clear message of the prospects involved. Eventually, the authors developed a scaled-up process (63 kg!), corroborating the consistency of the small-scale results.

As a final example, which may have a bearing on the future development of enantioselective enzymatic catalysis in organic media, the report by Lee and coworkers on the CaLB-catalyzed enantioselective acylation of racemic alkyl lactates with vinyl alkanoates to provide alkyl (R)-lactate provides an interesting case [137]. After investigating several process conditions, including the effects of the organic solvent, the alkyl chain length of the alkyl lactates and vinyl alkanoates and the temperature on the *ee*-values and the reaction rate, they found that a highly efficient resolution could be developed by running the reaction in neat substrates. Eventually, the CaLB-catalyzed acylation and subsequent work-up successfully provided both butyl (R)-O-butanoyllactate and butyl (S)-lactate in excellent yields (48%) and enantioselectivities (>99.5% *ee*) on a 100-gram scale.

In a similar approach, Kasture and coworkers describe the use of neat substrate (ethyl acetate both as alcohol donor and as the reaction medium) in the preparation of chirally pure S-(–)-1,4-benzodioxan-2-carboxylate, an important drug intermediate used in the synthesis of doxazosin mesylate, from racemic 1,4-benzodioxan-2-carboxylic acid [138]. Again, CaLB catalyzed the transesterification reaction with good enantioselectivity ($E = 160$) and acceptable enantiomeric excess (>95%) and chemical yield (50%).

Finally, Sanfilippo and coworkers describe the enzymatic kinetic resolution of atropisomeric (±)-3,3′-bis(hydroxymethyl)-2,2′-bipyridine N,N-dioxide by enantioselective esterifcation in an unusual medium of 2-propanol/vinyl acetate (20:80) [139]. Lipase from *Mucor miehei* (immobilized lipase preparation, Lipozyme®) was found to give good enantioselectivity with an (*aS*)-enantiopreference in the axial recognition and allowed efficient preparation of both enantioforms with *ee* > 98%. Despite the fact that the propanol reacts with the acyl donor, this did not diminish its positive effects on the solubility of the bipyridyl substrate.

Considering the beneficial effect of the use of neat substrates on the reduction of the environmental impact of industrial solvents, it might seem that further examples of this approach in kinetic resolution processes may be expected in the near future.

2.9
Concluding Remarks

Despite the large number of reports of successful efforts to increase the *E*-value of enzymes by managing the medium composition, proper understanding of the factors that are involved is still largely lacking. The situation is somewhat reminiscent of the state of affairs in the study of enzymatic catalysis during the first half of the previous century. Whereas the chemical events during catalysis could be understood in the context of known chemistry, the magic of the extraordinary rate enhancement by enzymes had to be left to "conformational effects". Although the quantitative description of these effects is still a matter of ongoing research, a fair amount of insight is now available. From the work reviewed in this chapter, it would appear that the current understanding of the role of organic solvents on enzyme enantioselectivity finds itself in a similar predicament, with "flexibility" taking the role of "conformational changes".

Somewhat surprisingly, preliminary conclusions appear to suggest that enantioselectivity is highest under conditions (solvents, cosolvents, additives) where a certain degree of flexibility, as compared to the enzyme in aqueous buffer, is restored or maintained [122, 123, 125, 140, 141]. At first sight, this would appear to be rather counterintuitive, considering that one might expect a more rigid active site to lead to stricter chiral discrimination. Support for the latter view may be obtained from the well-documented observation that the majority of enzymes show an inverse relationship between E-value and temperature, [142] suggesting that the more rigid low-temperature conformations have the better enantioselective properties. Other explanations have, however, been proposed [143]. In this respect, an early report by Phillips and coworkers stating that both temperature and DMSO increase the enantioselectivity in the pig liver esterase-catalyzed hydrolysis of methyl alkyl dimethylmalonates is of interest [144].

Referring to the analysis of kinetic barriers that contribute to the differential rates, outlined above, it would appear that enantioselectivity may not reside in any single part of the kinetic scheme. In fact, analysis of the distribution of entropic and enthalpic contributions to the enantioselectivity of CaLB in mixtures of organic solvents showed that the gradual replacement of one organic solvent with the other may lead to a switch from one "state" to another [145]. More information on the response of individual steps in the kinetic scheme to a change of solvent is needed to arrive at a conclusive answer.

From a practical point of view, however, the situation certainly looks brighter. In any given case, small-scale experiments using a limited set of organic solvents and/or low-molecular additives, may lead to the formulation of favorable conditions. From the examples provided in this chapter, this certainly appears to be a worthwhile strategy.

References

1 Albery, W. J., Knowles, J. R. (1976) Biochemistry 15, 5631–5640.
2 Benner, S. A. (1989) Chem. Rev. 89, 789–806.
3 Fersht, A. (1999) Structure and mechanism in protein science: A guide to enzyme catalysis and protein folding, Freeman, New York.

References

4 Collins, A. N., Sheldrake, G. N., Crosby, J. (1992) *Chirality in industry*, John Wiley Sons, Chichester, UK.

5 Sheldon, R. A. (1993) *Chirotechnology*, Marcel Dekker, Inc., New York.

6 Singer, S. J. (1962) *Advances in Protein Chemistry* **17**, 1–68.

7 Halling, P., Kvittingen, L. (1999) *TIBTECH* **17**, 343–344.

8 Klibanov, A. M. (2000) *TIBTech* **18**, 84–85.

9 Klibanov, A. M., Samokhin, G. P., Martinek, K., Berezin, I. V. (1977) *Biotechnol. Bioeng.* **19**, 1351–1361.

10 Zaks, A., Klibanov, A. M. (1984) *Science* **224**, 1249–1251.

11 Zaks, A., Klibanov, A. M. (1985) *Proc. Natl. Acad. Sci. USA* **82**, 3192–3196.

12 Carrea, G., Riva, S. (2000) *Angew. Chem. Int. Ed.* **39**, 2226–2254.

13 Koskinen, A. M. P., Klibanov, A. M. (1996) *Enzymatic reactions in organic media*, 1st ed., Blackie Academic Professional, London.

14 Schmid, A., Dordick, J. S., Hauer, B., Kiener, A., Wubbolts, M., Witholt, B. (2001) *Nature* **409**, 258–268.

15 Klibanov, A. M. (1990) *Acc. Chem. Res.* **23**, 114–120.

16 Klibanov, A. M. (2001) *Nature* **409**, 241–246.

17 Carrea, G., Ottolina, G., Riva, S., Secundo, F. (1992) in Progress in Biotechnology, 8 (Biocatalyis in Non-conventional Media, edited by J. Tramper et al.), Elsevier Science Publishers B.V., pp 111–119.

18 Hirose, Y., Kariya, K., Sasaki, I., Kurono, Y., Ebiike, H., Achiwa, K. (1992) *Tetrahedron Lett.* **33**, 7157–60.

19 Kitaguchi, H., Fitzpatrick, P. A., Huber, J. E., Klibanov, A. M. (1989) *J. Am. Chem. Soc.* **111**, 3094–3095.

20 Tawaki, S., Klibanov, A. M. (1992) *J. Am. Chem. Soc.* **114**, 1882–1884.

21 van Tol, J. B. A., Schouten, I. J. C., Jongejan, J. A., Duine, J. A. (1990b) in 5th European Congress on Biotechnology (Christiansen, C., Munck, L., Villadsen, J., Eds.) pp 268, Munksgaard Intl. Publisher, Copenhagen.

22 Wescott, C. R., Klibanov, A. M. (1994) *Biochim. Biophys. Acta* **1206**, 1–9.

23 Wu, S. H., Chu, F. Y., Wang, K. T. (1991) *Bioorg. Med. Chem. Lett.* **1**, 339–42.

24 Cleland, W. W. (1963) *Biochim. Biophys. Acta* **67**, 104–137.

25 NC-IUB. (1982) *Eur. J. Biochem.* **128**, 281–291.

26 Barnsley, E. A. (1993) *Biochem. J.* **291**, 323–328.

27 Chen, C. S., Fujimoto, Y., Girdaukas, G., Sih, S. J. (1982) *J. Am. Chem. Soc.* **104**, 7294–7299.

28 Jongejan, J. A., van Tol, J. B. A., Geerlof, A., Duine, J. A. (1990) in 5th European Congress on Biotechnology (Christiansen, C., Munck, L., Villadsen, J., Eds.) pp 268, Munksgaard Intl. Publisher, Copenhagen.

29 van Tol, J. B. A., Jongejan, J. A., Geerlof, A., Duine, J. A. (1991) *Recl. Trav. Chim. Pays-Bas* **110**, 255–262.

30 Straathof, A. J. J., Jongejan, J. A. (1997) *Enzyme Microbiol. Technol.* **21**, 559–571.

31 Straathof, A. J. J., Rakels, J. L. L., Heijnen, J. J. (1992) *Biocatalysis* **7**, 13–27.

32 Faber, K. (2000) *Biotransformations in organic chemistry*, 4th ed., Springer, Berlin.

33 Kazlauskas, R. J. (2005) *Nature* **436**, August 25.

34 Reetz, M. T., Wilensek, S., Zha, D., Jaeger, K.-E. (2001) *Angew. Chem. Int. Ed.* **40**, 3589–3591.

35 Kawanami, Y., Honnma, A., Ohta, K., Matsumoto, N. (2005) *Tetrahedron* **61**, 693–697.

36 Kawanami, Y., Itoh, K. (2005) *Chem. Lett.* **34**, 682–683.

37 Mezzetti, A., Keith, C., Kazlauskas, R. J. (2003) *Tetrahedron: Asymmetry* **14**, 3917–3924.

38 Shang, C.-S., Hsu, C.-S. (2003) *Biotechnol. Lett.* **25**, 413–416.

39 Watanabe, K., Koshiba, T., Yasufuku, Y., Miyazawa, T., Ueji, S.-I. (2001a) *Bioorg. Chem.* **29**, 65–76.

40 Khmelnitsky, Y. L., Welch, S. H., Clark, D. S., Dordick, J. S. (1994) *J. Am. Chem. Soc.* **116**, 2647–2648.

41 Ueji, S.-I., Mori, S.-I., Yumoto, H., Hiroshima, N., Ebara, Y. (2005) *Chem. Lett.* **34**, 110–111.

42 Secundo, F., Phillips, R. S. (1996) *Enzyme Microbiol. Technol.* **19**, 487–492.

43 Vidinha, P., Harper, N., Micaelo, N. M., Laurenco, N. M. T., Gomes da Silva, M. D. R., Cabral, J. M. S., Afonso, C. A. M., Soares, C. M., Barreiros, S. (2003) *Biotechnol. Bioeng.* **85**, 442–449.

44 Wang, P.-Y., Tsai, S.-W. (2005) *Enzyme Microbiol. Technol.* **37**, 266–271.

45 Pham, V. T., Phillips, R. S., Ljungdahl, L. G. (1989) *J. Am. Chem. soc.* **111**, 1935–1936.

46 Phillips, R. S., Zheng, C., Pham, V. T., Andrade, F. A. C., Andrade, M. A. C. (1994) *Biocatalysis* **10**, 77–86.

47 Sakai, T., Kawabata, I., Kishimoto, T., Ema, T., Utaka, M. (1997) *J. Org. Chem.* **62**, 4906–4907.

48 Sakai, T., Kishimoto, T., Tanaka, Y., Ema, T., Utaka, M. (1998) *Tetrahedron Lett.* **39**, 7881–7884.

49 Vänttinen, E., Kanerva, L. T. (1997) *Tetrahedron: Asymmetry* **8**, 923–933.

50 Fantin, G., Fogagnolo, M., Guerzoni, M. E., Lanciotti, R., Medici, A., Pedrini, P., Rossi, D. (1996) *Tetrahedron: Asymmetry* **7**, 2879–2887.

51 Kamat, S. V., Beckman, E. J., Russell, A. J. (1993a) *J. Am. Chem. Soc.* **115**, 8845–8846.

52 Kim, M.-J., Kim, H. M., Kim, D., Ahn, Y., Park, J. (2004) *Green Chem.* **6**, 471–474.

53 Overbeeke, P. L. A., Koops, B. C., Verheij, H. M., Slotboom, A. J., Egmond, M. R., Jongejan, J. A., Heijnen, J. J. (2000c) *Biocatalysis and Biotransformations* **18**, 59–77.

54 Secundo, F., Carrea, G. (2002) *J. Mol. Catal. B: Enzymatic* **19–20**, 93–102.

55 Ueji, S.-I., Tanaka, H., Hanaoka, T., Ueda, A., Watanabe, K., Kaihatsu, K., Ebara, Y. (2001) *Chem. Lett.*, 1066–1067.

56 Overbeeke, P. L. A., Govardhan, C., Khalef, N., Jongejan, J. A., Heijnen, J. J. (2000a) *J. Mol. Catal. B: Enzymatic* **10**, 385–393.

57 van Langen, L. M., Selassa, R. P., Van Rantwijk, F., Sheldon, R. A. (2005) *Org. Lett.* **7**, 327–329.

58 Maccarrone, M., Bari, M., Battista, N., Finazzi-Agro, A. (2001) *Biophys. Chem.* **90**, 97–101.

59 Fitzpatrick, P. A., Klibanov, A. M. (1991) *J. Am. Chem. Soc.* **113**, 3166–71.

60 Terradas, F., Teston-Henry, M., Fitzpatrick, P. A., Klibanov, A. M. (1993) *J. Am. Chem. Soc.* **115**, 390–396.

61 Ke, T., Wescott, C. R., Klibanov, A. M. (1996) *J. Am. Chem. Soc.* **118**, 3366–3374.

62 Mahmoudian, M., Baines, B. S., Dawson, M. J., Lawrence, G. C. (1992) *Enzyme Microbiol. Technol.* **14**, 911–916.

63 Wescott, C. R., Noritomi, H., Klibanov, A. M. (1996) *J. Am. Chem. Soc.* **118**, 10365–10370.

64 Sakurai, T., Margolin, A. L., Russell, A. J., Klibanov, A. M. (1988) *J. Am. Chem. Soc.* **110**, 7236–7237.

65 Kamat, S. V., Iwaskewycs, B., Beckman, E. J., Russell, A. J. (1993b) *Proc. Natl. Acad. Sci. USA* **90**, 2940–2944.

66 Rogalska, E., Ransac, S., Verger, R. (1993) *J. Biol. Chem.* **268**, 792–4.

67 Chaudhary, A. K., Kamat, S. V., Beckman, E. J., Nurok, D., Kleyle, R. M., Hajdu, P., Russell, A. J. (1996) *J. Am. Chem. Soc.* **118**, 12891–12901.

68 Berglund, P. (2001) *Biomolecular Eng.* **18**, 13–22.

69 Gubicza, L., Kelemen-Horvath, I. (1996) *Med. Fac. Landbouww. Univ. Gent* **61**, 1361–1365.

70 Dordick, J. S. (1992) *Biotechnol. Prog.* **8**, 259–267.

71 Mishra, P., Griebenow, K., Klibanov, A. M. (1996) *Biotechnol. Bioeng.* **52**, 609–614.

72 Noritomi, H., Almarsson, O., Barletta, G. L., Klibanov, A. M. (1996) *Biotechnol. Bioeng.* **51**, 95–99.

73 Wescott, C. R., Klibanov, A. M. (1993) *J. Am. Chem. Soc.* **115**, 10362–10363.

74 Parida, S., Dordick, J. S. (1991) *J. Am. Chem. Soc.* **113**, 2253–9.

75 Ottosson, J., Fransson, L., King, J. W., Hult, K. (2002) *Biochim. Biophys. Acta* **1594**, 325–334.

76 Ducret, A., Trani, M., Lortie, R. (1998) *Enzyme Microb. Technol.* **22**, 212–216.

77 Ueji, S.-I., Taniguchi, T., Okamoto, T., Watanabe, K., Ebara, Y., Ohta, H. (2003) *Bull. Chem. Soc. Jpn.* **76**, 399–403.

78 Lee, S. B. (1995) *J. Ferment. Bioeng.* **80**, 141–147.

79 Laane, C., Boeren, S., Vos, K., Veeger, C. (1987) *Biotechnol. Bioeng.* **30**, 81–87.

80 Carrea, G., Ottolina, G., Riva, S. (1995) *TIBTECH* **13**, 63–70.
81 Secundo, F., Riva, S., Carrea, G. (1992) *Tetrahedron: Asymmetry* **3**, 267–80.
82 Kanerva, L. T., Vihanto, J., Halme, M. H., Loponen, J. M., Euranto, E. K. (1990) *Acta Chem. Scand.* **44**, 1032–1035.
83 Miyazawa, T., Kurita, S., Ueji, S., Yamada, T., Kuwata, S. (1992) *Biotechnol. Lett.* **14**, 941–946.
84 Lou, W.-Y., Zong, M.-H. (2006) *Chirality* **18**, 814–821.
85 Ulbert, O., Frater, T., Belafi-Bako, K., Gubicza, L. (2004) *J. Mol. Catal. B: Enzymatic* **31**, 39–45.
86 van Tol, J. B. A., Stevens, R. M. M., Veldhuizen, W. J., Jongejan, J. A., Duine, J. A. (1995c) *Biotechnol. Bioeng.* **47**, 71–81.
87 Wolff, A., Straathof, A. J. J., Jongejan, J. A., Heijnen, J. J. (1997) *Biocatalysis and Biotransformations* **15**, 175–184.
88 van Tol, J. B. A., Kraayveld, D. E., Jongejan, J. A., Duine, J. A. (1995b) *Biocatalysis and Biotransformations* **12**, 119–136.
89 Fryszkowska, A., Komar, M., Koszelewski, D., Ostaszewski, R. (2006) *Tetrahedron: Asymmetry* **17**, 961–966.
90 Luque, S., Ke, T., Klibanov, A. M. (1998) *Biocatal. Biotransform.* **16**, 233–248.
91 Jones, J. B., Mehes, M. M. (1979) *Can. J. Chem.* **57**, 2245–2248.
92 Rakels, J. L. L., Caillat, P., Straathof, A. J. J., Heijnen, J. J. (1994) *Biotechnol. Prog.* **10**, 403–409.
93 Jongejan, J. A., van Tol, J. B. A., Duine, J. A. (1994) *Chimica Oggi*, 15–24.
94 Janssen, A. E. M., Padt, A. van der, Sonsbeek, H. M. Van, Rite, K. Van 't. (1993) *Biotechnol. Bioeng.* **41**, 95–103.
95 Wolff, A., Straathof, A. J. J., Heijnen, J. J. (1994) *Biocatalysis* **11**, 249–261.
96 Bovara, R., Carrea, G., Ottolina, G., Riva, S. (1993) *Biotechnol. Lett.* **15**, 169–174.
97 Chua, L. S., Sarmidi, M. R. (2006) *J. Mol. Catal. B: Enzymatic* **38**, 551–556.
98 Högberg, H. E., Edlund, H., Berglund, P., Hedenström, E. (1993) *Tetrahedron: Asymmetry* **4**, 2123–2126.
99 Ors, M., Morcuende, A., Jimenez-Vacas, M. I., Valverde, S., Herradon, B. (1996) *Synlett* **5**, 449–451.
100 Persichetti, R. A., Lalonde, J. J., Govardhan, C. P., Khalaf, N. K., Margolin, A. L. (1996) *Tetrahedron Lett.* **37**, 6507–6510.
101 Palomo, J. M., Fernandez-Lorente, G., Mateo, C., Fuentes, M., Guisan, J. M., Fernandez-Lafuente, R. (2002) *Tetrahedron: Asymmetry* **13**, 2653–2659.
102 Palomo, J. M., Segura, R. L., Mateo, C., Terreni, M., Guisan, J. M., Fernandez-Lafuente, R. (2005) *Tetrahedron: Asymmetry* **16**, 869–874.
103 van Tol, J. B. A., Jongejan, J. A., Duine, J. A. (1995a) *Biocatalysis and Biotransformations* **12**, 99–117.
104 Truhlar, D. G., Garrett, B. C., Klippenstein, S. J. (1996) *J. Phys. Chem.* **100**, 12771–12800.
105 Truhlar, D. G., Hase, W. L., Hynes, J. T. (1983) *J. Phys. Chem.* **87**, 2664–2682.
106 Rotticci, D., Norin, T., Hult, K. (2000) *Org. Lett.* **2**, 1373–1376.
107 Anthonsen, T., Jongejan, J. A. (1997) *Methods Enzymol.* **286**, 473–495.
108 Dai, L., Klibanov, A. M. (2000) *Biotechnol. Bioeng.* **70**, 353–357.
109 Klibanov, A. M. (2003) *Curr. Opin. Biotechnol.* **14**, 427–431.
110 Cowan, D. A. (1997) *Comp. Biochem. Physiol.* **118A**, 429–438.
111 Keinan, E., Hafeli, E. K., Seth, K. K., Lamed, R. (1986) *J. Am. Chem. Soc.* **108**, 162–169.
112 Simpson, H., Cowan, D. A. (1997) *Pept. Prot. Lett.* **4**, 25–32.
113 De Gonzalo, G., Ottolina, G., Zambianchi, F., Fraaije, M. W. Carrea, G. (2006) *J. Mol. Catal. B: Enzymatic* **39**, 91–97.
114 Nakamura, K., Kinoshita, M., Ohno, A. (1995) *Tetrahedron* **51**, 8799–8808.
115 Kinoshita, M., Ohno, A. (1996) *Tetrahedron* **52**, 5397–5406.
116 Mattos, C., Ringe, D. (2001) *Curr. Opin. Struct. Biol.* **11**, 761–764.
117 Yun, H., Kim, J., Kinnera, K., Kim, B.-G. (2005) *Biotechnol. Bioeng.* **93**, 390–395.
118 Nakamura, K., Kondo, S.-I., Kawai, Y., Ohno, A. (1993) *Tetrahedron: Asymmetry* **4**, 1253–1254.

119 Odell, B., Earlam, J. (1985) *J. Chem. Soc., Chem. Commun.*, **359**.

120 Broos, C. J., Martin, M. N., Rouwenhorst, I., Verboom, W., Reinhoudt, D. N. (1991) *Recl. Trav. Chim. Pays-Bas* **110**, 255–258.

121 Mine, Y., Zhang, L., Fukanaga, K., Sugimura, Y. (2005) *Biotechnol. Lett.* **27**, 383–388.

122 Griebenow, K., Laureano, Y. D., Santos, A. M., Montanez Clemente, I., Rodriguez, L., Vidal, M. W., Barletta, G. (1999) *J. Am. Chem. Soc.* **121**, 8157–8163.

123 Santos, A. M., Vidal, M., Pacheco, V., Frontera, J., Baez, C., Ornellas, O., Barletta, G., Griebenow, K. (2001) *Biotechnol. Bioeng.* **74**, 295–308.

124 Montanez-Clemente, I., Alvira, E., Macias, M., Ferrer, A., Fonceca, M., Rodriguez, J., Gonzalez, A., Barletta, G. (2002) *Biotechnol. Bioeng.* **78**, 53–59.

125 Broos, J., Visser, A. J. W. G., Engbersen, J. F. J., Verboom, W., Hoek, A. V., Reinhoudt, D. N. (1995) *J. Am. Chem. Soc.* **117**, 12657–12663.

126 Watanabe, K., Uno, T., Koshiba, T., Okamoto, T., Ueji, S.-I. (2004) *Bull. Chem. Soc. Jpn* **77**, 543–548.

127 Zheng, Y.-J., Ornstein, R. L. (1996) *Biopolymers* **38**, 791–799.

128 Sharpless, K. B. (1988) *Janssen Chimica Acta* **6**, 3–6.

129 Klunder, J. M., Ko, S. Y., Sharpless, K. B. (1986) *J. Org. Chem.* **51**, 3710–3712.

130 Gao, Y., Hanson, R. M., Klunder, J. M., Ko, S. Y., Masamune, H., Sharpless, K. B. (1987) *J. Am. Chem. Soc.* **109**, 5765–5780.

131 Ladner, W. E., Whitesides, G. M. (1984) *J. Am. Chem. Soc.* **106**, 7250–1.

132 Philippi, M. C., Jongejan, J. A., Duine, J. A. (1987) in *Proc. 4th European Congress on Biotechnology 1987* (Neijssel, O. M., van der Meer, R. R., Luyben, K. C. A. M., Eds.) pp 281–284, Elsevier Science Publishers, A'dam.

133 van Tol, J. B. A., Jongejan, J. A., Duine, J. A. (1990a) in *5th European Congress on Biotechnology* (Christiansen, C., Munck, L., Villadsen, J., Eds.) pp 267, Munksgaard Intl. Publisher, Copenhagen.

134 van Tol, J. B. A., Jongejan, J. A., Duine, J. A. (1992) *Prog. Biotechnol.* **8**, 237–43.

135 Ramos Tombo, G. M., Schär, H. P., Fernandez, X., Busquets, I., Ghisalba, O. (1986) *Tetrahedron Lett.* **27**, 5707–5710.

136 Yazbeck, D., Derrick, A., Panesar, M., Deese, A., Gujral, A., Tao, J. (2006) *Org. Proc. Res. Dev.* **10**, 655–660.

137 Lee, Y. S., Hong, J. H., Jeon, N. Y., Won, K., Kim, B. T. (2004) *Organic Process Research Development* **8**, 948–951.

138 Kasture, S. M., Varma, R., Kalkote, U. R., Nene, S., Kulkarni, B. D. (2005) *Biochem. Eng. J.* **27**, 66–71.

139 Sanfilippo, C., D'Antona, N., Nicolosi, G. (2006) *Tetrahedron: Asymmetry* **17**, 12–14.

140 Watanabe, K., Ueji, S.-I. (2001) *J. Chem. Soc. Perkin* **1**, 1386–1390.

141 Watanabe, K., Yoshida, T., Ueji, S.-I. (2001b) *Chem. Commun.*, 1260–1261.

142 Overbeeke, P. L. A., Ottosson, J., Hult, K., Jongejan, J. A., Duine, J. A. (1999) *Biocatalysis and Biotransformations* **17**, 61–79.

143 Cainelli, G., Galletti, P., Giacombini, D., Gualandi, A., Quintavalla, A. (2003) *Helv. Chim. Acta* **86**, 3548–3559.

144 Andrade, M. A. C., Andrade, F. A. C., Phillips, R. S. (1991) *Bioorganic and Medicinal Chem. Lett.* **1**, 373–376.

145 Overbeeke, P. L. A., Jongejan, J. A., Heijnen, J. J. (2000b) *Biotechnol. Bioeng.* **70**, 278–290.

3
Activating Enzymes for Use in Organic Solvents
Anne L. Serdakowski, Jonathan S. Dordick

3.1
Introduction

Enzymes are highly selective catalysts that perform intricate chemistries at ambient temperatures and pressures. While water is the solvent of life, it is a poor solvent for most synthetic organic reactions, primarily because of the low aqueous solubility of uncharged organic compounds and the high reactivity of water as a nucleophile. From a processing standpoint, the high boiling point and low vapor pressure of water results in expensive dewatering steps to isolate the product(s) from an aqueous-based biotransformation. Finally, unwanted side reactions such as hydrolysis, racemization, polymerization, and decomposition often occur in water, which limit many of the reactions of interest in organic transformations. Because of these drawbacks, synthetic applications of enzymes have turned to the use of nonaqueous solvents to replace water as a bulk medium. Examples of biocatalysis in organic solvents are now well known, as is evident in other chapters in this book.

Potential Advantages of Employing Enzymes in Organic Media [1]
1. Increased solubility of nonpolar substrates
2. Thermodynamic equilibria favor synthesis over hydrolysis
3. Suppression of water-dependent side reactions (e.g., hydrolysis of acid anhydrides and halides, polymerization of quinones)
4. Alteration of substrate specificity
5. Ease of enzyme recovery via filtration or centrifugation
6. Ease of immobilization, e.g., via simple adsorption onto nonporous surfaces: enzymes cannot desorb from these surfaces in nonaqueous media
7. Ease of product recovery from low-boiling, high-vapor-pressure solvents
8. Often enhanced thermostability

3 Activating Enzymes for Use in Organic Solvents

9. Potential for enzymes to be used directly in a chemical process, eliminating microbial contamination

Despite these advantages, native enzymes almost universally exhibit very low activities in organic solvents – often 4–5 orders of magnitude lower than in aqueous solutions. This loss in catalytic activity may be attributed to several factors, including a decrease in the polarity of the enzyme's microenvironment, the loss of critical water residues from the enzyme's surface, the decreased conformational mobility of the enzyme's structure, ground-state stabilization of hydrophobic substrates, and deactivation during the preparation of the biocatalyst for use in nonaqueous media, e.g., lyophilization [2]. These and other factors are highlighted in Figure 3.1.

Herein we provide a general overview of enzymatic catalysis in organic solvents, with a focus on the role of water and solvent on enzyme structure and function, and how an increased knowledge of this role has led to methods to activate enzymes for optimal use in organic media.

3.2
Water – A Unique and Necessary Solvent for Enzymatic Catalysis

3.2.1
Challenges for Enzymatic Catalysis *in Water*

Water is a unique solvent, which also happens to be ubiquitous to life on Earth. Paradoxical as it may seem, any discussion on enzymatic catalysis in organic sol-

Organic solvents decrease enzyme activity

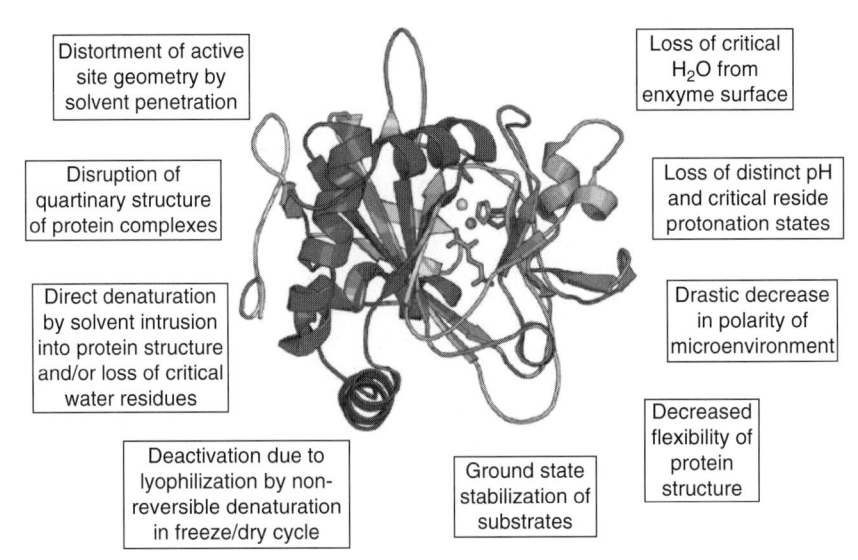

- Distortment of active site geometry by solvent penetration
- Disruption of quartinary structure of protein complexes
- Direct denaturation by solvent intrusion into protein structure and/or loss of critical water residues
- Deactivation due to lyophilization by non-reversible denaturation in freeze/dry cycle
- Ground state stabilization of substrates
- Loss of critical H$_2$O from enxyme surface
- Loss of distinct pH and critical reside protonation states
- Drastic decrease in polarity of microenvironment
- Decreased flexibility of protein structure

Figure 3.1 Factors contributing to the loss in enzymatic activity in organic media [2, 3].

vents, and in particular enzyme structure and function, should put water first and foremost. Even more intriguing is that while water is critical for life, and hence biological function, the physicochemical properties of this solvent are harsh and present challenges for enzyme function. For example, bulk water is highly concentrated (55 M), and therefore, highly nucleophilic. As a result, labile functional groups such as nucleic acids and amino acids, polysaccharides and carbohydrates, and oxidized intermediates of metabolic pathways can be negatively affected by water's hydrolytic properties [4]. This high concentration also creates very large thermodynamic barriers to synthetic reactions that oppose hydrolysis, and which require vast amounts of ATP to overcome such unfavorable thermodynamics. Obvious examples include DNA replication and protein synthesis, which consume a large fraction of the ATP requirement of the cell. Biology has evolved elaborate architectures that control water within cells. In addition, nature has designed enzyme structures to position (and hence control) water within their active sites. In the process, enzymes have evolved to both require and organize water to maximize function. Using enzymes in organic solvents, then, would appear to require the breakdown of this evolutionary organization that makes enzymes extraordinary catalysts. Fortunately, while water is indeed important, it is not necessary to have bulk water. Indeed, nature has provided us with several examples of this.

3.2.2
Enzymes do Function Without Water as a Bulk Solvent – Lessons from Extreme Halophiles

Bulk water has a water activity of unity ($a_w = 1$). The presence of high concentrations of small solutes can dramatically reduce this value. For example, within the Archaea are microorganisms that have adapted to survival in nearly saturated salt solutions ($a_w = 0.75$). Hence, bacteria such as *Halobacterium* can function without bulk water, which demonstrates that their enzymes also function in lower water activity. How does this occur? At low concentrations of salt (<0.2 M), charge screening of the negative amino acid residues on halophilic proteins by the salt acts to stabilize the protein [5]. At higher salt concentrations, the weakly hydrophobic core of these proteins is stabilized via a salting-out phenomenon [6]. Anionic surface residues appear to act as multiple kosmotropes, which favors high-density water clustering [7], similar to what is observed with polyphosphates in non-halophiles [8]. The weakly chaotropic potassium ion is the main entity in the intracellular halophilic milieu, and its pairing with the kosmotropic anionic surface residues results in a maximum stabilization of biological macromolecules and assists in solubilizing the proteins in the cell [9]. Thus, halophiles have adapted to and flourish in a "nonaqueous" environment.

In fact, halophilic enzymes have been shown to function in low salt environments, where the salt is replaced by a suitable organic solvent; typically those that are kosmotropic. For example, an extracellular protease from *H. halobium*, which requires 4 M NaCl for optimal function, is highly active in as little as 0.2 M NaCl in the presence of 40% (v/v) dimethylsulfoxide (DMSO) [10]. The behavior of salts

3 Activating Enzymes for Use in Organic Solvents

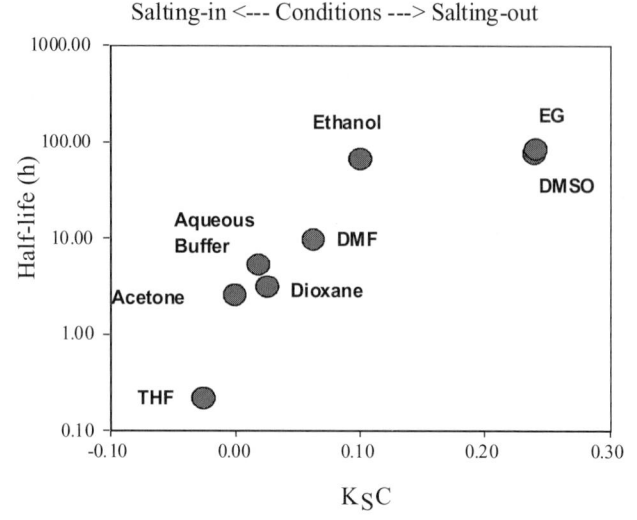

Figure 3.2 Replacement of high salt with organic solvents – stability of the extracellular protease from *Halobacterium halobium* in 0.36 M NaCl in water or various organic solvents. The K_SC value represents the Sechenow constant multiplied by the concentration of the solvent, which provides a measure of the salting-out (kosmotropic) capacity of the solution [10].

and solvents, in this case, is complementary; as shown in Figure 3.2 there is a strong correlation between the high salting-out nature of the organic solvent (as reflected in the value of K_SC) and the increased stability of the enzyme. Thus, in addition to salts, the salting-out characteristics of organic solvents can be used to stabilize the halophilic protease, and in some cases eliminate the need for a high-salt medium. More relevant to the contents of this chapter, these results provide strong evidence that organic solvents possess properties that mimic natural solutes critical for some for enzymes that naturally function in low-water environments.

3.2.3
Behavior of Enzymes in the Absence of Bulk Water

Molecular dynamics simulations have shown that the 3D structure of the serine protease subtilisin remains intact in solvents ranging from the highly polar acetonitrile to the highly nonpolar octane. In the latter, but not the former, there is a clear partitioning of the bulk solvent away from the enzyme, thereby allowing it to retain its native active site environment, including the location of structural water molecules and the location of solvent molecules that do not penetrate into the active site [11]. The localizing of water and organic solvent molecules in a nonaqueous environment is a direct extension of the evolutionary plasticity of nature in the design of enzyme structure. Even in the absence of a bulk water

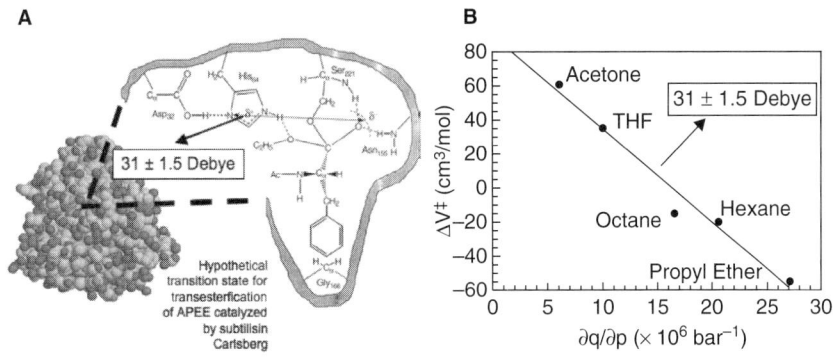

3.2 Water – A Unique and Necessary Solvent for Enzymatic Catalysis

Figure 3.3 (a) Transition state of subtilisin. (b) Plotting the activation volumes calculated from the pressure dependence on catalytic efficiency as a function of the pressure dependence on the Kirkwood parameter yields an estimate of the enzyme's transition state dipole moment. The continuity of the plot indicates that solvent does not disrupt the enzyme's strong dipole, thereby enabling catalysis to ensue [12].

solution, enzymes retain an ability to "store" structural water and repel organic solvents, all while enabling the enzyme transition state to remain structurally intact and functional.

To elucidate the influence of the solvent on an enzymic transition state, we in collaboration with Douglas Clark's group at the University of California, Berkeley, performed high-pressure kinetic studies [12]. Activation volumes were obtained, which were then applied in an electrostatic model of Kirkwood to calculate a lower limit of subtilisin's transition-state dipole moment (Figure 3.3). The resulting dipole moment of 31 ± 1.5 Debye was extraordinarily high and indicated that charge separation in the transition state remained intact in the nonaqueous environment. Furthermore, the dipole moment was unchanged in solvents ranging from water to octane, which indicated that organic solvent molecules do not penetrate within the dipole region of the transition state, and instead lie in the interior of the protein.

3.2.4
Removing Water from Enzymes – the Effect of Lyophilization on Enzyme Structure and Function

An extremely common practice when preparing enzymes for use in organic solvents is to remove the water via freeze drying to yield "dry" enzyme powders [13]. Lyophilization consists of three components: initial freezing, primary drying, and secondary drying [14–16]. During the freezing step the enzyme is immobilized in an ice, which fixes its structure [17]. Primary drying separates the ice from the protein by sublimation, and secondary drying removes a large fraction of the bound water from the protein. At the end of the primary drying stage, the water content remaining is about 10–15% (w/w), and at the end of the entire process it is reduced further to about 3–5% (w/w) [18]. Thus, following lyophilization the water content on a typical enzyme is far lower than that needed for monolayer coverage (ca. 30%, w/w).

Enzymes can still show activity after such extreme dehydration [19]. Rupley et al. used heat capacity and IR spectroscopy to show that hen lysozyme showed catalytic activity even before a monolayer of water was present on the enzyme. Later work [20, 21] showed that water forms (and, in the case of lyophilization, removed) on the enzyme in a highly ordered and defined manner. When water molecules come into contact with an enzyme, they bind first to the charged and polar amino acids. This specific hydration is important when considering the removal of water during the sublimation stage; when the enzyme is dried to the point when these water molecules are removed, the side chains of the amino acids interact with each other, thereby locking the protein's structure. These new interactions cause the enzyme to lose flexibility, affecting its catalytic activity. It is hypothesized that this rigidity aids the stability of enzyme molecules in 100°C anhydrous organic solvents, even though the same enzyme loses activity when heated in aqueous solutions [22].

One would imagine that the rapid freezing and subsequent sublimation of the water from an enzyme during lyophilization would destroy the delicate secondary structure of the protein; however, that conclusion has been controversial. A comprehensive analysis of the effects of freeze drying on protein structure was conducted by Dong et al. [23]. To examine the effect lyophilization had on enzyme secondary structure, this study compared the secondary structures of two enzymes; α-chymotrypsin and subtilisin Carlsberg. α-Chymotrypsin had a disordered secondary structure after lyophilization, but upon rehydration the protein was fully soluble and refolded to its natural conformation. Similar results were found with the subtilisin Carlsberg, but the perturbation of the secondary structure following lyophilization was more severe than that of the α-chymotrypsin. Additional studies using Raman [24, 25] and solid-state NMR [26], as well as hydrogen isotope-exchange/high-resolution NMR [27] agree with these findings, providing evidence supporting lyophilization-induced reversible denaturation.

More recent studies using FTIR have targeted the individual secondary structural elements, α-helix and β-sheet. Prestrelski et al. [28, 29] examined the second derivatives of vibrational spectra in the amide I region (1600–1720 cm^{-1}) and assigned correlation coefficients that reflected the overall (yet qualitative) changes in the secondary structure of the protein following lyophilization. Griebenow et al. [30] studied the amide III region (1220–1330 cm^{-1}) and developed a quantitative analysis of the changes in the secondary structural elements of a variety of lyophilized proteins. Following lyophilization, bovine pancreatic trypsin inhibitor, RNase A, chymotrypsinogen A, porcine insulin, horse myoglobin, and cytochrome *c* all showed an increased fraction of β-sheet, with a concomitant drop in the fractions of α-helices and unordered regions, relative to the native solution structures. This study conclusively showed that lyophilization has a disordering effect on the secondary structural elements of a protein. Moreover, the method of preparing an enzyme for use in dehydrated environments strongly influences its function.

Indeed, this is the basis behind many of the methods used to activate enzymes for use in organic solvents, as we will describe below.

3.3
Enzyme Activation in Nonaqueous Media

As described above (Figure 3.1), many factors appear to adversely affect the catalytic function of enzymes in organic solvents. Nonetheless, the situation is not hopeless, and the literature is replete with examples of the activation of enzymes through relatively simple methodologies, often with striking results. Because most enzymes are used as heterogeneous catalysts in organic solvents, the vast majority of techniques developed to date to overcome poor enzyme activity in nonaqueous media involve modifications to the enzyme through lyophilization. However, some activating approaches have resulted from the *solubilization* of enzymes in the organic solvent. These techniques, while beyond the scope of the current chapter, nonetheless provide additional routes to the design and use of highly active enzyme preparations. The highly diverse array of methods to activate enzymes in organic media is summarized in Table 3.1, and many of these examples are discussed below.

3.3.1
Addition of Water and Water Mimics

Enzymes are very sensitive to small changes in the water content of a predominantly nonaqueous medium. This affect was studied extensively by Affleck et al., who measured changes in activity, conformation, and active site polarity of subtilisin Carlsberg in anhydrous THF as small amounts of water were added [54]. In the absence of added water, the enzyme's structure and relative catalytic activity was very low in THF. The water associated with the enzyme [ca. 9% (w/w)], was strongly bound to ionizable groups on the protein [55]. As water was added to the organic solvent, the active-site polarity and flexibility increased, as well as the transesterification activity (Figure 3.4). A critical point was reached at about 15% (w/w) water (5–20 µl H_2O/ml THF), in which the polarity of the active site leveled off (Figure 3.4b) and the transesterification activity began to decline (Figure 3.4a). Enzyme flexibility continued to increase with added water, causing conformational changes, even below monolayer coverage hydration (estimated to be ca. 35% for subtilisin). These conformational changes are believed to cause the observed decrease in transesterification activity.

In these studies, water has been shown to increase both flexibility and active-site polarity, increasing enzyme activity. It would seem likely that other solvents with the same characteristic as water in these regards (the ability to form multiple hydrogen bonds with the enzyme) would have similar effects. This is indeed the case with a variety of hydrogen bond-forming solvents [31, 63] which act as "water

Table 3.1 Methods for enzyme activation for use in non-aqueous media.

Additive	Enzyme	Solvent	Activation	Reference
Addition of water and water mimics				
Addition of 4% H_2O	Thermolysin	tert-Amyl alcohol	3500	31
Addition of 1% H_2O + 9% formamide	Thermolysin	"	3800	31
Addition of 1% H_2O + 9% THF	Thermolysin	"	25	31
Addition of 40% Formamide	subtilisin Carlsberg	Acetonitrile	155	32
Lyophilization in the presence of excipients				
Lyoprotectants				
Trehalose	subtilisin Carlsberg	Hexane	4	33
Sucrose	subtilisin Carlsberg	"	14	33
Methyl-β-cyclodextrin	subtilisin Carlsberg	1,4-Dioxane	100	34
Sorbitol	α-chymotrypsin	Pyridine	19	35
Sorbitol	α-chymotrypsin	Diisopropyl ether	2	36
Tris	Horseradish Peroxidase	97% Acetone	27	37
Sorbitol	Protease (*Aspergillus oryzae*)	Pyridine	64	35
Sucrose	Protease (*Aspergillus oryzae*)	"	38	35
Xylitol	Protease (*Aspergillus oryzae*)	"	32	35
Trehalose	Protease (*Aspergillus oryzae*)	"	26	35
Mannitol	Protease (*Aspergillus oryzae*)	"	20	35
Polymers				
Poly(vinyl pyrrolidone)	α-chymotrypsin	Isooctane	1500	38
Poly(ethylene glycol) + Trehalose	subtilisin Carlsberg	n-Octane	70	39
Trans-1,2-cyclohexanol and poly(ethylene glycol)	Soybean peroxidase	Acetone	800	40
Poly(ethylene glycol)	Soybean peroxidase	"	35	37
Poly(ethylene glycol)	Horseradish peroxidase	97% Acetone	7	37
Poly(ethylene glycol)	Protease (*Aspergillus oryzae*)	Pyridine	8	35
Substrates/Inhibitors				
Thymidine	subtilisin Carlsberg	Tetrahydrofuran	50	41
N-Ac-Tyr-NH_2	subtilisin Carlsberg	Octane	55	42
Salts				
KCl (98%, v/v)	Lipase (*Humicola lanuginosa*)	Hexane	46	43
KCl (98%, v/v)	Lipase (*Mucor javanicus*)	Toluene	5	44
KCl (98%, w/w)	Penicillin amidase	Hexane	755	45
KCl (98%, w/w)	Penicillin amidase	Toluene	354	45
KCl (98%, w/w)	Penicillin amidase	Ethyl acetate	214	45
KAc (98%, w/w)	Penicillin amidase	Hexane	16900	46
CsCl (98%, w/w)	Penicillin amidase	"	25400	46
KCl (98%, v/v)	subtilisin Carlsberg	"	1920	47

Table 3.1 Continued

Additive	Enzyme	Solvent	Activation	Reference
Protein Engineering				
D60N + Q103R + N218S mutant	subtilisin E	85% Dimethylformamide	38	48
Gly$_{166}$ → Asn	subtilisin BPN'	Octane	370	49
Asp$_{248}$ Asn + Asn$_{218}$ → Ser	subtilisin E	80% Dimethylformamide	4	50
Immobilization				
Nanoporous Sol-gel glass	α-chymotrypsin	Hexane	110	51
Nylon-66 membrane	Laccase (*Trametes versicolor*)	Diethyl ether	52	52
Nylon-66 membrane	Laccase (*Trametes versicolor*)	Ethyl acetate	89	52
Nylon-66 membrane	Laccase (*Trametes versicolor*)	Methylene chloride	5	52
Solid State Buffers				
AMPSO/NaAMPSO	α-chymotrypsin	Acetonitrile	9	53
AMPSO/NaAMPSO	subtilisin Carlsberg	Tetrahydrofuran	24	53
HEPES/KHEPES	α-chymotrypsin	Acetonitrile	5	53
HEPES/KHEPES	subtilisin Carlsberg	Tetrahydrofuran	15	53

mimics". The effects of these water mimics can be seen in Figure 3.6; thermolysin's activity in *tert*-amyl alcohol varies with the co-solvent's ability to form hydrogen bonds with the enzyme [31].

These observations reveal that enzyme hydration below monolayer values has an effect on enzyme activity, and in some cases such activation can be dramatic. However, in all these examples, water was added and this may not be appropriate where nearly anhydrous reaction conditions are necessary.

3.3.2
Immobilization

Denaturation and the potential for diffusional limitations (due to the insolubility of native enzymes in organic solvents) are significant concerns that have influenced researchers in the nonaqueous enzymology field since its inception. One way to overcome both concerns is to employ the rather mature technique of attaching an enzyme to a solid support. This would aid in stabilization [64] and facilitate transport of the substrate to the enzyme to reduce potential diffusional limitations [65–67]. Examples of insoluble supports available for covalent enzyme attachment are glass [68], alumina [69], silica [69], and chitosan [70]. There are numerous advantages to these systems. First, the deposition of an enzyme onto a solid surface can increase the interfacial surface area between the protein and the solvent, thereby increasing the reaction rate [71–73]. Second, the chemical nature of the support can affect the partitioning of both the water and the substrate [74]. Hydrophilic beads have been shown to retain an aqueous microenvironment that aids

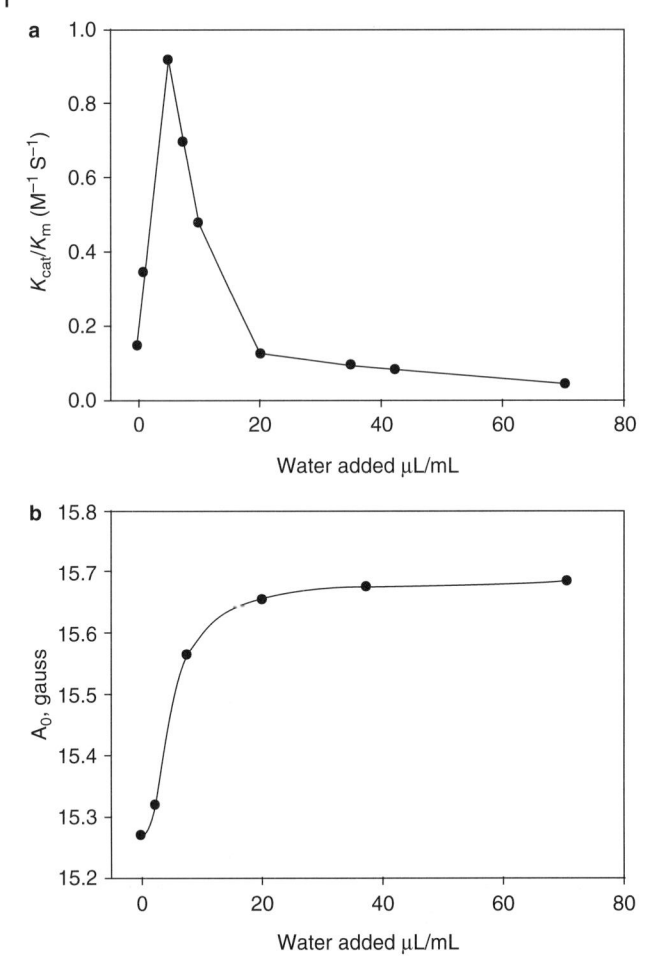

Figure 3.4 (a) Catalytic efficiencies of subtilisin Carlsberg in THF with different water contents. (b) Polarity of subtilisin's active site (as indicated by the hyperfine splitting constant, A_0, of the active-site bound spin label) with different water contents [54].

catalysis in nonaqueous media [65, 75, 76], leading to methods for choosing immobilization materials that compensate for water stripping by highly polar solvents [3, 77].

Immobilization of enzymes to solid supports can increase activity over a wide range of solvents [78]. As seen in Table 3.2, the transesterification of N-acetyl-L-phenylalanine ethyl ester (APEE) with 1-propanol by α-chymotrypsin (Scheme 3.2) immobilized to glass is 1–2 orders of magnitude higher than that of the free, lyophilized enzyme.

This universal activation does not appear to be due to the solvent's ability to change the water distribution around the enzyme. This can be deduced because

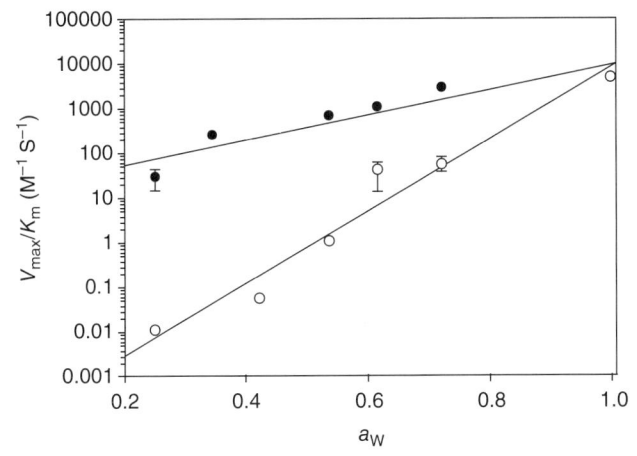

Figure 3.5 Catalytic efficiency of SBP-catalyzed oxidation of p-cresol in acetonitrile (●) and ethyl acetate (○) as a function of a_w. Enzymatic activity increases over 5 orders of magnitude as water is added to the system [60].

Scheme 3.1 SBP oxidation of p-cresol to form polyphenol product.

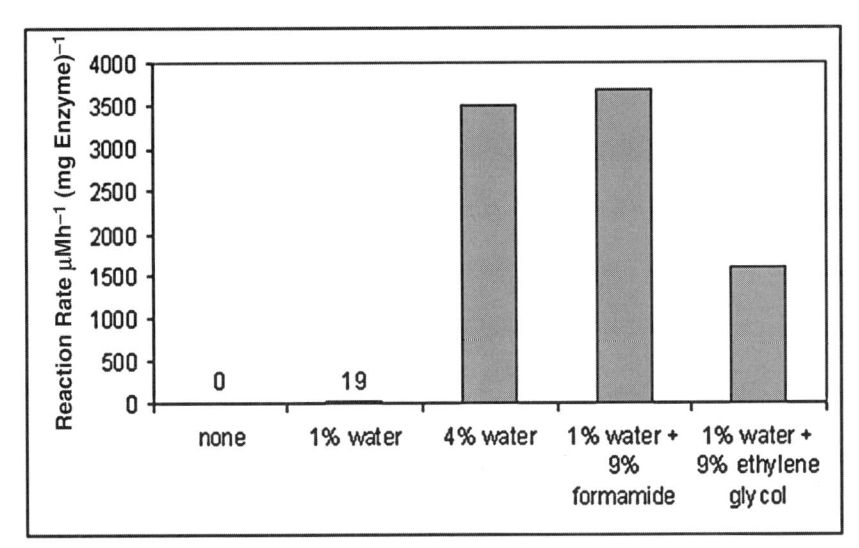

Figure 3.6 Initial rates of the reaction between Z-Gly-Gly-Phe and Phe-NH$_2$ catalyzed by thermolysin in tert-amyl alcohol containing various cosolvents [31].

Table 3.2 Values of $(k_{cat}/K_m)_{app}$ (M^{-1} min^{-1}) × 100 for freely suspended and glass-immobilized α-chymotrypsin in various solvents. In all solvent conditions, the enzyme is 1–2 orders of magnitude more active than the free, lyophilized enzyme [78].

Solvent	Freely suspended chymotrypsin	Immobilized chymotrypsin	Ratio[a]
n-Octane	221	19 120	87
Carbon tetrachloride	15	2 150	141
Toluene	7.7	557	72
Diethyl ether	33	1 370	42
Tetrahydrofuran	7.2	91	13
Acetonitrile	0	56	–

a Ratio defined as: $(k_{cat}/K_m)_{app,immobilized}/(k_{cat}/K_m)_{app,suspended}$.

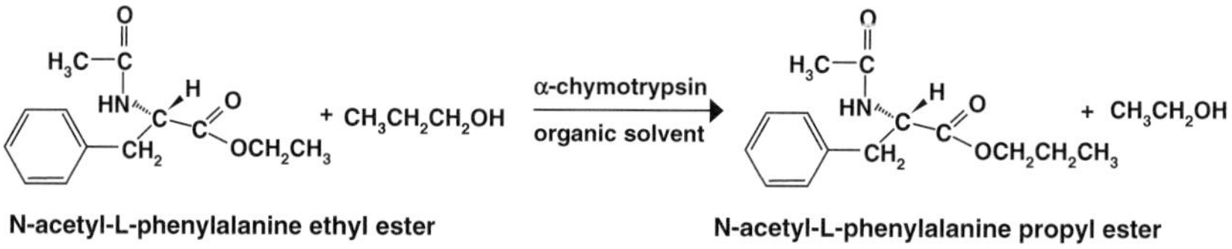

Scheme 3.2 Trans, of N-acetyl-L-phenylalanine ethyl ester with 1-PrOH by α-chymotrypsin to form N-acetyl-L-phenylalanine propyl ester.

the activation did not correlate with the polarity of the solvent, and Karl Fisher titration revealed that the water contents of the immobilized enzyme and the glass beads alone were similar [3, 78]. Electron Paramagnetic Resonance (EPR) spectroscopy coupled with computer simulations indicated that the active site of the immobilized enzyme was more flexible, and its polarity increased as water content was increased in a THF system as compared to the suspended enzyme [78].

3.3.3
Solid-State Buffers

Lyophilized enzymes have a pH memory, meaning that the activity of the enzyme in organic solvent parallels its pH-activity profile of the aqueous solution from which it was lyophilized [36, 79–81]. However, very often acidic or basic mixtures within a nonaqueous reaction mixture such as reactant, products, or impurities, can disrupt this delicate protonation state, leading to changes in catalytic activity. To counteract this potential problem, solid-state buffers have been developed to protect the enzyme's protonation state in the nonaqueous environment [53, 82]. These solid-state buffers contain pairs of crystalline solids that can be intercon-

Table 3.3 Variation of catalytic rate in the presence of different solid state buffers [53].

Solid-state buffer	Rate nmol mg^{-1} min^{-1} α-chymotrypsin in CH$_3$CN	Subtilisin in THF
None	11.9	2.2
AMPSO/NaAMPSO	108	52
HEPES/KHEPES	59	33.3
MOPS/NaMOPS	35.9	12.1

verted by exchange of H$^+$ and another ion, such as Cl$^-$. In this method, enzymes are immobilized to a solid support and then exposed to a given solid-state buffer pair, allowing the enzyme to exchange protons and counterions until an equilibrium protonation state is reached.

This method was employed for transesterification reactions with both α-chymotrypsin and subtilisin Carlsberg with a variety of H$^+$/Na$^+$ buffers [53]. With both enzymes (which differ widely in secondary and tertiary structures) and two polar solvents, acetonitrile and THF, the activating effect of the solid-state buffer was clearly evident (Table 3.3). The observation that a variety of buffer pairs show success in activating two dissimilar enzymes in synthetically useful solvents makes this method for activation promising and novel.

3.3.4
Lyophilization in the Presence of Excipients

It is well known that enzymes can be deactivated by lyophilization, often through the process of dehydration. Nonetheless, large increases in nonaqueous activity are observed when enzymes are lyophilized in the presence of lyoprotectant-based excipients [35, 83]. Lyoprotectants acts as a substitute for water molecules (which are removed in the lyophilization process) by forming H-bonds with the protein structure [13]. It has been shown that the absence of lyoprotectants during the freeze drying process leads to both a random enzyme structure and an increase in the β-sheet content of the protein [13]. Some examples of excipients that have shown this combined activation and structural protection are sugars (sucrose, trehalose, lactose, maltose, inulin), polyols (sorbitol, mannitol), amino acids (glycine, glutamine, histidine, leucine, threonine, arginine, lysine), and polymers [poly-(vinyl pyrroline), poly-(ethylene glycol)] [84]. Studies with FTIR have shown that lyophilization in the presence of lyoprotectants results in an increase in the fraction of protein in a native conformation as opposed to lyophilizing without the additive [28].

Lyoprotectants can affect enzyme stability in both stages of lyophilization: the freezing and the drying stages. In the freezing stage of lyophilization, ice crystals form and have been shown to be a cause of enzyme denaturation. Studies have shown that when added as a lyoprotectant, the amorphous polyol mannitol stabilizes the enzymes β-galactosidase and L-asparaginase. However, if the crystallinity of the mannitol increases because of annealing during the freezing stage, the stabilizing effect is completely lost [85]. Carbohydrates are a good example of an additive that targets the dehydration stage of lyophilization, preserving hydrogen bonds by acting as water substitutes, helping to maintain an active conformation of the enzyme during the loss of water [28, 86].

As a result, various excipients (from small molecules to polymers) have been added to the aqueous solution prior to lyophilization. Some of the most common are carbohydrates [86], polymers [87], organic buffer [81], competitive inhibitors [42], or nonbuffer salts [88].

3.3.4.1 Polymers

Polymers, such as poly-(ethylene glycol) (PEG), have been shown to activate enzymes for use in nonaqueous media when co-lyophilized with the enzyme prior to use [35, 37, 39]. Many polymers are classified as cryoprotectants, protecting the enzyme during the freezing stage of lyophilization [13]. These additives are thought to protect the enzyme from denaturation prior to the loss of bound water [37]. Other hypotheses have suggested that polymers such as PEG function as chaperones, helping a protein to fold into its native structure prior to lyophilization [89 91]. Both of these explanations suggest that polymers such as PEG interact directly with the protein molecule, an important characteristic examined by Mi et al. [92] when they used structural and binding assays to study the effects different molecular weight PEGs had on the cryoprotection of lactate dehydrogenase. Their findings suggested that there were extensive non-specific interactions between the PEG and the enzyme in a concentration-dependent manner, protecting the enzyme's native secondary structure during freeze drying, resulting in an increase in activity. This method of protection has been applied to enzymes in organic solvents [37], increasing activity multiple times over that of the free lyophilized enzyme.

3.3.4.2 Crown Ethers and Cyclodextrins

Crown ethers are heterocyclic chemical compounds that, in their simplest form, are cyclic oligomers of ethylene oxide. The essential repeating unit of any simple crown ether is ethyleneoxy, i.e., —CH_2CH_2O—, which repeats twice in dioxane and six times in 18-crown-6. Crown ethers can activate enzymes for use in organic solvents through two methods: (a) direct addition of 18-crown-6 to the reaction solvent [93], or (b) co-lyophilization of the enzyme with 18-crown-6, the latter being the most effective [94, 95].

Crown ethers may serve a dual role in their activating behavior. It is thought that crown ethers reduce the formation of inter- and intramolecular salt bridges in suspended enzymes [96]. It is known that intermolecular salt bridge formation leads to the clustering of proteins, and intramolecular salt bridges result in a lower enzyme activity because of conformation changes and reduced flexibility, and may block the active site. Since 18-crown-6 is an effective chelator of amino groups [97], it may form complexes with cationic lysine residues on the surface of an enzyme, thus screening the charge and preventing undesirable salt bridge formation.

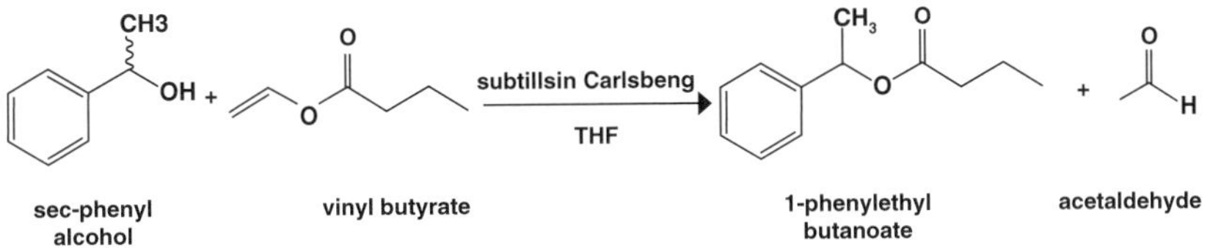

Scheme 3.3 Tran, of *sec*-phenethyl alcohol with vinyl butyrate by subtilisin Carlsberg in THF.

When 18-crown-6 was co-lyophilized with α-chymotrypsin, a 470-fold activation was seen over the free enzyme in the transesterification of APEE with 1-propanol in cyclohexane (Scheme 3.2) [96]. There was a low apparent specificity for the size and macrocyclic nature of the crown ether additives, suggesting that, during lyophilization, 18-crown-6 protects the overall native conformation and acts as a lyoprotectant. To examine this global effect, FTIR was used to examine the effect of crown ethers on the secondary structure of enzymes. In one study [98], subtilisin Carlsberg was shown to retain its secondary structure in 1,4-dioxane when lyophilized in a 1:1 ratio with 18-crown-6. In addition, examination of FTIR spectra from varying incubation temperatures indicated that an increase in crown ether content in the final enzyme preparation resulted in a decreased denaturation temperature in the solvent, indicating a more flexible protein structure.

Cyclodextrins (sometimes called cycloamyloses) make up a family of cyclic oligosaccharides composed of 5 or more α-D-glucopyranoside units linked 1 → 4, as in amylose (a fragment of starch), and have been used as activating macrocyclic excipients in nonaqueous biocatalysis. Griebenow et al. observed that lyophilization of methyl-β-cyclodextrin with subtilisin Carlsberg in a 6:1 weight ratio resulted in a 164-fold rate enhancement in THF for the transesterification rate of *sec*-phenethyl alcohol with vinyl butyrate (Scheme 3.3) [98].

3.4 Salt-Activated Enzymes

A particularly interesting, and extremely simple, method of activating enzymes for use in nonaqueous media is lyophilization in the presence of nonbuffer salts. In the process, a great deal about excipient-enzyme-water interactions has been learned along with an appreciation of how enzymes adjust to novel microenvironments while retaining their intrinsic catalytic properties.

Khmelnitsky et al. were the first to observe the activating effects salt showed on enzymes in the nonaqueous environment [88]. As shown in Figure 3.7, the transesterification activity of the serine protease subtilisin Carlsberg in anhydrous solvents is strongly dependent on the KCl content in a lyophilized enzyme preparation and increases sharply as the salt content is increased. This increase in activity was determined to be a result primarily of an increase in k_{cat} and not a decrease in K_m, as shown in (Table 3.4).

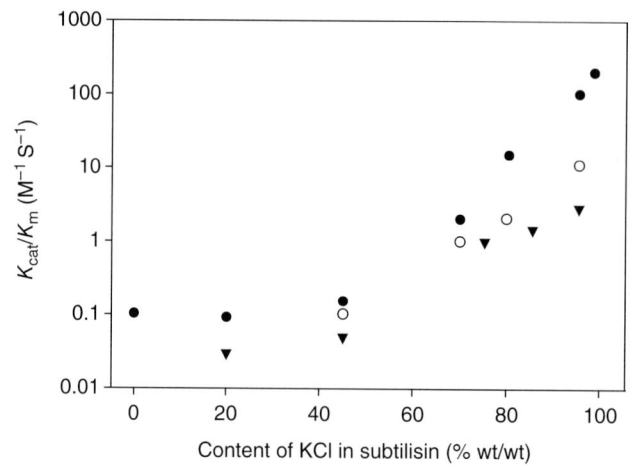

Figure 3.7 Catalytic activity of subtilisin in anhydrous organic solvents (● n-hexane, ○ diisopropyl ether, ▼ THF) as a function of the KCl content in the dry catalyst. The activity is expressed in terms of k_{cat}/K_m of the transesterification reaction between N-acetyl-L-phenylalanine ethyl ester and n-propanol, used in concentrations of 10 mM and 0.85 M, respectively [88].

Table 3.4 Effect of KCl as a salt matrix on subtilisin Carlsberg and chymotrypsin in anhydrous hexane[a] [88].

Enzyme	Salt content (%, w/w)	k_{cat} (s^{-1})	K_m (mM)	k_{cat}/K_m (M^{-1}s^{-1})
Subtilisin	0	0.027	260	0.104
	98	10.4	26.7	390
Chymotrypsin[b]	0	4.2×10^{-4}	33.0	0.013
	94	220×10^{-4}	33.0	0.67

a For the transesterification reaction of APEE with 0.85 M 1-propanol.
b The concentration of active sites of α-chymotrypsin in n-hexane was assumed to be approximately equal to that found in aqueous solutions (66% of the total protein). It should be noted that, even in the active fraction of active sites for α-chymotrypsin in the nearly pure enzymes, suspension was only 10% (the same as for subtilisin), and the fraction of active sites in the salt-rich catalyst was 66%, only a 6-fold enhancement in k_{cat} (and thus k_{cat}/K_m) would result.

3.4.1
Salt Activation is not due to a Relaxation of Diffusional Limitations

Could this salt activation phenomenon be a result of relaxed diffusional limitations in a concentrated salt/enzyme formulation as compared to the salt-free preparation? To answer this question, Bedell et al. [99] measured the initial rates of subtilisin Carlsberg-catalyzed transesterification of APEE with n-PrOH in hexane (Scheme 3.4) with two different enzyme preparations (Figure 3.8) (a) 98% (w/w)

3.4 Salt-Activated Enzymes

Scheme 3.4 Trans, of N-acetyl-L-phenylalanine ethyl ester with 1-PrOH by subtilisin Carlsberg to form N-acetyl-L-phenylalanine propyl ester.

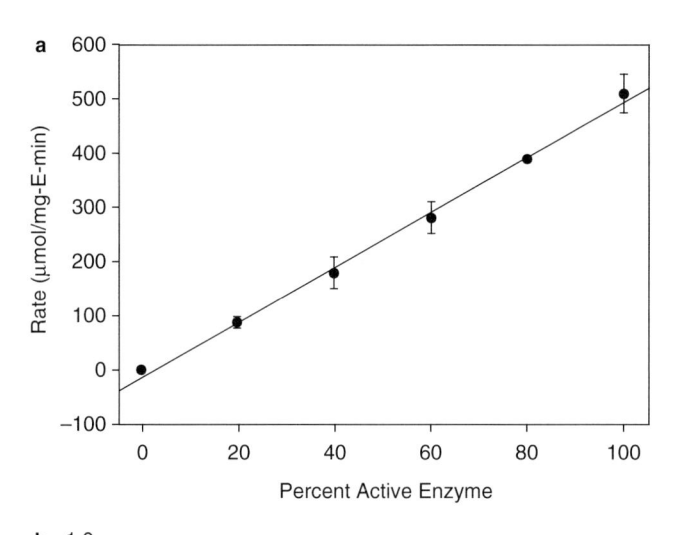

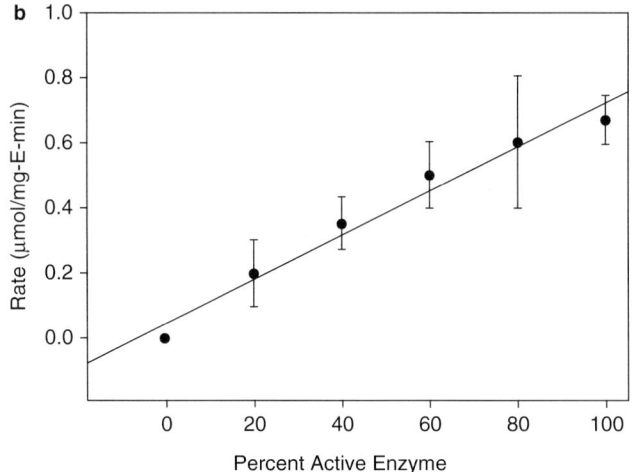

Figure 3.8 (a), (b) Initial rates of subtilisin Carlsberg-catalyzed transesterification of N-Ac-L-Phe-OEt with n-PrOH in hexane as a function of active enzyme in different salt-enzyme preparations. Catalyst composition (a) 98% (w/w) KCl, 1% (w/w) SC, and 1% phosphate buffer (average of three preparations), (b) 99% (w/w) subtilisin and 1% phosphate buffer (average of five preparations) [99].

KCl, 1% (w/w) SC, and 1% phosphate buffer; (b) 99% (w/w) SC and 1% phosphate buffer. These initial rates were plotted against the percentage of active enzyme in the lyophilized sample (the fraction of active enzyme is an adequate measure of enzyme concentration in these systems, because the total weight of enzyme per weight of catalyst is constant for each biocatalyst preparation).

This study employed conventional diffusion-reaction theory, showing that with diffusion-limited reactions the internal effectiveness factor of a heterogeneous catalyst is inversely related to the Thiele modulus. Using a standard definition of the Thiele modulus [100], the observed reaction rate of an immobilized-enzyme reaction will vary with the square root of the immobilized-enzyme concentration in a diffusion-limited system. In this case, a plot of the reaction rate versus the enzyme loading in the catalyst formulation will be nonlinear.

3.4.2
Mechanism of Salt Activation

Salts provide a wonderfully diverse group of small molecules that can bind to macromolecules, structure water, and influence the stability and activity of biological molecules. In addition, the physicochemical behavior of salts can be quantified, much like the behavior of organic solvents. As a result, quantitative analysis of enzyme structure and function in different microenvironments can be achieved. One of these quantitative parameters is the Jones-Dole B coefficient (JDB). Specifically, cations and anions can be categorized by the JDB, which measures the degree to which a salt increases or decreases the viscosity of water [101]. Ions can be grouped into two categories; kosmotropes or chaotropes. Kosmotropes are assigned positive JDB coefficients and exhibit strong interactions with water, increasing its surface tension. Conversely, chaotropes are expressed by negative JDB coefficients and exhibit weaker interactions with water than water does with itself, subsequently decreasing the surface tension of water [9].

One of the first attempts to quantify the behavior of salts on enzyme function in organic solvents was performed by Ru et al. [33]. Subtilisin Carlsberg was lyophilized with different salts (containing kosmotropic or chaotropic anions), and the reactivity of the enzyme (preparations containing 98%, w/w, of the salt, 1%, w/w, of the enzyme, and 1%, w/w, of phosphate buffer) in the transesterification of APEE with 1-PrOH in hexane was evaluated (Figure 3.9). The catalytic efficiency (k_{cat}/K_m) of the transesterification reaction increased as the kosmotropicity of the anion increased. It was suggested that highly kosmotropic salts act as both cryo- and lyoprotectants during freeze drying, giving a more robust catalyst resistant to the denaturing effect of the solvent than less kosmotropic, or chaotropic, salts.

Similar results were also found for penicillin amidase-catalyzed transesterification of phenoxyacetate methyl ester with 1-propanol in hexane (Scheme 3.5) [46]. The initial rates of various penicillin amidase preparations containing acetate or chloride salts and their 1:1 binary mixtures were measured, which resulted in the identification of formulations with various degrees of activation. For example, the

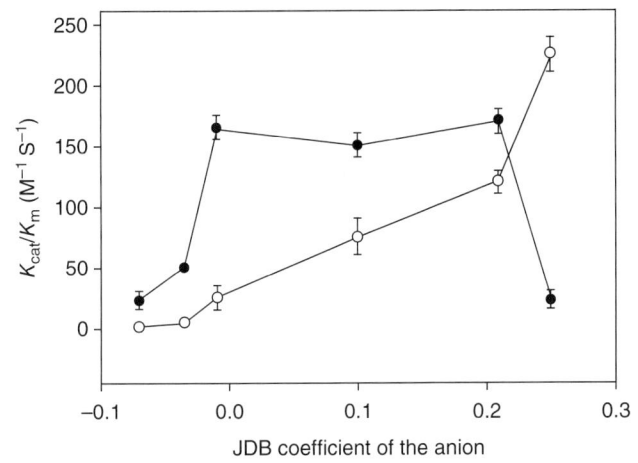

Figure 3.9 The catalytic efficiency, k_{cat}/K_m (M^{-1} s^{-1}), of subtilisin Carlsberg activated by various sodium (○) and potassium (●) salts. Values represent the average of four trials from lyophilized enzyme samples prepared independently, and the error bars represent one standard deviation from the mean [33]. The low activity for potassium acetate may have been due to the poor lyophilization of this salt, which undergoes melting on the freeze dryer.

Scheme 3.5 Transesterification of phenoxyacetate methyl ester with 1-PrOH in hexane by penicillin amidase.

V_{max}/K_m value of a potassium acetate preparation was 22-fold more active than a KCl preparation, and ~17 000-fold more active than the salt-free preparation. Moreover, a preparation containing equal amounts of potassium acetate and KCl was more active than either of the two salts alone, suggesting that the mixture of the two salts has unique properties that are distinct from the individual salts, and that the salt effect is not simply an additive function of composition.

These results were then correlated to the Jones-Dole B coefficient to investigate the dependence of enzyme activation on the kosmotropicity of the salt in a solvent such as hexane. Specifically, plotting enzyme activity as a function of the difference in JDB coefficients of the cations and anions of the salts, resulted in a clear trend towards increased enzyme activity when the difference between the kosmotropicity of the anion and the chaotropicity of the cation was increased (Figure 3.10) [46]. These results were consistent with those of Ru et al. [33], in that enzyme activity in salt-activated preparations in hexane positively correlates with increased kosmotropicity on the anion. As a result of the elucidation of the influence of the kosmotropic/chaotropic properties of salts on enzyme function, the role of water

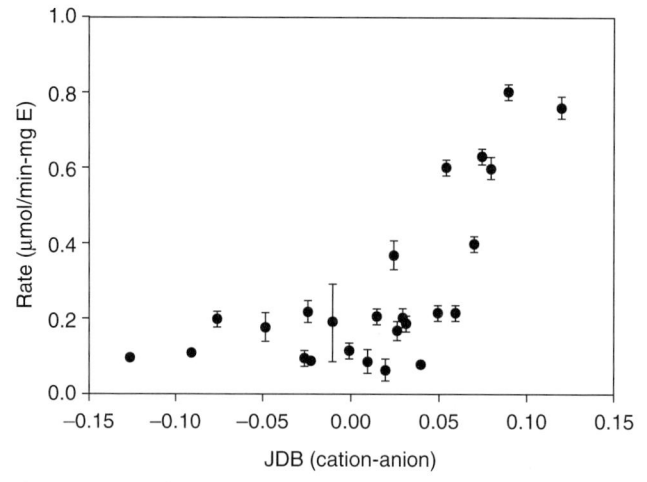

Figure 3.10 Correlation between the observed reactivity of penicillin amidase formulations and the difference in the Jones-Dole B coefficients of the kosmotropic anion and the chaotropic cation for single salt and binary salt mixtures [46].

on enzyme behavior in nonaqueous media has once again become a critical parameter that defines enzyme activity. Moreover, the molecular properties of protein-associated water remains a key factor in further understanding enzyme activity in nonaqueous media.

3.4.3
The Structural and Molecular Dynamics of Salt Activation

Solvent polarity is known to affect catalytic activity, yet consistent correlations between activity and solvent dielectric (ε) have not been observed [12, 102]. However, a striking correlation was found between the catalytic efficiency of salt-activated subtilisin Carlsberg and the mobility of water molecules (as determined using NMR relaxation techniques) associated with the enzyme in solvents of varying polarities (Figure 3.11) [103]. As the solvent polarity increased, the water mobility of the enzyme increased, yet the catalytic activity of the enzyme decreased. This is consistent with previous EPR and molecular dynamics (MD) studies, which indicated that enzyme flexibility increases with increasing solvent dielectric [104].

Enzyme flexibility is greater in solvents with high polarity because of weaker electrostatic interactions in these solvents [54, 104, 105]. The loss in enzyme activity seen in the NMR study described above may be attributed to the "water stripping" model; as water is stripped from the enzyme, locations in and on the enzyme previously inaccessible to the solvent may become accessible, thus permitting increased solvent-enzyme interactions [103]. As a result, enzyme structure may be disrupted (e.g., partially denatured), and catalytic activity is decreased. The partially denatured enzyme appears to exhibit greater flexibility as solvent polarity increases [106, 107].

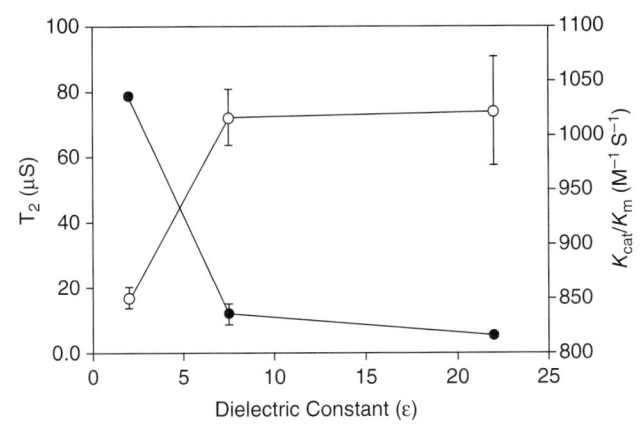

Figure 3.11 Catalytic efficiency, $(k_{cat}/K_m)_{app}$ (●), of salt-activated subtilisin Carlsberg in hexane, THF, and acetone in comparison with T_2 (transverse relaxation constant) (○) of mobile deuterons as a function of dielectric constant of solvent [103].

Eppler et al. [103] viewed these results as having a potential relationship to salt-activated enzyme preparations, particularly in relation to the mobility of enzyme-bound water. Specifically, the authors examined both water mobility [as measured by T_2-derived correlation times, $(\tau_c)_{D2O}$] and NaF-activated enzyme activity and observed a linear relationship. This suggests that the salt-activated enzymes contain a more mobile water population than salt-free enzymes, which facilitates a more aqueous-like local environment and dramatically increases enzyme activity through increased flexibility. Therefore, enzyme activation appears to correlate with the properties of enzyme-associated water. Once again, the physicochemical properties of water dictate enzyme structure, function, and dynamics. Hence, salt activation has proven to be a useful technique in activating enzymes for use in organic solvents and has provided a quantitative tool to better understand the role of water in enzymatic catalysis in dehydrated media.

3.5 Conclusions

We have focused this chapter on some of the basic principles behind enzymatic catalysis in nonaqueous enzymology, with specific attention placed on the deactivating effects of organic solvents and the re-activating effects of simple lyophilized excipients. Through increased structural and kinetic characterization and introduction of molecular dynamics and other computational tools, the study of enzymes in dehydrated environments provides a window into the broader understanding of enzymatic catalysis in both aqueous and nonaqueous milieu. Beyond organic solvents lie a range of other nonaqueous media, including room-temperature ionic liquids, vapors, and polymer melts. The gleanings of enzymatic catalysis in organic solvents, and the role of water in enzyme structure, function, and dynamics, will help us better design nonaqueous systems for commercial exploitation that involves all three of these media in addition to standard organic solvents. For example, combining enzyme design (e.g., to position water in regions of the protein to maximize protein dynamics) with excipient lyophilization may provide an intricate, and potentially more exploitable, understanding of enzyme function in distinctly nonaqueous environments. As a result, new routes to the preparation of highly active and stable biocatalysts will be developed for synthetic applications, sensors, functional biologically-based composites, or nanodevices that will ensure the bright future of biocatalysis.

References

1 Dordick, J. S. (1989) *Enzyme Microb. Technol.* **11**, 194–211.
2 Lee, M. Y., Dordick, J. S. (2002) *Curr. Opin. Biotechnol.* **13**, 376–384.
3 Gorman, L. S., Dordick, J. S. (1992) *Biotechnol. Bioeng.* **39**, 392–397.
4 Tester, J. W., Cline, J. A. (1999) *Corrosion* **55**, 1088–1100.
5 Soo-Hoo, T. S., Brown, A. D. (1967) *Biochim. Biophys. Acta* **135**, 164–166.
6 Lanyi, J. K. (1974) *Bacteriol. Rev.* **38**, 272–290.
7 Mershin, A., Kolomenski, A. A., Schuessler, H. A., Nanopoulos, D. V. (2004) *BioSystems* **77**, 73–85.
8 Guttman, H. S., Cayley, J. S., I, M., Anderson, C. F., Record, M. T. (1995) *Biochemistry* **34**, 1393–1404.
9 Collins, K. D. (1997) *Biophys. J.* **72**, 65–76.
10 Kim, J., Dordick, J. S. (1997) *Biotechnol. Bioeng.* **55**, 471–479.
11 Yang, L., Dordick, J. S., Garde, S. (2004) *Biophys. J.* **87**, 812–821.
12 Michels, P. C., Dordick, J. S., Clark, D. S. (1997) *J. Am. Chem. Soc.* **119**, 9331–9335.
13 Gupta, M. N., Roy, I. (2004) *Eur. J. Biochem.* **271**, 2575–2583.
14 Arakawa, T., Prestelski, S. J., Kenney, W. C., Carpenter, J. F. (2001) *Adv. Drug Deliv. Rev.* **46**, 307–326.
15 Fagain, C. O. (1996), ed. Doonan, S. (Humana Press, Totowa, NJ), pp. 323–327.
16 Roy, I., Gupta, M. N. (2004) *Biotechnol. Appl. Biochem.* **39**, 165–177.
17 Anchordoquy, T. J., Izutsu, K.-I., Randolph, T. W., Carpenter, J. F. (2001) *Arch. Biochem. Biophys.* **390**, 35–41.
18 Brulls, M., Folestad, S., Sparen, A., Rasmuson, A. (2003) *Pharm. Res.* **20**, 494–499.
19 Rupley, H. A., Gratton, E., Careri, G. (1983) *Trends Biochem. Sci.* **8**, 18–22.
20 Poole, P. L., Finney, J. L. (1983) *Int. J. Biol. Macromol.* **5**, 308–310.
21 Falconi, M., Brunelli, M., Pesce, A., Ferrario, M., Bolognesi, M., Desideri, A. (2003) *Proteins: Struct. Funct. Genet.* **51**, 607–615.
22 Klibanov, A. M., Ahern, T. J. (1987) in *Protein Engineering*, eds. Oxender, D. L., Fox, C. F. (Alan R. Liss, New York), pp. 213–218.
23 Dong, A., Meyer, J. D., Kendrick, B. S., Manning, M. C., Carpenter, J. F. (1996) *Arch. Biochem. Biophys.* **334**, 406–414.
24 Yu, N. T., Jo, B. H. (1973) *Arch. Biochem. Biophys.* **156**, 469–474.
25 Poole, P. L., Finney, J. L. (1983) *Biopolymers* **22**, 255–260.
26 Gregory, R. B., Gangoda, M., Gilpin, R. K., Su, W. (1993) *Biopolymers* **33**, 1871–1876.
27 Desai, U. R., Osterhout, J. J., Klibanov, A. M. (1994) *J. Am. Chem. Soc.* **116**, 9420–9422.

28 Prestrelski, S. J., Arakawa, T., Carpenter, J. F. (1993) *Arch. Biochem. Biophys.* **303**, 465–473.
29 Prestrelski, S. J., Tedeschi, N., Arakawa, T., Carpenter, J. F. (1993) *Biophys. J.* **65**, 661–671.
30 Griebenow, K., Klibanov, A. M. (1995) *Proc. Natl. Acad. Sci. USA* **92**, 10969–10976.
31 Kitaguchi, H., Klibanov, A. M. (1989) *J. Am. Chem. Soc.* **111**, 9272–9273.
32 Almarsson, O., Klibanov, A. M. (1995) *Biotechnol. Bioeng.* **49**, 87–92.
33 Ru, M. T., Hirokane, S. Y., Lo, A. S., Dordick, J. S., Reimer, J. A., Clark, D. S. (2000) *J. Am. Chem. Soc.* **122**, 1565–1571.
34 Montanez, C. I., Alvira, E., Macias, M., Ferrer, A., Fonceca, M., Rodriguez, J., Gonzalez, A., Barletta, G. (2002) *Biotechnol. Bioeng.* **78**, 53–59.
35 Dabulis, K., Klibanov, A. M. (1993) *Biotechnol. Bioeng.* **41**, 566–571.
36 Triantafyllou, A. O., Wehtje, E., Adlercreutz, P., Mattiasson, B. (1997) *Biotechnol. Bioeng.* **54**, 67–76.
37 Dai, L., Klibanov, A. M. (1999) *Proc. Natl. Acad. Sci. USA* **96**, 9475–9478.
38 Murakami, Y., Hoshi, R., Hirata, A. (2003) *J. Mol. Cat. B: Enzymatic* **22**, 79–88.
39 Roy, I., Sharma, A., Gupta, M. N. (2004) *Bioorg. Med. Chem. Lett.* **14**, 887–889.
40 Ozawa, S., Klibanov, A. M. (2000) *Biotechnol. Lett.* **22**, 1269–1272.
41 Rich, J. O., Dordick, J. S. (1997) *J. Am. Chem. Soc.* **119**, 3245–3252.
42 Russell, A. J., Klibanov, A. M. (1988) *J. Biol. Chem.* **263**, 11624–11626.
43 Persson, M., Mladenoska, I., Wehtje, E., Adlercreutz, P. (2002) *Enzyme Microb. Technol.* **31**, 833–841.
44 Altreuter, D. H., Dordick, J. S., Clark, D. S. (2002) *J. Am. Chem. Soc.* **124**, 1871–1876.
45 Lindsay, J. P., Clark, D. S., Dordick, J. S. (2002) *Enzyme Microb. Technol.* **31**, 193–197.
46 Lindsay, J. P., Clark, D. S., Dordick, J. S. (2004) *Biotechnol. Bioeng.* **85**, 553–560.
47 Ru, M. T., Wu, K. C., Lindsay, J. P., Dordick, J. S., Reimer, J. A., Clark, D. S. (2001) *Biotechnol. Bioeng.* **75**, 187–196.
48 Chen, K., Arnold, F. (1991) *Bio/Technology* **9**, 1073–1077.
49 Wangikar, P. P., Michels, P. C., Clark, D. S., Dordick, J. S. (1997) *J. Am. Chem. Soc.* **119**, 70–76.
50 Martinez, P., Van Dam, M. E., Robinson, A. C., Chen, K., Arnold, F. (1991) *Biotechnol. Bioeng.* **39**, 141–147.
51 Wang, P., Dai, S., Waezsada, S. D., Tsao, A. Y., Davison, B. H. (2001) *Biotechnol. Bioeng.* **74**, 249–253.
52 Ruiz, A. I., Malave, A. J., Felby, C., Griebenow, K. (2000) *Biotechnol. Lett.* **22**, 229–233.
53 Partridge, J., Halling, P. J., Moore, B. D. (2000) *J. Chem. Soc., Perkin Trans.* **2**, 465–471.
54 Affleck, R., Xu, Z. F., Suzawa, V., Focht, K., Clark, D. S., Dordick, J. S. (1992) *Proc. Natl. Acad. Sci.* **89**, 1100–1104.
55 Rupley, J. A., Yang, P.-H., Tollin, G. (1980) *ACS Symp. Ser.* **127**, 111–132.
56 Xu, Z. F., Affleck, R., Wangikar, P. P., Suzawa, V., Dordick, J. S., Clark, D. S. (1994) *Biotechnol. Bioeng.* **43**, 515–520.
57 Kazandjian, R. Z., Dordick, J. S., Klibanov, A. M. (1986) *Biotechnol. Bioeng.* **28**, 417–421.
58 Zaks, A., Klibanov, A. M. (1988) *J. Biol. Chem.* **263**, 8017–8021.
59 Laane, C., Boeren, S., Vos, K., Veeger, C. (1987) *Biotechnol. Bioeng.* **30**, 81–87.
60 Serdakowski, A. L., Munir, I. Z., Dordick, J. S. (2006) *J. Am. Chem. Soc.* **128**, 14272–14273.
61 Bell, G., Janssen, A. E. M., Halling, P. J. (1997) *Enzyme Microb. Technol.* **20**, 471–477.
62 Ohe, S. (1989) *Vapor-Equilibrium Data* (Elsevier, Kodansha, Tokyo).
63 Ray, A. (1971) *Nature* **231**, 313–315.
64 Klibanov, A. M. (1979) *Anal. Biochem.* **93**, 1–25.
65 Ingalls, R. G., Squires, R. G. (1975) *Biotechnol. Bioeng.* **17**, 1627–1637.
66 Aldercruetz, P. (1991) *Eur. J. Biochem.* **199**, 609–614.

67 Tanaka, A., Kawamoto, T. (1991) *Bioprocess Tech.* **14**, 183–208.

68 Wehtje, E., Aldercruetz, P., Mattiasson, B. (1992) *Biotechnol. Bioeng.* **41**, 171–178.

69 Hyndman, D., Flynn, G., Lever, G., Burrell, R. (1992) *Biotechnol. Bioeng.* **40**, 1319–1327.

70 Bindhu, L. V., Abramam, E. T. (2003) *J. Appl. Poly. Sci.* **88**, 1456–1464.

71 Klibanov, A. M. (1986) *Chemtech* **16**, 354–359.

72 Ryu, K., Stafford, D. R., Dordick, J. S. (1989) *ACS Symp. Ser. (Biocatal. Agric. Biotechnol.)*, 141–157.

73 Ryu, K., Dordick, J. S. (1989) *J. Am. Chem. Soc.* **111**, 8026–8027.

74 Clark, D. S. (1994) *Trends Biotechnol.* **12**, 439–443.

75 Tanaka, A., Kawamoto, T. (1991) in *Protein Immobilization: Fundamentals and Applications*, ed. Taylor, R. F. (Marcel Dekker, Inc, New York), pp. 183–208.

76 Reslow, M., Adlercreutz, P., Mattiasson, B. (1988) *Eur. J. Biochem.* **172**, 573–578.

77 Zaks, A., Klibanov, A. M. (1988) *J. Biol. Chem.* **263**, 3194–3201.

78 Suzawa, V. M., Khmelnitsky, Y. L., Yu, L., Giarto, L., Dordick, J. S., Clark, D. S. (1995) *J. Am. Chem. Soc.* **117**, 8435–8440.

79 Zaks, A., Klibanov, A. M. (1985) *Proc. Natl. Acad. Sci. USA* **82**, 3192–3196.

80 Yang, Z., Zacherl, D., Russell, A. J. (1993) *J. Am. Chem. Soc.* **115**, 12251–12257.

81 Skrika-Alexopoulos, E., Freedman, R. (1993) *Biotechnol. Bioeng.* **41**, 887–893.

82 Zacharis, E., Moore, B. D., Halling, P. J. (1997) *J. Am. Chem. Soc.* **119**, 12396–12397.

83 Triantafyllou, A. O., Wehtje, E., Adlercreutz, P., Mattiasson, B. (1995) *Biotechnol. Bioeng.* **45**, 406–414.

84 Ash, M., Ash, I. (2002) *Textbook of pharmaceutical additives* (Synapse Information Resources Inc, New York).

85 Izutsu, K., Yoshioka, S. (1994) *Chem. Pharm. Bull. (Tokyo)* **42**, 5–8.

86 Yamane, T., Ichiryu, T., Nagata, M., Ueno, A., Shimizu, S. (1990) *Biotechnol. Bioeng.* **36**, 1063–1069.

87 Otamiri, M., Adlercreutz, P., Mattiasson, B. (1991) *Biotechnol. Appl. Biochem.* **13**, 54–64.

88 Khmelnitsky, Y. L., Welch, S. H., Clark, D. S., Dordick, J. S. (1994) *Biotechnol. Bioeng.* **58**, 654–657.

89 Hartl, F. U. (1996) *Nature* **381**, 571–580.

90 Kurganov, B. I., Topchieva, I. N. (1998) *Biochemistry (Mosc).* **63**, 413–419.

91 Rozema, D., Gellman, S. H. (1990) *B. J. Biol. Chem.* **271**, 3478–3487.

92 Mi, Y., Wood, G., Thoma, L. (2004) *A.A.P.S. J.* **6**, 1–10.

93 Reinhoudt, D. N., Eendebak, A. M., Nijenhuis, W. F., Verboom, W., Kloosterman, M., Schoemaker, H. E. (1989) *J. Chem. Soc., Chem. Commun.* **1**, 399–400.

94 Broos, J. (1995) *J. Am. Chem. Soc.* **2**, 255–256.

95 van Unen, D. J., Engbersen, J. F., Reinhoudt, D. N. (1998) *Biotechnol. Bioeng.* **59**, 553–556.

96 van Unen, D.-J., Engbersen, J., Reinhoudt, D. (2002) *Biotechnol. Bioeng.* **77**, 248–255.

97 Gokel, G. (1991) *Crown ethers and cryptands* (Royal Society of Chemistry, Cambridge, UK).

98 Griebenow, K., Vidal, M., Baez, C., Santos, A. M., Barletta, G. (2001) *J. Am. Chem. Soc.* **123**, 5380–5381.

99 Bedell, B. A., Mozhaev, V. V., Clark, D. S., Dordick, J. S. (1998) *Biotechnol. Bioeng.* **58**, 654–657.

100 Blanch, H. W., Clark, D. S. (1996) *Biochemical Engineering* (Marcel Dekker, New York).

101 Jones, G., Dole, M. (1929) *J. Am. Chem. Soc.* **51**, 2950–2964.

102 Kim, J., Clark, D. S., Dordick, J. S. (2000) *Biotechnol. Bioeng.* **67**, 112–116.

103 Eppler, R. K., Komor, R. S., Huynh, J., Dordick, J. S., Reimer, J. A., Clark, D. S. (2006) *Proc. Natl. Acad. Sci.* **103**, 5706–5710.

104 Affleck, R., Haynes, C. A., Clark, D. S. (1992) *Proc. Natl. Acad. Sci.* **89**, 5167–5170.

105 Broos, J., Visser, A., Engbersen, J., Verboom, W., van Hoek, A., Reinhoudt, D. (1995) *J. Am. Chem. Soc.* **117**, 12657–12663.

106 Burke, P. A., Griffin, R. G., Klibanov, A. M. (1993) *Biotechnol. Bioeng.* **42**, 87–94.

107 Burke, P. A., Griffin, R. G., Klibanov, A. M. (1992) *J. Biol. Chem.* **267**, 20057–20064.

Part Two Biocatalysis in Neat Organic Solvents –
Synthetic Applications

4
Exploiting Enantioselectivity of Hydrolases in Organic Solvents

Hans-Erik Högberg

4.1
Introduction

The natural function of most hydrolases *in vivo* is to hydrolyze natural compounds, e.g., acid derivatives to the free acids and a leaving group such as an alcohol or an amine. In organic solvents it is possible to run such reactions in the reverse direction, allowing for synthesis of esters and amides. In most applications, biologically active compounds are needed in enantiomerically pure forms for use as, e.g., drugs and agrochemicals. Therefore the ability to prepare enantiomerically pure compounds has become a key issue in organic chemistry. Being proteins built from naturally occurring chiral, enantiopure amino acids, hydrolases are also enantiomerically pure, chiral polymers. Thus, when reacting with a chiral, racemic substrate, they will react faster with one enantiomer of the substrate than the other. This enantioselectivity has formed the basis for the widespread use of hydrolytic enzymes for the synthesis of enantiomerically pure compounds using this type of kinetic resolution. Another approach to chiral, enantiomerically pure compounds is desymmetrization of prochiral compounds, which also can be achieved by hydrolases. The use of hydrolases for the preparation of enantiomerically pure compounds has been extensively reviewed [1]. This chapter describes some of the applications of hydrolases in organic solvents for the purpose of obtaining enantiomerically enriched organic compounds.

4.1.1
Enantioselectivity

Hydrolases reacting with a racemic substrate will normally display a preference for reacting faster with one enantiomer of the substrate e.g., the R-enantiomer than with the S-one (Scheme 4.1). The initial relative rate of reaction for the enantiomers of a racemic substrate, when the conversion is zero, is called the enantiomeric ratio or the E-value and is defined as the ratio between the specificity constants (k_{cat}/K_m) for the two competing enantiomers. When the enantiomeric excesses of one of either the substrate or the product, ee_s and ee_p [2], respectively,

Organic Synthesis with Enzymes in Non-Aqueous Media. Edited by Giacomo Carrea and Sergio Riva
Copyright © 2008 WILEY-VCH Verlag GmbH & Co. KGaA, Weinheim
ISBN: 978-3-527-31846-9

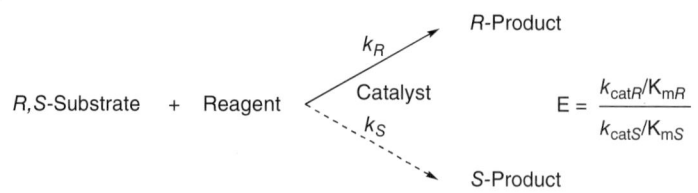

Scheme 4.1 Enantioselective kinetic resolution of a racemate. k = rate constants for the individual enantiomers of the substrate, E = enantiomeric ratio, i.e., the ratio between the specificity constants k_{cat}/K_m for the fast and slow reacting enantiomer. If a racemate is used as substrate, then these concentrations are equal at the start (i.e. zero conversion), and hence $E = k_R/k_S$.

and the conversion are known, the E-value of the reaction can be calculated from Equations 1–3 for irreversible reactions and Equations 4 and 5 for reversible ones [3, 4]. Normally the E-value is independent of conversion and remains constant for a given set of reaction conditions (for examples of exceptions see Ref. [5]). The E-value does, however, depend on the enzyme used and its preparation, the substrate, the solvent, the reaction temperature and other reaction conditions.

For irreversible reactions:

$$E = \frac{\ln[1-c(1+ee_p)]}{\ln[1-c(1-ee_p)]} \tag{1}$$

$$E = \frac{\ln[1-c(1+ee_s)]}{\ln[1-c(1-ee_s)]} \tag{2}$$

$$E = \frac{\ln\frac{ee_p(1-ee_s)}{(ee_p+ee_s)}}{\ln\frac{ee_p(1+ee_s)}{(ee_p+ee_s)}} \tag{3}$$

where c is conversion of substrate (0 to 1) and ee_p and ee_s are enantiomeric excesses (0 to 1) of product (P) and remaining substrate (S), respectively.

For reversible reactions:

$$E = \frac{\ln[1-(1+K)c(1+ee_p)]}{\ln[1-(1+K)c(1-ee_p)]} \tag{4}$$

$$E = \frac{\ln\{1-(1+K)[c+ee_s(1-c)]\}}{\ln\{1-(1+K)[c-ee_s(1-c)]\}} \tag{5}$$

where c, ee_p, and ee_s are the same as for Equation 1, Equation 2, and Equation 3, $K = (1-c_{eq})/c_{eq}$, and c_{eq} = conversion at equilibrium.

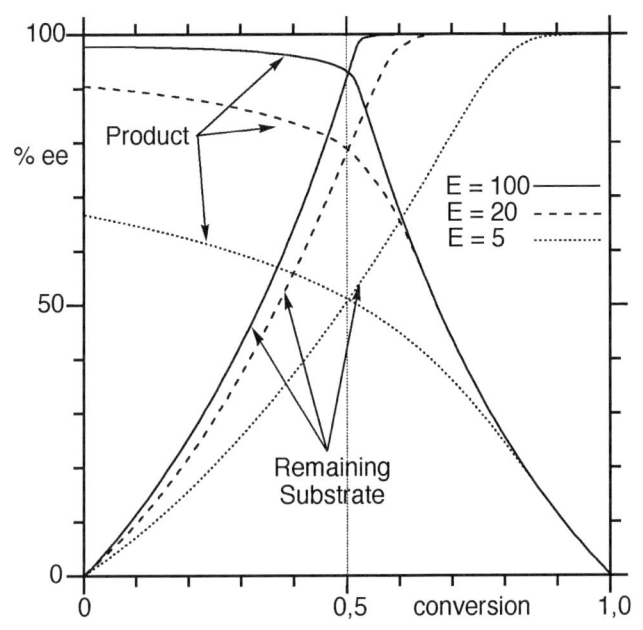

Figure 4.1 Irreversible reaction.

Graphs of the course of typical reactions are shown in Figures 4.1 and 4.2 for irreversible and reversible reactions, respectively. Note that at low conversions (roughly below 40% conversion) the product curves for a given E-value are virtually identical for both the irreversible and the reversible cases (Figure 4.1 and Figure 4.2, respectively).

For obtaining both the product and the remaining substrate in high enantiomeric excess in one reaction step, the E-value needs to be high, usually around or over 100. If the E-value is lower, isolation of the products and substrates followed by renewed reactions with these can be used in an iterative manner until sufficiently enantiomerically pure products are obtained.

Irreversible reactions. It is evident from Figure 4.1, that for an irreversible reaction it is possible to obtain the remaining substrate in very high enantiomeric excess even when the reaction has a rather low E-value, provided that the reaction is allowed to proceed to sufficiently high conversions. The product, however, cannot be obtained in a higher *ee* than that which is determined by the E-value. Thus at E = 100 and at the start of the reaction, the first product formed has a maximum *ee* of 98%, and the *ee* then decreases with increasing conversion. In this case, the remaining substrate can be obtained in 100% *ee* after approximately 55% conversion.

Reversible reactions. Figure 4.2 shows that in reversible reactions neither the product nor the remaining substrate can be obtained in a very high *ee*. Thus, the reversibility lowers the *ee*-value of the remaining substrate at high conversions. The problems associated with reversibility will be discussed in Section 4.1.3.

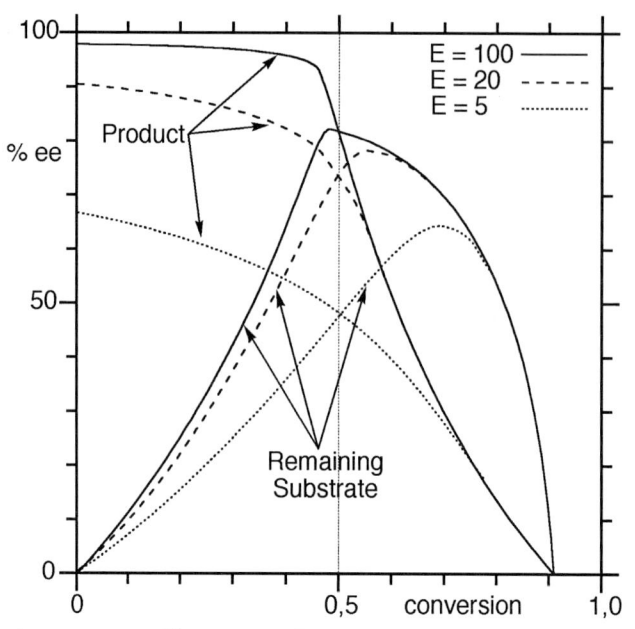

Figure 4.2 Reversible reaction with $c_{eq} = 0.9$.

Limitations of kinetic resolutions. If one is interested in preparing a single enantiomer, the major drawback of hydrolase-catalyzed kinetic resolutions is that only 50% yield of the desired enantiomer is possible from a racemic starting material. The introduction of dynamic kinetic resolution (DKR, see Chapter 6) provides one solution to this problem. Another one is to invert the configuration of the product or the remaining substrate after the hydrolase-mediated resolution, and examples of this approach are discussed below (Section 4.2.1.2).

4.1.2
Desymmetrization

Desymmetrization of symmetric compounds involves breaking the symmetry in a substrate, i.e. elimination of one or several elements of symmetry (Scheme 4.2). Thus, it is an asymmetric synthetic transformation and can therefore give a quantitative yield.

Hence, when enantiopure compounds are needed, desymmetrization constitutes a useful alternative to kinetic resolution of racemates. Hydrolases are useful for such transformations, in both hydrolytic and acylation reactions [6]. *Meso*-compounds have been used extensively in such reactions. The success of such a reaction depends on one of the *pro-R* or *pro-S* groups reacting much faster than the other. If the monoderivatized product reacts further, the second step of course gives the doubly reacted *meso*-product. If the second step favors the minor one of

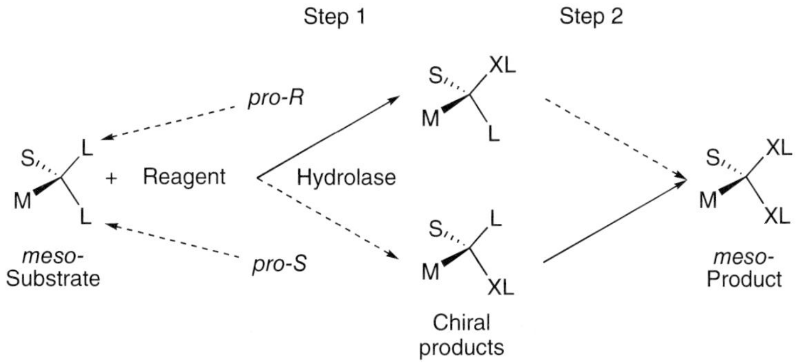

Scheme 4.2 Desymmetrization of a *meso*-substrate.

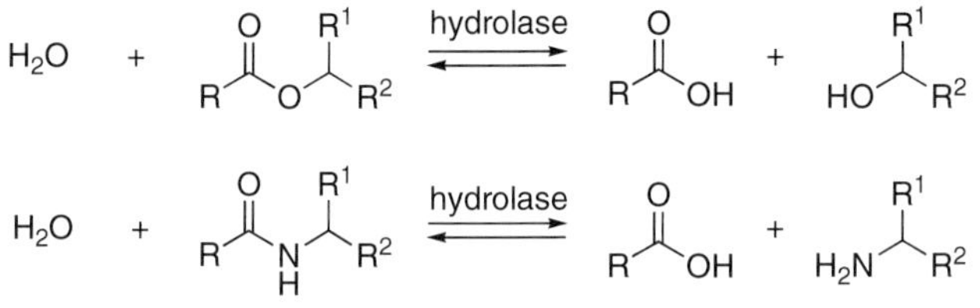

Scheme 4.3 Hydrolase-catalyzed reactions.

the chiral products, the major one can be obtained in a high *ee* even if the selectivity in the first step is fairly low.

4.1.3
The Reversibility Problem

Two typical hydrolase-catalyzed reactions are shown in Scheme 4.3. It is important to note that these reactions are reversible, and in water the equilibrium of course favors hydrolysis. However, the use of hydrophobic organic solvents allows the acylation to give e.g., esters and amides using hydrolases as catalysts. Many of these enzymes are commercially available and can be used in hydrophobic organic solvents as received, and because they are insoluble in the reaction medium they can easily be recovered by filtration and used again.

Even in hydrophobic solvents the water formed will eventually establish an equilibrium. Therefore, when high *ee*s are desired for both the products and the remaining substrates, reversibility (Scheme 4.3: compare Figure 4.1 and Figure 4.2) in hydrolase-catalyzed resolutions causes problems, and several methods to circumvent these problems have been developed. The simplest option is to shift the equilibrium by employing an excess of one of the reagents, e.g., by using it as the solvent, provided that the enzyme tolerates this. Water can also be removed from the reaction mixture by addition of drying agents or by azeotropic distillation.

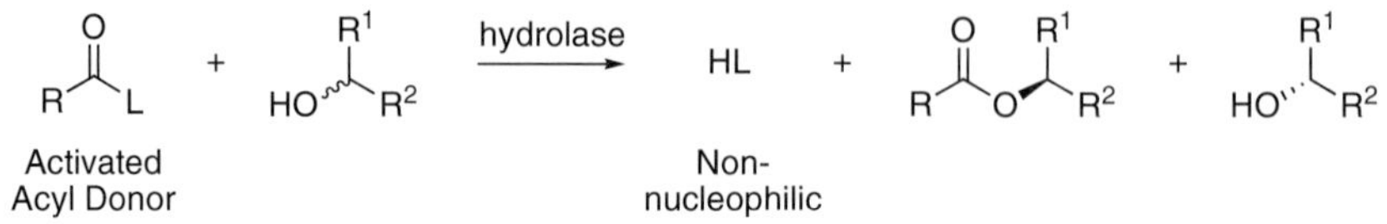

Scheme 4.4 Irreversible hydrolase-catalyzed acylation with an activated acyl donor.

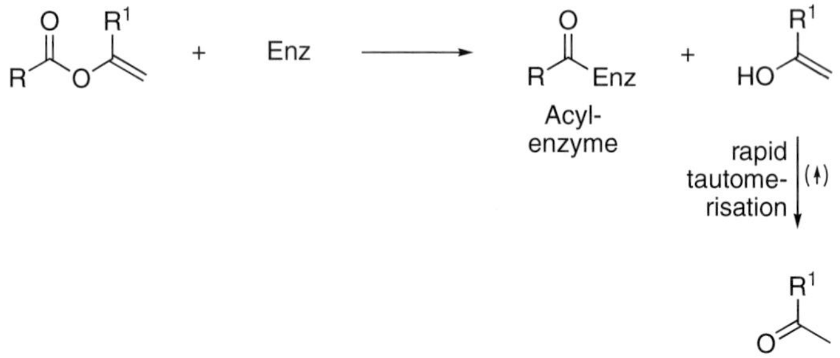

Scheme 4.5 Vinyl esters as acyl donors.

However many hydrolases do not tolerate such conditions. Thus, these options rarely give the desired result. Therefore a number of acid derivatives have been used as substitutes for acids as acyl donors in hydrolase-catalyzed reactions. These acyl donors are designed so that the protonated leaving group LH is too weak a nucleophile to attack the acylated enzyme (for example, ester formation, Scheme 4.4). Thus, the reversibility is strongly hindered or completely stopped. If the E-value is high, this process allows us to acquire both the product and the remaining substrate of high ee.

Many such activated acyl derivatives have been developed, and the field has been reviewed [7–9]. The most commonly used irreversible acyl donors are various types of vinyl esters. During the acylation of the enzyme, vinyl alcohols are liberated, which rapidly tautomerize to non-nucleophilic carbonyl compounds (Scheme 4.5). The acyl-enzyme then reacts with the racemic nucleophile (e.g., an alcohol or amine). Many vinyl esters and isopropenyl acetate are commercially available, and others can be made from vinyl and isopropenyl acetate by Lewis acid- or palladium-catalyzed reactions with acids [10–12] or from transition metal-catalyzed additions to acetylenes [13–15]. If ethoxyacetylene is used in such reactions, R^1 in the resulting acyl donor will be OEt (Scheme 4.5), and hence the end product from the acyl donor leaving group will be the innocuous ethyl acetate [16]. Other frequently used acylation agents that act as more or less irreversible acyl donors are the easily prepared 2,2,2-trifluoro- and 2,2,2-trichloro-ethyl esters [17–23]. Less frequently used are oxime esters and cyanomethyl ester [7]. S-ethyl thioesters such as the thiooctanoate has also been used, and here the ethanethiol formed is allowed to evaporate to displace the equilibrium [24, 25]. Some anhydrides can also serve as irreversible acyl donors.

Some hydrolases, especially the lipases from *Candida rugosa* (CRL) and *Geotrichum candidum* are sensitive to the acetaldehyde produced from reactions with

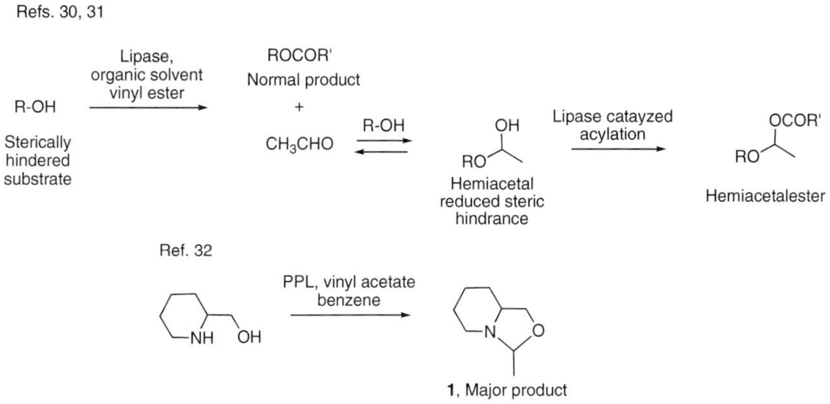

Scheme 4.6 Anomalous products from reactions with vinyl esters.

vinyl esters [26]. In spite of this, they have been used successfully with vinyl esters under controlled conditions [27–29].

Although vinyl esters are widely used as acyl donors, it is worth noting that when these acyl donors are employed with some sterically demanding substrates, some undesired products may be formed, namely hemiacetals and their esters. The acetaldehyde liberated from the acylating agent reacts with the substrate alcohol yielding a hemiacetal, which is acylated by the enzyme to yield the hemiacetal ester (Scheme 4.6) [30, 31]. Because the hemiacetals and their esters are decomposed to the substrate alcohol during workup, they may easily escape notice, and thus erroneous conclusions regarding the E-value of the reaction can be drawn. Another example of acetaldehyde reacting with a substrate is the attempt to resolve a 2-aminoalcohol by porcine pancreas lipase (PPL)-catalyzed acylation with vinyl acetate. Here the product formed is, instead of the expected ester, a cyclic one, the oxazolidine **1**, which is formed via acetaldehyde reacting with the substrate (Scheme 4.6) [32].

4.1.4
Determining and Optimizing Enantioselectivity

Enantiomeric purity. In order to assess the efficiency of an enantioselective hydrolase-catalyzed reaction, it is imperative that one can accurately measure at least the conversion and the enantiomeric excesses of either the substrate or the product (see equations Equation 1, Equation 2, and Equation 3). Although optical rotation is sometimes used to assess enantiomeric excess, it is not recommended. Much better alternatives are various chromatographic methods. For volatile compounds, capillary gas chromatography on a chiral liquid phase is probably the most convenient method. Numerous commercial suppliers offer a large variety of columns with different chiral liquid phases. Hence it is often easy to find suitable conditions for enantioselective GC-separations that yield *ee*-values in excess of

99%, which can usually be determined with a high precision (±0.05%) [33, 34]. For nonvolatile compounds, HPLC on a chiral phase is the method of choice.

Solvent and water content. When reactions are performed in organic solvents [1a], the solvent of choice is often a hydrophobic solvent with logP > 1.5. However, despite a massive amount of work on this matter, no general rules can be formulated. Usually one has to scan a number of solvents to find the one that gives the best E for the individual case. The water activity of the reaction system can influence E quite substantially, but no general rules can be formulated, and usually one has to find the optimal conditions for the individual substrate. It is important to note, however, that if the water content is high in, e.g., an acylation reaction system, water can compete with the desired nucleophile and give an acid that can disturb the efficiency of the resolution.

Genetic engineering. The X-ray structures are known for many hydrolases, allowing for modeling of the substrate in the active site as well as structurally based, random or rational protein mutation to magnify or invert enantioselectivity. An example of the latter is provided by the rational design of a mutant of *Candida antarctica* lipase B (CALB), which, instead of the wild-type *R*-selectivity, displayed *S*-selectivity toward 1-phenylethanol [35]. Several other hydrolases have also been redesigned by protein engineering to invert or enhance their enantioselectivity [36–40].

4.2
Enantioselective Reactions in Organic Solvents

The exploitation of hydrolases as catalysts in organic solvents is described in the following sections. The overview is by no means comprehensive, and the selection of examples is subjective. The names of the hydrolases will be those used by the authors, despite the fact that some of the names have been changed since the cited work was published.

4.2.1
Reactions of Alcohols

Alcohols, especially secondary ones, are the most popular substrates for hydrolase-catalyzed resolutions, but also racemic primary alcohols may be resolved. Tertiary ones are unreactive to most hydrolases, but exceptions exist.

4.2.1.1 Primary Alcohols
Whereas resolutions of secondary alcohols by hydrolase-catalyzed acylation in organic solvents often proceed with high E-values (see Section 4.2.1.2), primary alcohols are much more difficult to resolve. There are, however, many examples of successful resolutions of primary alcohols, and a selection of these will be described here.

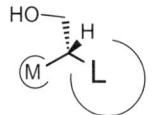

Figure 4.3 Preferential acylation mode of lipases for primary alcohols with a large substituent (**L**) in a large hydrophobic pocket and a medium-sized substituent (**M**) in a smaller pocket, provided no oxygen atoms are located at the stereogenic center.

Although quite reliable empirical rules exist for the enantioselectivity of hydrolases for secondary alcohols (see Section 4.2.1.2), such rules are not as developed for primary alcohols, partly because many hydrolases often show low enantioselectivity. With some exceptions, lipases from *Pseudomonas sp.* and porcine pancreas lipase (PPL) often display sufficient selectivity for practical use. The model described in Figure 4.3 has been developed for *Pseudomonas cepacia* lipase (reclassified as *Burkholderia cepacia*), and, provided that no oxygen is attached to the stereogenic center, it works well for this lipase in many cases [41]. However, as soon as primary alcohols are resolved by enzyme catalysis, independent proof of configuration for a previously unknown product is recommended.

Hydrolase-catalyzed resolution of simple 2-methylalkanols by acylation with vinyl acetate in organic solvents proceeds with low to moderate E-values (Table 4.1) [42, 43].

Lipases from *Pseudomonas sp.* [Amano PS and *Pseudomonas fluorescens* lipase (PFL)] are useful. Provided that the conversion is high enough, the remaining (*R*)-2-methylalkanols *R*-**2** can be obtained almost enantiomerically pure. The (*S*)-2-methylalkanyl acetates can only be obtained in modest enantiomeric purity at conversions ≈ 40%.

3-Aryl-2-methylpropan-1-ols can be successfully resolved using catalysis by a lipase with rather high E-values. This is exemplified by the reactions of some alcohols where the aryl group is a thienyl moiety. Thus, some racemic 2-methyl-3-thienylpropanols were treated with Amano PS and vinyl acetate in *t*-butylmethyl ether (TBME) with initial water activity $a_w = 0.32$ to furnish the acetates **3-6** shown in high *ee*s and good E-values (Scheme 4.7) [44]. Because the thiophene moiety can be hydrogenated by Raney nickel to furnish a tetramethylene chain [-$(CH_2)_4$-], this offers a method for the preparation of long-chain 2-methylalkanols of higher enantiomeric purity than that provided by direct enzyme-catalyzed resolution of the latter (Table 4.1).

E-values of similar magnitudes can also be obtained with other 3-aryl-2-methylpropan-1-ols {Aryl = phenyl, Amano PS/vinyl acetate/chloroform, E = 116 [44]; aryl = 1- or 2-naphthyl, PFL/vinyl acetate/chloroform, E = 150 or 45, respectively [45]; aryl = 2-furyl Amano PS/vinyl acetate/chloroform E = 105 [44]; aryl = 2-methoxyphenyl, Amano PS/vinyl acetate/TBME, E = 105 [44]}.

Table 4.1 Lipase-catalyzed resolution of 2-methylalkanols (**2**) with vinyl acetate in dichloromethane or chloroform at room temperature.

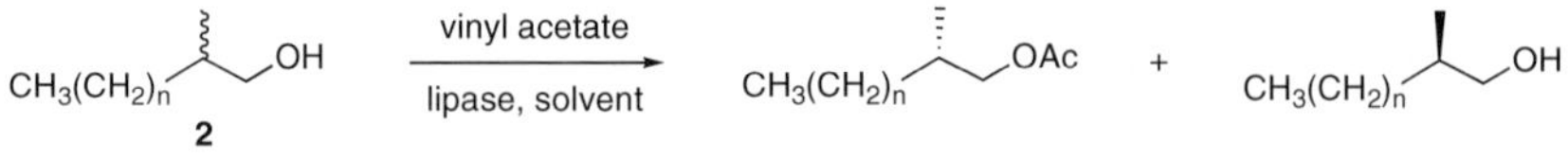

$$CH_3(CH_2)_n\diagdown OH \xrightarrow[\text{lipase, solvent}]{\text{vinyl acetate}} CH_3(CH_2)_n\diagdown OAc + CH_3(CH_2)_n\diagdown OH$$
2

			Substrate 2				
n	Enzyme	Solvent	%Conv	ee$_{prod}$	ee$_{substr}$	E	Ref.
2	Amano PS	CH$_2$Cl$_2$	80.2	28.7	98	5.9	42
2	Amano PS	CH$_2$Cl$_2$	74.6	33.7	99	8.7	42
5	Amano PS	CH$_2$Cl$_2$	78.0	27.4	96.2	5.7	42
5	PFL	CHCl$_3$	39.5	60.0		5.8	43
7	Amano PS	CH$_2$Cl$_2$	70.0	42.4	98.1	9.9	42
7	Amano PS	CHCl$_3$	40.6	73.4		10.7	43
9	PFL	CHCl$_3$	39.4	74.4		10.9	43
11	PFL	CHCl$_3$	42.7	72.4		10.6	43

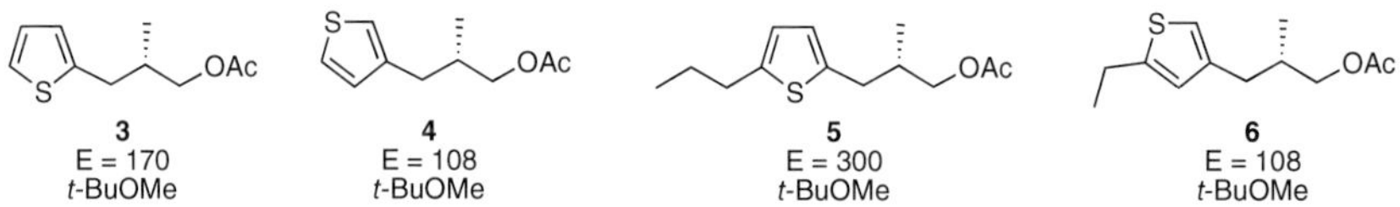

3
E = 170
t-BuOMe

4
E = 108
t-BuOMe

5
E = 300
t-BuOMe

6
E = 108
t-BuOMe

Scheme 4.7 Acylation products obtained from some 2-methyl-3-thienylpropanols.

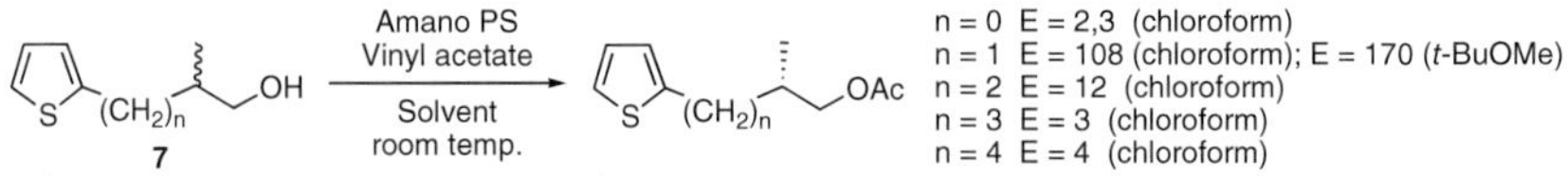

Amano PS
Vinyl acetate
Solvent
room temp.

7

n = 0 E = 2,3 (chloroform)
n = 1 E = 108 (chloroform); E = 170 (t-BuOMe)
n = 2 E = 12 (chloroform)
n = 3 E = 3 (chloroform)
n = 4 E = 4 (chloroform)

Scheme 4.8 Preferentially formed acylation products obtained from some 2-methyl-ω-thienylalkan-1-ols (**7**).

It is, however, important that the aryl group is placed at the correct position in the chain. Moving it either further away from or closer to the stereogenic center as shown with the substrates of type **7** gives much lower E-values (Scheme 4.8) [44].

Although lipases from *Pseudomonas* are usually the catalysts of choice for primary alcohols, 2-(2-furyl)-propan-1-ol (Scheme 4.8: **7** n = 0 with O instead of S) actually gives a higher E (E = 20) with *Candida antarctica* lipase B (CALB) than it does with *Pseudomonas sp.* lipase (PSL) (E = 2) on acylation with vinyl acetate in pentane [78].

A number of racemic 2-aryloxy-1-propanols **8** have been resolved by *Pseudomonas sp.* lipase (Amano AK)-catalyzed acylation in diisopropyl ether (DIPE) at 25 °C (Scheme 4.9) [46]. Because an oxygen atom is directly attached to the stereogenic center, the selection rule described in Figure 4.3 does not hold.

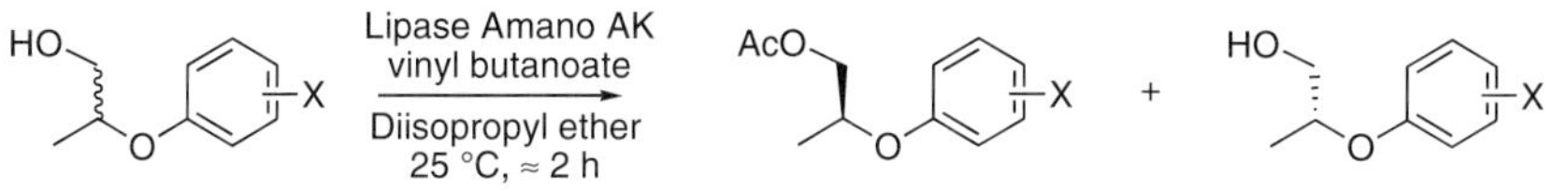

4.2 Enantioselective Reactions in Organic Solvents

Scheme 4.9 Resolution of some 2-aryloxy-1-propanols.

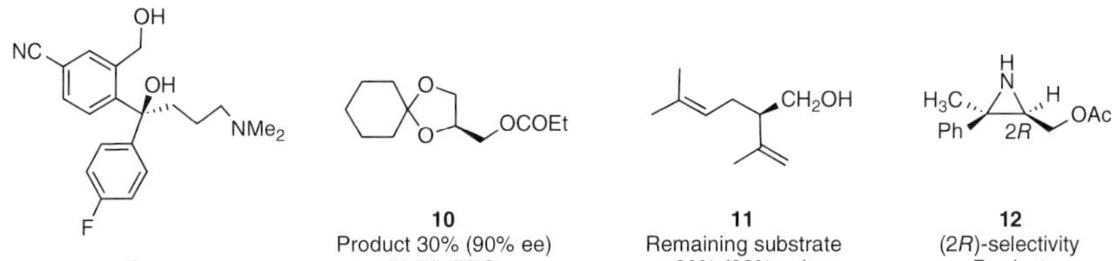

Compound **8**, Lipase Amano AK, vinyl butanoate, Diisopropyl ether, 25 °C, ≈ 2 h:

X = H, E = 35
X = 4-F, E = 34
X = 2-Cl, E = 48
X = 3-Cl, E = 55
X = 4-Cl, E = 58
X = 2,4-Cl₂, E = 19
X = 4-Et, E = 19
X = 4-i-Pr, E = 15

Scheme 4.10 Some examples of resolutions of primary alcohols.

9 Remaining substrate at 56% conv. (>99% ee) E = 59

10 Product 30% (90% ee) PLE/MPEG vinyl propionate Water saturated toluene E = 28

11 Remaining substrate 20% (98% ee) Hog pancreas lipase Succinic anhydride Diethyl ether, room temp, 48 h E ≈ 12

12 (2R)-selectivity Product 30% conv. (96% ee) PS-C II, vinyl acetate Acetone -20 °C, 29 h E = 73

CALB and vinyl acetate enantioselectively acylate a racemic precursor to the antidepressant Citalopram in toluene at 30 °C, furnishing the desired (S)-enantiomer **9** as the remaining substrate in a high *ee* (Scheme 4.10) [47]. It is worth noting that in this case the stereogenic center is four bonds removed from the reacting center. Other examples of primary alcohols that have been resolved with moderate success are also shown in Scheme 4.10. Thus pig liver esterase (PLE) co-lyophilized with methoxypolyethyleneglycol (MPEG) catalyzed the acylation by vinyl propionate of the racemic 1,2-cyclohexanone ketal of glycerol to give **10** [48]. Racemic lavandulol rac-**11** has been resolved by using hog pancreas lipase and succinic anhydride as acyl donor [49]. In contrast, racemic lavandulol, when treated with CALB and acetic acid as acyl donor in hexane, furnished the opposite enantiopreference, with E = 19 at 25 °C [50]. An aziridinemethanol has been successfully resolved at −20 °C using lipase PS-C II immobilized on thyonite to give **12** [51]. Under similar conditions the diastereomer where the phenyl group and methyl group have switched positions proceed with the opposite (2S)-selectivity at −40 °C (E = 55) [51].

In cases where the E-value is rather low, repeated resolutions may be required to supply the desired enantiomers in high *ee*s. The synthesis of both enantiomers of the marine natural product laurene **13** provides an illustrative example. Here, the racemic cyclopentylcarbinol **12** is acylated with vinyl acetate in hexane catalyzed by CRL (E ≈ 35) to provide the remaining substrate in high *ee*. Chemical hydrolysis followed by repeated CRL-catalyzed acylation and then chemical hydrolysis provides the other enantiomer of the alcohol (Scheme 4.11) [52].

In a similar way, resolution of the racemic cyclohomogeraniol **14** (Scheme 4.12) proceeds with a low enantiomeric ratio (E ≈ 4) on acylation with vinyl acetate

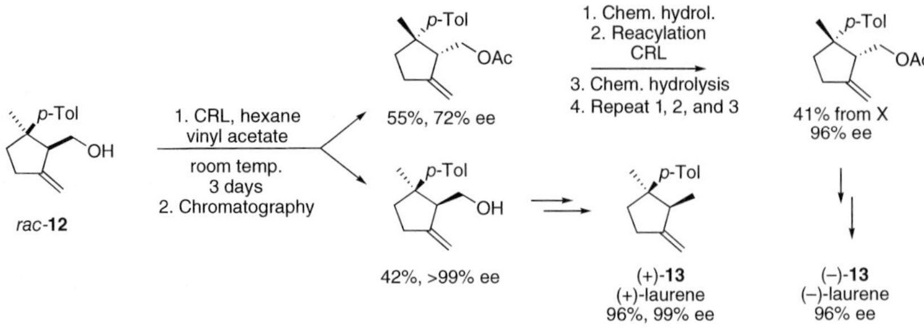

Scheme 4.11 Enantioselective synthesis of (+)- and (−)-laurene (**13**).

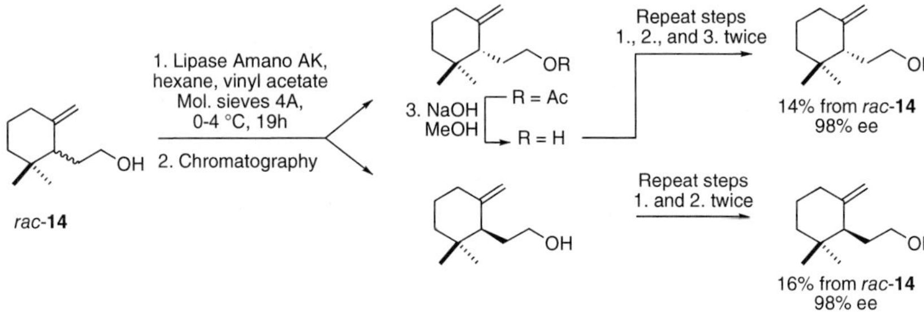

Scheme 4.12 Iterative resolution of the cyclohomogeraniol *rac*-**14**.

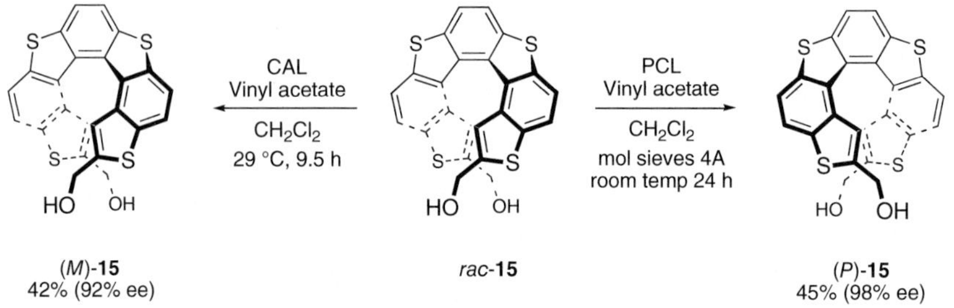

Scheme 4.13 Resolution of a heterohelicene.

catalyzed by Amano AK lipase in hexane. Hence, by chemical hydrolysis of the produced ester and repeated separate acylations of the resulting alcohol and the remaining alcohol from the first step gave increased *ees*. Repetition of the latter procedures once again furnished after chemical hydrolysis both enantiomers of the cyclohomogeraniol **14** in high *ees*, albeit in low yields [53].

An interesting case is the resolution of a heterohelicenediol. The very bulky racemic substrate *rac*-**15** is resolved by lipase-catalyzed acylation with vinyl acetate in dichloromethane. Oddly enough, *Candida antarctica* lipase (CAL) and *Pseudomonas cepacia* lipase (PCL) display opposite enantiopreferences [54]. In Scheme 4.13, only the remaining substrates are shown and not the other products, the mono- and diacetates.

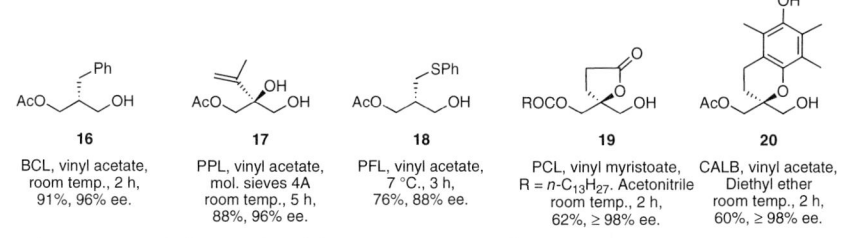

Scheme 4.14 Hydrolase-catalyzed desymmetrization products obtained from the corresponding prochiral diols.

Compound **16**: BCL, vinyl acetate, room temp., 2 h, 91%, 96% ee.

Compound **17**: PPL, vinyl acetate, mol. sieves 4A, room temp., 5 h, 88%, 96% ee.

Compound **18**: PFL, vinyl acetate, 7 °C., 3 h, 76%, 88% ee.

Compound **19**: PCL, vinyl myristoate, R = n-$C_{13}H_{27}$. Acetonitrile, room temp., 2 h, 62%, ≥ 98% ee.

Compound **20**: CALB, vinyl acetate, Diethyl ether, room temp., 2 h, 60%, ≥ 98% ee.

Compound **21**: PFL, vinyl acetate, Chloroform, room temp., ≈ 10 days, 50%, 79% ee.

Compound **22**: CALB, vinyl acetate, room temp., 7 h, 59%, 92% ee.

Scheme 4.15 Desymmetrization products of some phosphorus containing prochiral diols.

Desymmetrization of primary diols provides an efficient route to enantiopure compounds. Examples of such resolutions are numerous, and a few illustrative examples will be given here. Many 1,3-diols can be efficiently desymmetrized by hydrolase-catalyzed acylation. One must always keep in mind that acyl migration can occur in such monoesters of 1,3-diols, but in many cases this is not a problem. Thus, in vinyl acetate in the presence of BCL (*Burkholderia cepacia* lipase) racemic 2-benzyl-1,3-propanediol provides a good yield of enantiomerically enriched monoacetate **16** (Scheme 4.14, product shown) [55]. Similarly the 2-isopropenylglycerol with vinyl acetate and PPL furnished the monoester **17** [56]. PFL in vinyl acetate furnishes the monoacetate **18** from the corresponding diol [57]. The butyrolactone monoacetate **19** can be obtained in a fair yield and high *ee* using vinyl myristoate and PCL as catalyst. [58]. The monoacetate **20** is obtained upon treatment of the corresponding diol with vinyl acetate and CALB [59].

Desymmetrization of some phosphorus derivatives has been achieved (Scheme 4.15). Thus, the monoacetate **21** can be obtained from the corresponding diol by PFL-catalyzed acylation with vinyl acetate in chloroform [60]. The monoacetate **22** can be obtained from the corresponding bis(hydroxymethyl)phenylphosphine borane by acylation of the latter in vinyl acetate mediated by CALB [61].

Hydrolase-catalyzed desymmetrizations of *meso*-diols have also been exploited (Scheme 4.16). Thus, the monoacetate **23** is produced in excellent yield and *ee* by acylation with vinyl acetate of the corresponding *meso*-diol catalyzed by CRL in hexane [62]. In a similar way the monoacetate **24** is produced from the *meso*-tetrol by acylation in vinyl acetate catalyzed by PPL [63]. The *meso*-piperidine derivative

4 Exploiting Enantioselectivity of Hydrolases in Organic Solvents

Scheme 4.16 Desymmetrizations of primary *meso*-diols.

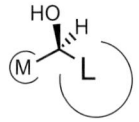

23	24	25	26
CRL, vinyl acetate, Hexane, Mol. sieves 3A, room temp., 4 h, 94%, 97% ee.	PPL, vinyl acetate, room temp., 24 h, 76%, >98% ee.	*Candida antarctica* lipase, vinyl acetate, room temp., 3 h, 80%, 95% ee.	Lipase Amano AK, vinyl acetate, benzene 37 °C, 5 days, 68%, 95% ee. R = *p*-methoxybenzyl

Figure 4.4 Preferential acylation mode of lipases for secondary alcohols with large substituent (**L**) in large hydrophobic pocket and medium-sized substituent (**M**) in smaller pocket as described by Kazlauskas' rule [72].

25, intermediate in enantioselective synthesis of an indolizidine alkaloid, has been prepared by *Candida antarctica* lipase-catalyzed acylation in vinyl acetate at room temperature [64]. The oxacyclononene derivative **26**, a potential building block for ciguatoxin synthesis, has been prepared from the corresponding *meso*-diol by lipase Amano AK (from *Pseudomonas sp.*) catalyzed acylation in benzene [65].

4.2.1.2 Secondary Alcohols

Resolutions of secondary alcohols by hydrolase-catalyzed acylation in organic solvents have become one of the most popular methods for preparing enantioenriched secondary alcohols and their esters. Compared with primary alcohols (see Section 4.2.1.1), secondary alcohols often give higher E-values. There are numerous examples of successful resolutions of secondary alcohols, and a few will be described here (for reviews see Ref. [1]).

As mentioned earlier (Section 4.2.1.1), empirical rules for the enantioselectivity of hydrolases have been developed. It is important to keep in mind that these rules do not work for all substrates. Most rules are based on pockets, which indicate how the steric bulk of the substituents in the substrate fit into the environment of the active site. Thus, such rules have been suggested for pig liver esterase(PLE) [66], the protease subtilisin [66–68], and certain lipases [69–71]. For secondary alcohols, most lipases follow the simple rule of Kazlauskas, which was developed for *Pseudomonas cepacia*, and which is depicted in Figure 4.4 [72]. This model implies that the fast-reacting enantiomers binds to the active site as described in Figure 4.4, whereas the slowly reacting one is not able to achieve a comfortable fit, because it will require the large substituent **L** to fit into the smaller pocket. In contrast to lipases, subtilisin displays opposite enantioselectivity toward secondary alcohols [68].

It is worth noting here that with two enzymes displaying opposite enantioselectivity it is possible to produce both enantiomers of the ester products. If the remaining alcohols can be continuously and rapidly racemized during the much slower acylation reaction, either the R- or S-esters can be obtained in high yields (>>50%) from reactions catalyzed by two hydrolases that display opposite enantiopreference. The combined process of racemization and simultaneous resolution, dynamic kinetic resolution (DKR), is described in Chapter 6.

If the difference in bulk in the secondary alcohol substrates between the groups **M** and **L** is large (Figure 4.4), the E-value in a hydrolase-catalyzed reaction can be expected to be large. Therefore, there are numerous examples of hydrolase-catalyzed acylations in organic solvents of 1-arylethanols. According to Kaslauskas' rule, lipases have been used to supply R-1-arylethyl esters and S-1-arylethanols, whereas the corresponding S-esters and R-alcohols can often be obtained from subtilisin-catalyzed acylations.

1-Arylalkanols have been one of the most studied groups of compounds, partly because of the different spatial requirements of the substituents, which facilitate the resolution, and such compounds have served well in model studies. Thus, some 1-arylethanols can be very efficiently resolved in hexane using CALB-catalyzed (Novozym 435) acylation with vinyl acetate to give the esters **26-31** in very high ees and with excellent enantioselectivity (E > 200) (Scheme 4.17, top half) [73]. When the same reactions were performed in supercritical carbon dioxide, the reactions proceeded faster and with similar excellent E-values [73].

In contrast to the R-preference displayed by CALB, quite impressive S-preference toward several 1-aryl-1-ethanols can be achieved by using a protease, the commercially available subtilisin Carlsberg, as catalyst, and isopropenyl pentanoate as the acyl donor in THF in the presence of sodium carbonate [74]. To achieve a successful resolution, the protease has to be treated with a mixture of two surfactants, octyl-β-D-glycopyranoside and Brij 56 [the monocetyl ether of polyoxyethylene (10)], 4/1/1 by weight, at pH 7.2, and then lyophilized before use. The pentanoates **32–37** are obtained with fair enantioselectivity (E ≥ 47) (Scheme 4.17, bottom half) [74]. When the kinetic resolution conditions were established, efficient conditions for dynamic kinetic resolutions (DKR) were also established (see Chapter 6) [74].

Many 1-heteroaryl-1-alkanols are also easily resolved by hydrolase-catalyzed reactions in organic solvents. Thus, some racemic 1-(2-pyridyl)-1-alkanols and isoquinonyl-1-ethanols are easily resolved by CALB (Novozym 435) and vinyl acetate in diisopropyl ether to give the acetates **38–44** (Scheme 4.18) [75].

Some 1-(2-thienyl)-1-alkanols of type **45** can be efficiently resolved by acylation with vinyl butanoate catalyzed by CALB (Novozym 435) in various solvents (Scheme 4.19) [76].

Some 1-(2-furyl)-1-alkanols have also been resolved by hydrolase-catalyzed acylations (Scheme 4.20). Thus 1-(2-furyl)-1-ethanol (**46**) is efficiently resolved by acylation with vinyl acetate catalyzed either by Lipozyme IM or PPL [77]. Resolution with a more complex acyl donor, ethoxyvinyl methyl fumarate, catalyzed by Lipase LIP (from *Pseudomonas aeruginosa*) has also been achieved [95]. The

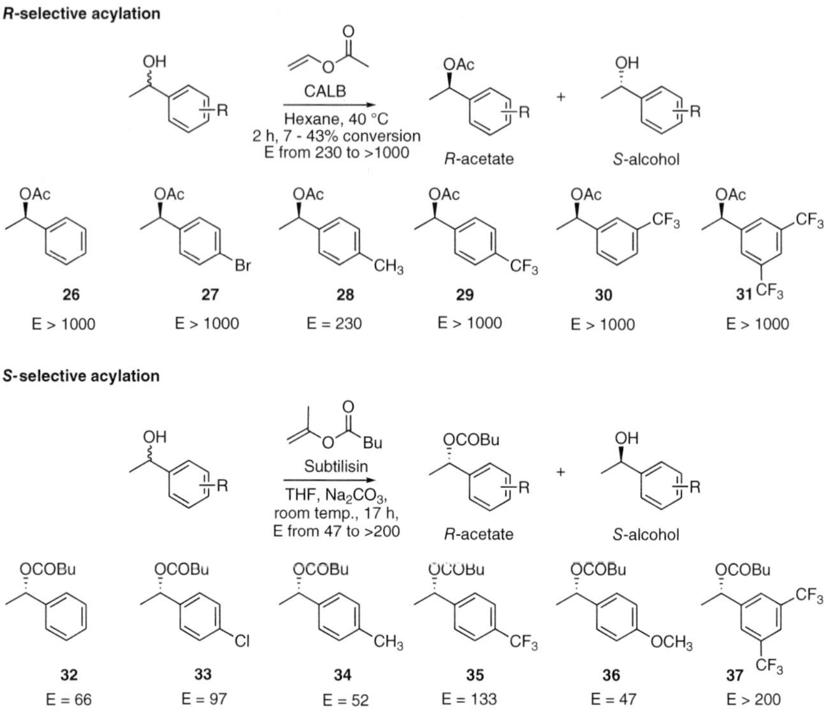

R-selective acylation

Scheme 4.17 Kinetic resolution of some 1-arylethanols. Preferentially formed esters **26–37** are shown.

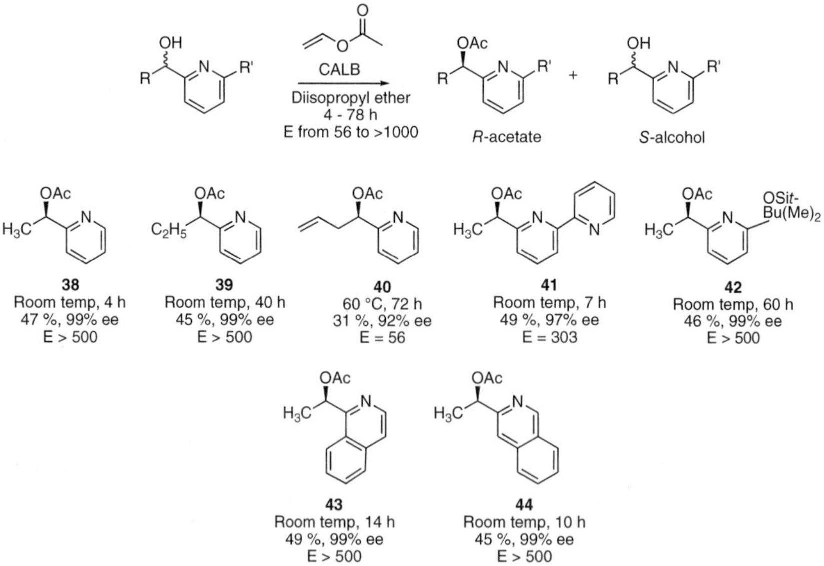

Compounds (R-selective acylation, CALB, Hexane, 40 °C, 2 h, 7 - 43% conversion, E from 230 to >1000):

- **26**: E > 1000
- **27** (4-Br): E > 1000
- **28** (4-CH₃): E = 230
- **29** (4-CF₃): E > 1000
- **30** (3-CF₃): E > 1000
- **31** (3,5-CF₃): E > 1000

S-selective acylation (Subtilisin, THF, Na₂CO₃, room temp., 17 h, E from 47 to >200):

- **32**: E = 66
- **33** (4-Cl): E = 97
- **34** (4-CH₃): E = 52
- **35** (4-CF₃): E = 133
- **36** (4-OCH₃): E = 47
- **37** (3,5-CF₃): E > 200

Scheme 4.18 Resolution of some 2-pyridyl-1-alkanols and quinolyl-1ethanols providing the acetates **38–42** and **43, 44**, respectively.

(CALB, Diisopropyl ether, 4 - 78 h, E from 56 to >1000)

- **38**: Room temp, 4 h, 47 %, 99% ee, E > 500
- **39** (C₂H₅): Room temp, 40 h, 45 %, 99% ee, E > 500
- **40** (allyl): 60 °C, 72 h, 31 %, 92% ee, E = 56
- **41**: Room temp, 7 h, 49 %, 97% ee, E = 303
- **42** (OSit-Bu(Me)₂): Room temp, 60 h, 46 %, 99% ee, E > 500
- **43**: Room temp, 14 h, 49 %, 99% ee, E > 500
- **44**: Room temp, 10 h, 45 %, 99% ee, E > 500

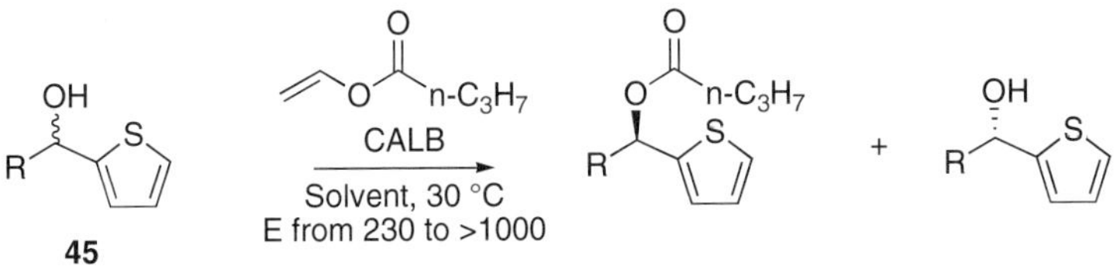

Scheme 4.19 Kinetic resolution of 1-(2-thienyl)-1-alkanols of type **45**.

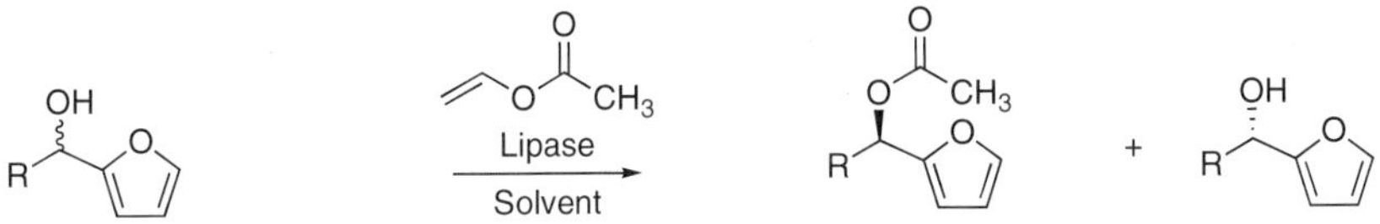

R = CH$_3$ (Hexane) E > 200
R = C$_2$H$_5$ (Hexane) E > 500
R = n-C$_3$H$_7$ (Benzene) E > 500
R = n-C$_4$H$_9$ (Toluene) E > 200

46: R = CH$_3$ (CCl$_4$, Lipozyme IM) E = 87
THF, PPL) E = 266

47: R = CH$_2$CH=CH$_2$ (pentane, CALB) E = 75

Scheme 4.20 Kinetic resolution of some 2-furylcarbinols **46** [77] and **47** [78].

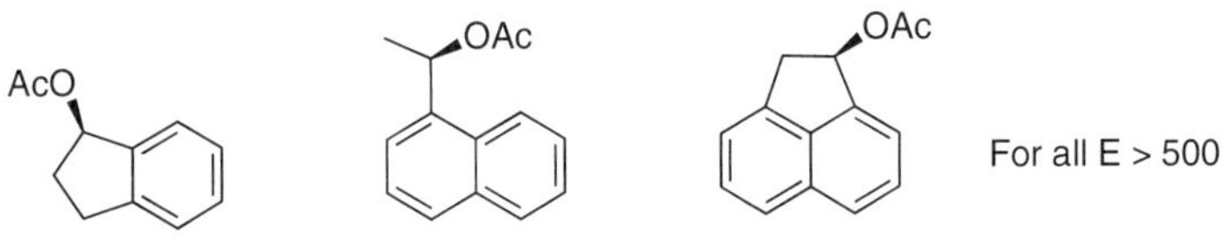

For all E > 500

Scheme 4.21 Kinetic resolution of some complex arylalkylcarbinols.

bulkier 2-furylallylcarbinol **47** can be resolved by vinyl acetate acylation catalyzed either by CALB or *Alcaligenes sp.* lipase (ASL) in pentane with E 75 or 80, respectively, but when ASL is used, hydrolysis of the racemic acetate is more practical [78].

Kinetic resolution of some more complex arylalkylcarbinols have also been achieved by acylation with isopropenyl acetate catalyzed by CALB (Scheme 4.21) [79]. The reactions are performed in diethyl ether at room temperature for 24–30 h. With an appropriate amount of CALB, the reactions proceed with excellent E, providing the acetates **48–50**.

There are numerous reports of successful resolutions of other types of secondary alcohols. A few representative examples will be discussed here.

With the exception of 2-butanol (E = 9), simple 2-alkanols can be efficiently resolved by using CALB-catalyzed (Novozym 435) acylation in hexane using S-ethyl thiooctanoate or vinyl butyrate as acyl donor to produce the esters **51–59**

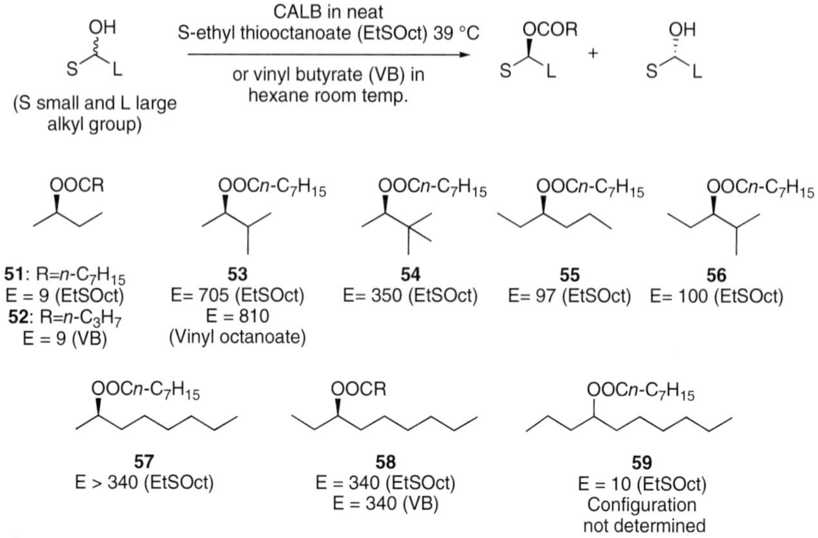

Scheme 4.22 CALB-catalyzed acylation of some secondary alkanols.

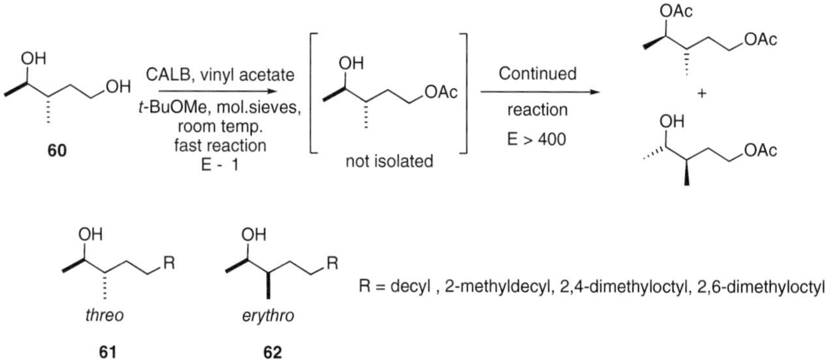

Scheme 4.23 CALB-catalyzed resolution of *anti*-3-methyl-1,4-pentanediol (**60**) and alcohols **61** and **62**.

(Scheme 4.22) [80]. The use of long-chain acyl donors for the acylation of 3-methyl-2-butanol was found to increase E. Thus CALB-catalyzed acylation of this alcohol with vinyl propionate, butyrate, hexanoate, and octanoate gave E = 470, 390, 720, and 810, respectively [81].

The resolution of *anti*-3-methyl-1,4-pentanediol (**60**) proceeds via rapid Novozym 435 (CALB) catalyzed monoacylation with vinyl acetate in TBME under dry conditions followed by a slower enantioselective acylation of the secondary alcohol moiety (Scheme 4.23) [82]. Similarly, diastereomerically pure *threo-* and *erythro*-3-methylalkan-2-ols of type **61** and **62**, respectively, have been successfully acylated with 2R-preference ("E" > 100 to >1000) using vinyl acetate, propionate or butyrate

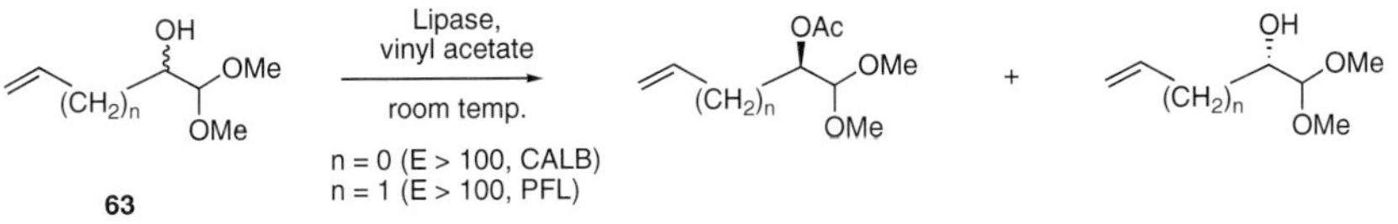

Scheme 4.24 Resolution of some 2-hydroxyacetals.

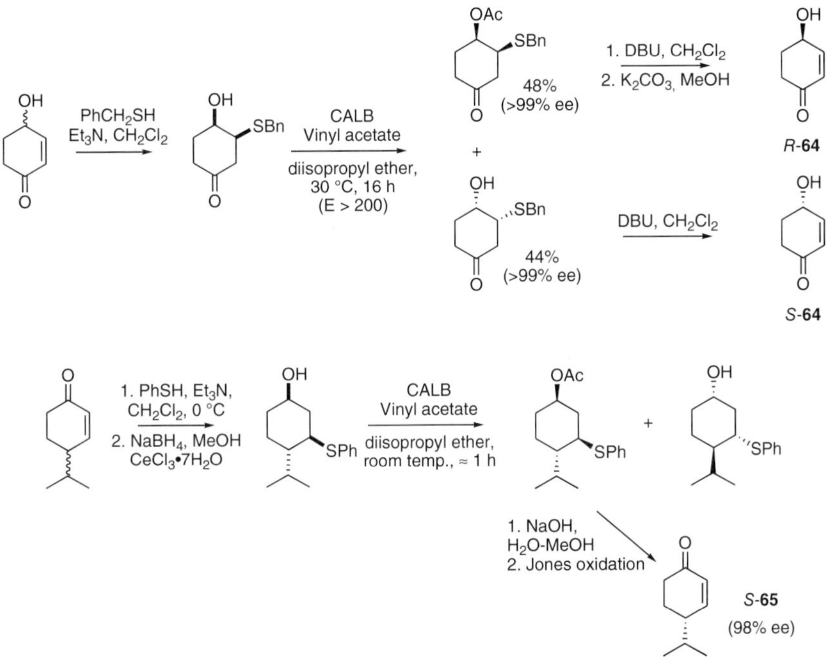

Scheme 4.25 Strategy for the resolution of cyclohex-2-enols and cyclohex-2-enones.

as acyl donor in hexane at room temperature and lipase Amano PS as such or immobilized on diatomite [83].

Two 2-hydroxyaldehydes protected as acetals (**63**) have been resolved by lipase-catalyzed acylation with vinyl acetate as reagent and solvent (Scheme 4.24) [84]. For the vinyl derivative (n = 0) Chirazyme L2 (CALB) gives the best E, whereas *Pseudomonas fluorescens* lipase (PFL) provides the best E for n = 1 [84].

Because of the minimal steric difference between the substituents surrounding the alcohol moiety in simple 2-cyclohexenols, these are not easy to resolve by hydrolase-catalyzed reactions. However if a sterically bulky group can be temporarily introduced and then removed, such resolution will be facilitated. One example of this approach is the preparation of enantiomerically enriched 4-hydroxycyclohex-2-enone (**64**) (Scheme 4.25) [85]. A similar approach was used for the preparation of enantiomerically enriched cryptone (**65**) (Scheme 4.25) [86].

The enzyme-mediated preparation of the four stereoisomers of the odorant Jasmal has been described [87]. Thus, the two racemic diastereomeric alcohols **66**

Scheme 4.26 Lipase PS-catalyzed acylation for resolution of the racemic cis- and trans-isomers of 4-acetoxy-3-pentyltetrahydropyran (Jasmal).

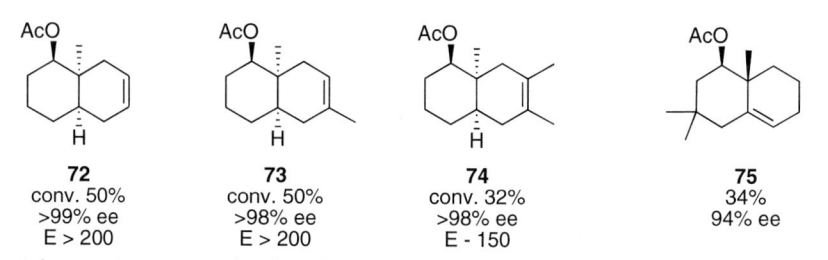

72
conv. 50%
>99% ee
E > 200

73
conv. 50%
>98% ee
E > 200

74
conv. 32%
>98% ee
E - 150

75
34%
94% ee

Scheme 4.27 Lipase-catalyzed resolution of some octahydronaphthalenols.

and **67**, the alcohol precursors of Jasmal, were separated by chromatography and individually subjected to Lipase PS and vinyl acetate in TBME. The product mixtures were separated, and the remaining alcohol was acetylated. Thus, all four stereoisomers of Jasmal **68–71** were prepared in high *ee* (Scheme 4.26) [87].

Successful resolutions of hydrogenated naphthalenes by acylation in organic solvent have been described. Thus, CALB (Novozym 435) catalyzes the acylation with vinyl acetate at 40 °C in hexane of three *cis*-fused octahydronaphthalenols to furnish after 48 h the acetylated products **72–74** and the remaining alcohol in high *ee*s and with excellent E (Scheme 4.27, products **72-74** shown) [88]. Similarly, a key intermediate in the synthesis of natural products of the marasmane and lactarane

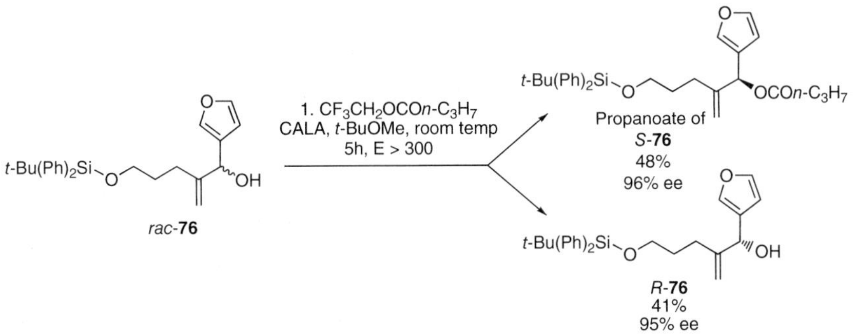

Scheme 4.28 CALA-catalyzed resolution of a sterically demanding vinylfurylcarbinol.

type, another racemic octahydronaphthalenol, was enantioselectively acylated with vinyl acetate catalyzed by *Candida rugosa* lipase (CRL) in diisopropyl ether at 45 °C for 22 h (Scheme 4.27, product **75** shown) [89]. In the latter case, the more commonly used lipases CALB or lipase PS gave undesired products later identified as acetates of hemiacetals.

The rather complex furylvinylcarbinol derivative **76** shown in Scheme 4.28 was required in enantiopure form as a key intermediate in the synthesis of the natural product cneorin. The carbinol moiety is heavily substituted with sterically demanding groups. Therefore attempts to resolve the furylvinylcarbinol with CALB or lipase PS-II led to very slow reactions. However, the rarely used enzyme *Candida antarctica* lipase A (CALA), which is known to act on sterically hindered substrates offers an alternative. Thus acylation of the furylvinylcarbinol **76** with 2,2,2-trifluoroethyl butanoate catalyzed by CALA (immobilized on celite with sucrose at pH 7.9) furnished the enantiomerically enriched propanoate of *S*-**76** and *R*-**76** (Scheme 4.28) [90]. Small-scale experiments gave E > 300.

Hydrolase-catalyzed acylation can be used to purify a diastereo- and enantiomerically enriched product. For example dimethylzinc addition to the racemic aldehyde **77** furnishes the racemic phenylsulfanylbutanol **78** (Scheme 4.29) in a 95/5 (2R^*,3R^*)/(2R^*,3S^*)-ratio. When this is treated with Chirazyme L2 (CALB) and vinyl acetate in heptane it is resolved with a high E-value (>400) [91]. However the diastereomeric ratio in the remaining substrate and produced ester is virtually unchanged. To circumvent the problematic contamination with the undesired diastereomers, enantiomerically enriched aldehyde **77** was reacted with dimethylzinc to furnish one major stereoisomer of **78** contaminated with a small amount of a mixture of the other three (Scheme 4.29). Because the two major contaminants had the opposite configuration at position 2 relative to the major product, these contaminants were efficiently removed from the major product and the trace by-product by treatment with the 2R-selective Chirazyme L2 (CALB) and vinyl acetate in heptane to furnish virtually diastereo- and enantiomerically pure acetate (2R,3R)-**79** or the alcohol (2S,3S)-**78** (Scheme 4.29) [91].

One of the limitations of kinetic resolutions is the maximum obtainable yield of 50% of one pure enantiomer. One way of circumventing this limitation is of

Schemem 4.29 Lipase-catalyzed purification by removal of undesired stereoisomers from the product. Chirazyme L2 (CALB), vinyl acetate, n-heptane, mol. sieves 3A, 25 °C.

course continuous and rapid inversion of the substrate, i.e. dynamic kinetic resolution (DKR), which is discussed in detail in Chapter 6. Another way is to react the substrate to approximately 50% conversion and then invert the configuration of the remaining substrate. After an initial kinetic resolution, the Mitsunobu reaction can be used to invert the remaining secondary alcohols under simultaneous acylation, and therefore it can be used in an essentially one-pot procedure for the production of a single enantiomerically highly enriched ester [92]. Thus CALB-catalyzed acylation with vinyl butanoate of some 3-chloro-1-aryl-1-propanols **80a** or **b** in hexane or toluene furnished the R-esters **81** and **82**, respectively, and the remaining S-alcohols S-**80** (Scheme 4.30, top) [93]. After removal of the enzyme and solvent, the product mixture was dissolved in ether and treated with butanoic acid, triphenylphosphine, and diethyl diazodicarboxylate (Mitsunobu conditions), which led to exclusive conversion of the remaining alcohols S-**80a** or S-**80b** to give the R-butanoate esters **81** and **82**, respectively. Thus a hydrolase-catalyzed acylation and a subsequent Mitsunobu reaction of the remaining substrate yield the same product, the R-butanoate ester. A similar approach was used to prepare the marine natural product dictyoprolene (Scheme 4.30) [94]. Thus the racemic 1,5-undecadien-3-ol **83** was acylated in hexane with vinyl acetate catalyzed by lipase Amano PS. The reaction was stopped at 51% conversion by removal of the biocatalyst. After concentration and treatment under Mitsunobu conditions (anhydrous THF, acetic acid, PPh$_3$, diisopropyl diazodicarboxylate), the remaining substrate furnished the ester S-**84** along with the same compound produced by the Amano

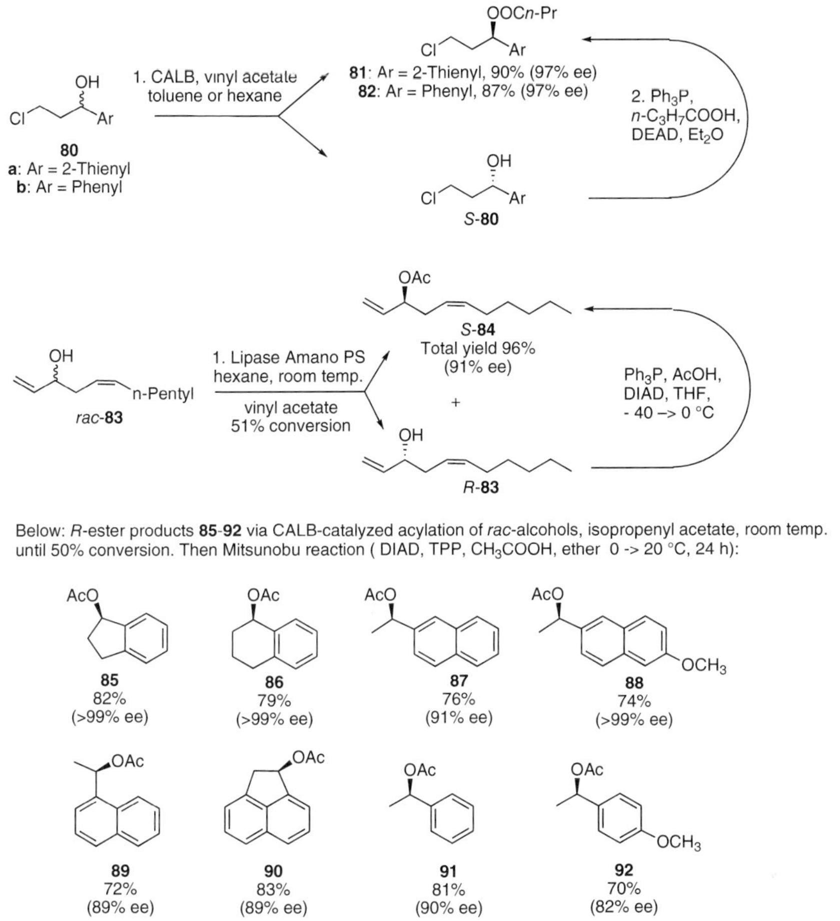

Scheme 4.30

Scheme 4.30 Combined kinetic resolution and Mitsunobu inversion.

Top reaction: Compound **80** (a: Ar = 2-Thienyl; b: Ar = Phenyl), Cl-CH$_2$-CH(OH)-Ar.
1. CALB, vinyl acetate, toluene or hexane → **81**: Ar = 2-Thienyl, 90% (97% ee); **82**: Ar = Phenyl, 87% (97% ee) (as OOCn-Pr esters). 2. Ph$_3$P, n-C$_3$H$_7$COOH, DEAD, Et$_2$O → S-**80**.

Middle reaction: rac-**83** (n-Pentyl allylic alcohol). 1. Lipase Amano PS, hexane, room temp., vinyl acetate, 51% conversion → S-**84** (OAc), Total yield 96% (91% ee) + R-**83** (OH). Then Ph$_3$P, AcOH, DIAD, THF, −40 → 0 °C.

Below: R-ester products **85–92** via CALB-catalyzed acylation of rac-alcohols, isopropenyl acetate, room temp. until 50% conversion. Then Mitsunobu reaction (DIAD, TPP, CH$_3$COOH, ether 0 → 20 °C, 24 h):

85 82% (>99% ee) — AcO-indane
86 79% (>99% ee) — OAc-tetrahydronaphthalene
87 76% (91% ee) — AcO-naphthyl
88 74% (>99% ee) — AcO-methoxynaphthyl, OCH$_3$
89 72% (89% ee) — OAc-naphthyl
90 83% (89% ee) — OAc-acenaphthyl
91 81% (90% ee) — OAc-phenyl
92 70% (82% ee) — OAc-(4-OCH$_3$)phenyl

PS-catalyzed reaction [94]. A number of arylalkylcarbinyl esters **85–92** were also prepared in 70–83% yield (82–>99% ee) via a similar approach (Scheme 4.30, bottom) [79].

Hydrolase-catalyzed domino reactions incorporating a resolution and a subsequent cycloaddition reaction have been described [95–97]. This constitutes an attractive approach to complex synthetic intermediates. For example, the 1-(3-methyl-2-furyl)]propanol rac-**93** reacts with ethoxyvinyl methyl fumarate (**94**) catalyzed by Lipase LIP (from *Pseudomonas aeruginosa*) to furnish a dienophilic fumarate ester, which spontaneously undergoes an intramolecular Diels-Alder reaction with the furan moiety furnishing exclusively the *syn*-adduct, the oxabicyclohexene **95** in excellent ee along with the remaining alcohol S-**96** (Scheme 4.31) [95]. A similar approach has been used for a procedure that includes a series of domino reactions that includes dynamic kinetic resolution of the 3-vinylcyclohex-

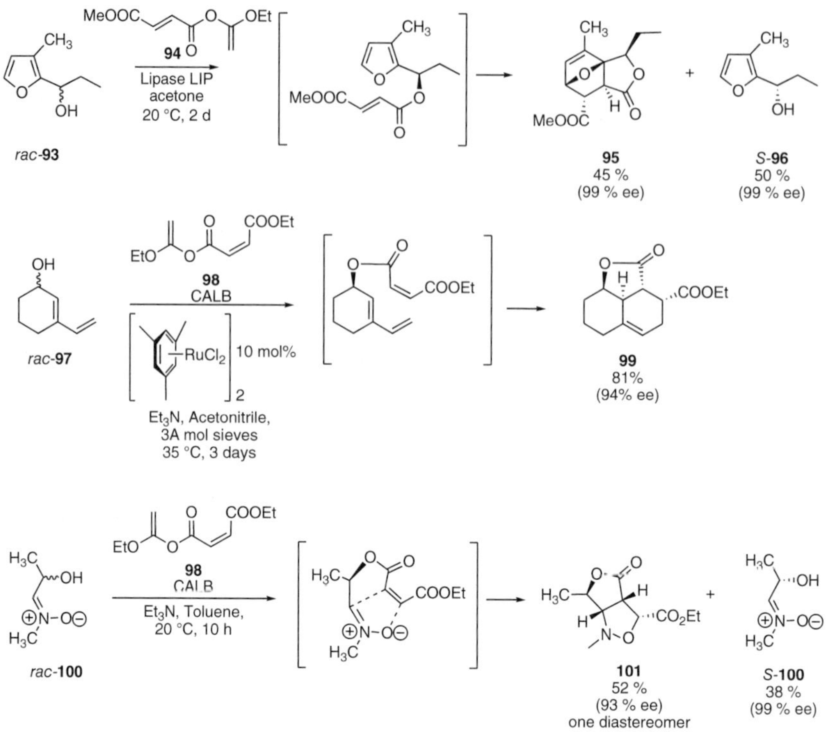

Scheme 4.31 Domino hydrolase-catalyzed/cycloaddition reactions.

2-en-1-ol rac-**97** by reaction (CALB catalysis) with **98** followed by a Diels-Alder reaction of the resulting maleate ester to furnish the highly enantiomerically enriched decalin **99** (Scheme 4.31) [96]. A domino hydrolase-catalyzed acylation/1,3-dipolar cycloaddition has also been described (Scheme 4.31) [97]. The reaction of the acyl donor, 1-ethoxyvinyl ethyl maleate (**98**), with a racemic 2-hydroxypropylnitrone rac-**100** catalyzed by CALB gives a nitrone maleate ester, which readily undergoes intramolecular 1,3-dipolar cyclooaddition to yield a single diastereomer of the adduct **101** in a high ee along with the remaining enantiomerically pure starting nitrone S-**100** [97].

Hydrolase-catalyzed resolution combined with a subsequent ring-closing metathesis (RCM) reaction is a fairly new development that constitutes a promising strategy for forming α,β-unsaturated lactones [98, 99]. Thus CALB and vinyl acrylate in hexane has been used to resolve (E = 65) the vinylhomoallylalcohol rac-**102** (Scheme 4.32). After chromatographic separation, the enantiomerically enriched acrylate R-**103** was treated with Grubbs' catalyst in dichloromethane to furnish the desired hexenolide, the natural product (+)-goniothalamin R-**104** in (96% ee) [98]. Acylation with acryloyl chloride of the remaining alcohol S-**102** followed by RCM furnished (−)-goniothalamin S-**104** [98]. A similar strategy was used to furnish highly enantiomerically enriched butenolides, e.g., the butenolides

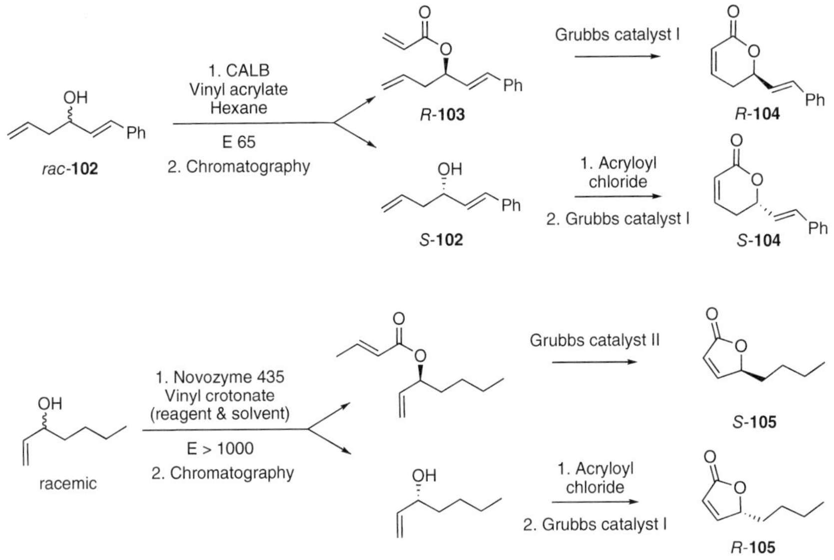

Scheme 4.32 Combined resolution and RCM to give α,β-unsaturated lactones.

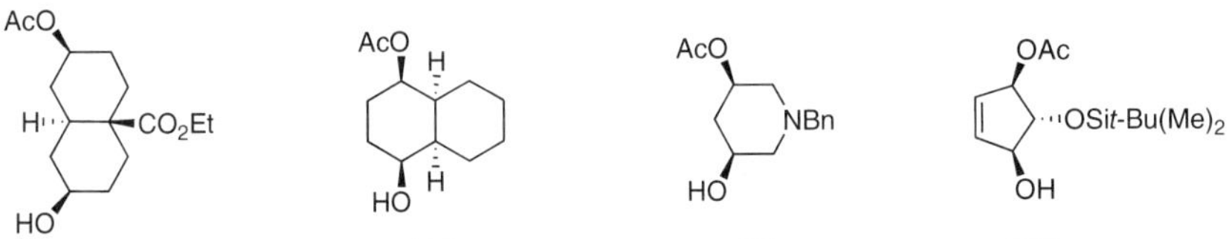

106 96% (>99% ee)

107 85% (>98% ee)

108 91% (>99% ee)

109 99% (>99% ee)

Scheme 4.33 Some examples of products obtained by hydrolase-catalyzed desymmetrizations of *meso*-diols (products shown).

S-**105** and R-**105** (>98 % ee, Scheme 4.32), useful intermediates for the preparation of the enantiomers of quercus or whiskey-lactone [99].

As pointed out in Section 4.1.2, desymmetrizations of *meso*-compounds provides a possibility to obtain a quantitative yield of enantiopure compounds. Some examples of desymmetrizations of secondary *meso*-diols using hydrolases and an acyl donor in an organic solvent are given below. Thus, the racemic *meso*-diol corresponding to the *trans*-decalin monoester **106** (Scheme 4.33) can be efficiently desymmetrized by lipase AK and vinyl acetate in diisopropyl ether at 33 °C for 12 h to give the monoester **106** [100]. The *cis*-decalin **107** is obtained by CALB-catalyzed reaction of the *rac-meso*-diol in neat vinyl acetate at 40 °C for 1 day [101]. A *meso-N*-benzylpiperidinediol reacts with isopropenyl acetate catalyzed by lipase PS-C Amano II in toluene at room temperature for 4 h to give the monoacetate **108** [102]. Lipase Amano AK and vinyl acetate at 33 °C for 13 h reacted with a *meso*-cyclopentenediol to give the monoacetate **109** (Scheme 4.33) [103].

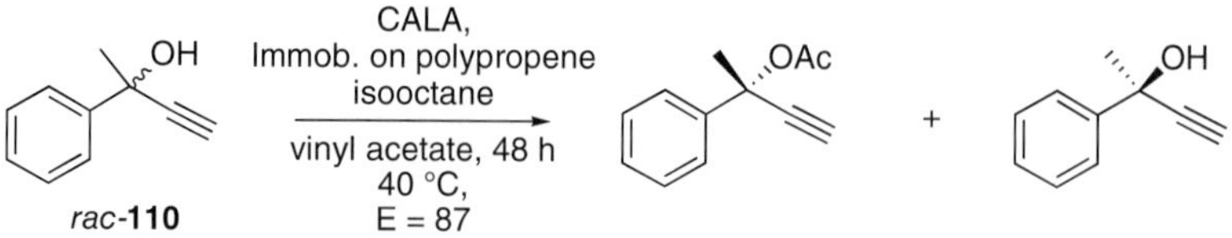

Scheme 4.34 Resolution of a tertiary alcohol.

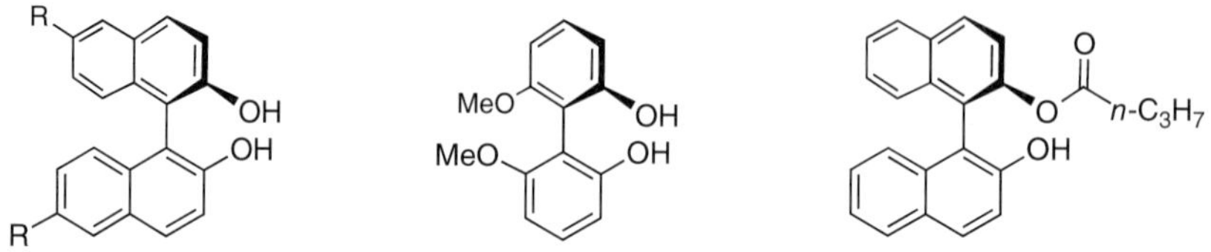

111: R = H; E = 85
112: R = Br; E = 80
113: R = OMe; E = 14

114
E > 100

115
E = 91

Scheme 4.35 Lipase-catalyzed acylation 111–114 and butanolysis 115 of phenolic biaryls. Preferentially reacted enantiomers are shown.

As described above, the resolution of many types of secondary alcohols by hydrolase-catalyzed acylation in an organic solvent is usually possible after screening for a selective lipase and optimization of the reaction conditions.

4.2.1.3 Tertiary Alcohols

Tertiary alcohols are very unreactive toward hydrolases. There are, however, exceptions. Some enzymes have a certain amino acid motif located in the oxyanion binding pocket that allows the docking of space-demanding alcohols such as tertiary ones into the acylated enzyme [104]. One such hydrolase is *Candida antarctica* lipase A (CALA), which has been found to catalyze the acylation of the tertiary 2-phenyl-3-butyn-2-ol *rac*-110 by vinyl acetate in organic solvents. Thus efficient resolution of 110 was achieved in isooctane at room temperature (Scheme 4.34) [105].

4.2.1.4 Resolution of Dihydroxybiaryls

Chiral enantiopure 2,2′-dihydroxybiaryls are important as chiral ligands and are also structural motifs occurring in some natural products. Hydrolase-catalyzed resolution by acylation in organic solvents of some dihydroxybiaryls has been successfully achieved. Thus, the racemic binaphthols 111–113 have been resolved by mono acylation with vinyl acetate in *t*-butylmethyl ether (TBME) at 45 °C catalyzed by *Pseudomonas* sp. lipase (Scheme 4.35) [106]. In a similar way the 2,2′-dihydroxybiphenyl 114 can be acylated with vinyl acetate catalyzed by PSL immobilized on celite in TBME at 45 °C (Scheme 4.35) [107]. Butanolysis of the racemic monobutyrate of binaphthol *rac*-115 catalyzed by CALB in toluene at 80 °C for 72 h gives (*R*)-binaphthol (93% ee) at ca. 50% conversion [108].

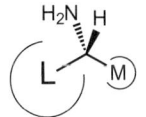

Figure 4.5 Empirical selection rule for lipases for preferential acylation of amines. Subtilisin prefers the opposite configuration.

4.2.2 Reaction of Amines

Enantiopure amines are important industrial building blocks and are used as resolving agents, chiral auxiliaries, and intermediates for pharmaceuticals and agrochemicals. Hydrolase-catalyzed resolution is an important tool for their preparation [109]. A quite reliable empirical rule exists for the enantioselectivity of hydrolases for amines: most lipases catalyze aminolysis of esters with the preference described in Figure 4.5, but subtilisin reacts with opposite enantiopreference. A great number of chiral racemic amines have been successfully resolved by acylation in organic solvents. It is important to remember that amines may react spontaneously with esters. Therefore, the reaction conditions must be adjusted to reduce the spontaneous reaction to ensure that the enantiomeric excess of the product is as high as possible. Therefore, the acyl donor should have a moderate reactivity to ensure that the relative rate of the uncatalyzed reaction is slow relative to the hydrolase-catalyzed one. A proper choice of solvent is crucial, and the spontaneous reaction is often faster in non-polar solvents than in polar ones.

4.2.2.1 Alkylamines

Several chiral racemic alkylamines have been successfully resolved using hydrolase-catalyzed acylation reactions with esters as acyl donors. A few examples are described here (Table 4.2).

4.2.2.2 Arylalkylamines

Hydrolase-catalyzed acylation of 1-arylethylamines has been studied extensively, and the reactions usually proceed with excellent E-values, providing easy access to both enantiomers of the amine after chemical hydrolysis. Some examples are listed in Table 4.3.

Other arylalkylamines have been resolved with quite impressive E-values using CALB (Scheme 4.36). Often the solvent and the acyl donor have to be carefully chosen to achieve an optimal E.

Examples of other amines that have been rather efficiently resolved are given in Scheme 4.37. Thus, amphetamine (**122a**) and methoxylated derivatives thereof (**122b–d**) have been resolved at 28 °C catalyzed by CALB using ethyl acetate as the acyl donor and solvent [122]. Under the same conditions, the acylation of

Table 4.2 Hydrolase-catalyzed enantioselective resolution by acylation of alkylamines. Preferentially acylated enantiomers shown. In some cases E-values are based on calculations from the literature data. Adapted from Ref. [109].

Lipase: (R)-116, structure with NH_2 and CH_3 on R

Subtilisin: (S)-116, structure with NH_2 and CH_3 on R

- a: R = C_2H_5
- b: R = n-C_3H_7
- c: R = n-C_4H_9
- d: R = t-C_4H_9
- e: R = n-C_5H_{11}
- f: R = n-C_6H_{13}
- g: R = n-C_9H_{19}
- h: R = CH_3OCH_2
- i: R = $cyclo$-C_6H_{11}

Amine	Hydrolase	Acyl donor	Solvent	T °C	E	Amide	Ref.
116a	CALB	Methyl methacrylate	THF	30	55	R	110
116b	CALB	EtOAc	neat		>200	R	111
116c	CALB	EtOAc	neat	21	≈20	R	112
116d	CALB	Isopropyl acetate	DME	rt	>700	R	113
116e	Subtilisin	Trifluoroethyl butyrate	3-Methyl-3-pentanol	45	6	S	114
116e	CALB	1-Phenylethyl acetate	Dioxane		>200	R	115
116f	CALB	Ethyl acetate	neat	21	>60	R	112
116g	CALB	Ethyl acetate	neat	21	>50	R	112
116h	CALB	Ethyl methoxyacetate	neat	rt	58	R	116
116i	CALB	Dibenzyl carbonate	Toluene	rt	17	R	117
116i	Subtilisin	Trifluoroethyl butyrate	3-Methyl-3-pentanol	30	23	S	114

trans-phenylcyclopentylamine (**123**) proceeds with an excellent E-value but the corresponding *cis*-isomer gives a low E (16) and reacts much more slowly [123]. By running the reaction with the *cis*-isomer in TBME at 28 °C and changing the acyl donor to *rac-cis*-2-phenylcyclopentyl methoxyacetate, one can obtain a very high E (ca. 1000) [123]. When treated with CALB in dioxane at 30 °C in the presence of an equimolar amount of dimethyl malonate, the *trans*-cyclopentanediamine **124** can be efficiently resolved to give the enantiomerically pure (*R,R*)-bis(amidoester) and the remaining (*S,S*)-diamine (87% *ee*) [124]. The reaction is a sequential kinetic resolution: in the first step, with E = 21, the first malonyl group is put onto **124**. Because the bulk is increased in the second acylation step with **125** as the substrate, the E-value is much higher [124].

The sterically rather demanding racemic binaphthylalkylamines **126** and **127** can be efficiently resolved in diisopropyl ether using *Pseudomonas aeruginosa* lipase (LIP) as catalyst and trifluoroethyl butyrate as acylating agent for **126** and ethyl butyrate for **127** (Scheme 4.38) [125].

Table 4.3 Hydrolase-catalyzed enantioselective resolution by acylation of arylalkylamines. Preferentially acylated enantiomers shown. In some cases E-values are based on calculations from the data given in the references. Adapted from [109].

Lipase: (R)-117, R-CH(NH₂)-CH₃
Subtilisin: (S)-117, R-CH(NH₂)-CH₃

a: R = Ph
b: R = 4-ClC₆H₄
c: R = 2-Pyridyl
d: R = 2-Furyl
e: R = 2-Thienyl
f: R = 1-Naphtyl
g: R = isoquinolinyl

Amine	Hydrolase	Acyl donor	Solvent	T °C	E	Amide	Ref.
117a	CALB	Ethyl acetate	neat		110	R	[111]
117a	CALB	i-Propyl methoxyacetate	TBME		>1000	R	[118]
117a	CALB	Dibenzyl carbonate	Toluene	rt	>300	R	[117]
117a	Subtilisin	Trifluroethyl butyrate	3-Methyl-3-pentanol	30	19	S	[114]
117b	CALB	Methyl methoxyacetate	t-Amylmethyl ether	40	>400	R	[119]
117c	CALB	Ethyl acetate	neat	30	66	R	[120]
117d	CALB	Ethyl acetate	neat	30	>100	R	[120]
117e	CALB	Ethyl acetate	Dioxane	30	>100	R	[120]
117f	CALB	i-Propyl acetate	DME	rt	650	R	[113]
117f	Subtilisin	Trifluoroethyl butyrate	3-Methyl-3-pentanol	30	150	S	[114]
117g	CALB	Ethyl acetate	DIPE	60	120	R	[121]

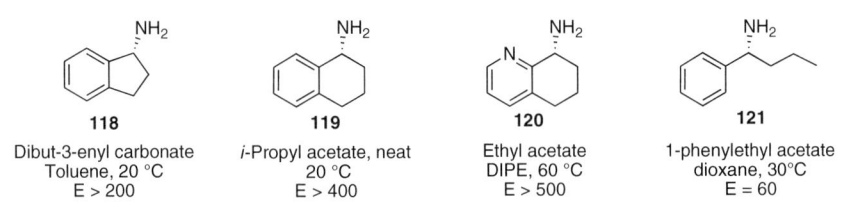

118
Dibut-3-enyl carbonate
Toluene, 20 °C
E > 200

119
i-Propyl acetate, neat
20 °C
E > 400

120
Ethyl acetate
DIPE, 60 °C
E > 500

121
1-phenylethyl acetate
dioxane, 30°C
E = 60

Scheme 4.36 CALB-catalyzed acylation of arylalkylamines **118** [117], **119** [113], **120** [121], and **121** [115]. The most reactive enantiomers toward acylation are shown.

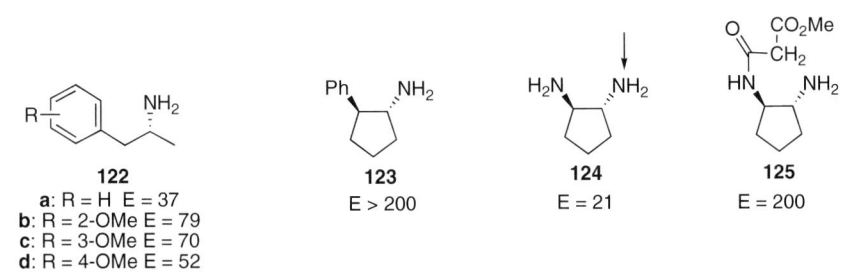

122
a: R = H E = 37
b: R = 2-OMe E = 79
c: R = 3-OMe E = 70
d: R = 4-OMe E = 52

123
E > 200

124
E = 21

125
E = 200

Scheme 4.37 Preferentially acylated enantiomers of some chiral amines.

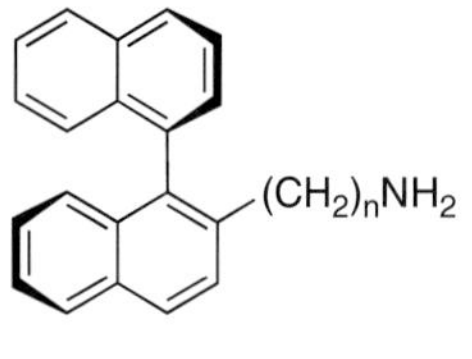

126: n = 1, E = 45
127: n = 2, E > 450

Scheme 4.38 Resolution of binaphthylalkylamines. Preferentially acylated enantiomers shown.

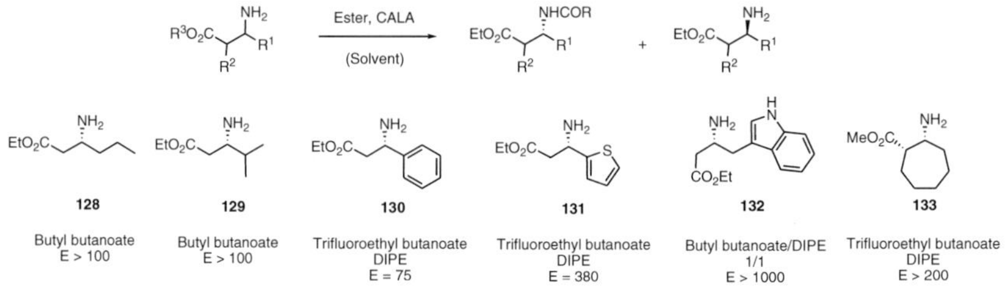

Compound	128	129	130	131	132	133
Acyl donor	Butyl butanoate	Butyl butanoate	Trifluoroethyl butanoate DIPE	Trifluoroethyl butanoate DIPE	Butyl butanoate/DIPE 1/1	Trifluoroethyl butanoate DIPE
E	E > 100	E > 100	E = 75	E = 380	E > 1000	E > 200

Scheme 4.39 Candida antarctica lipase A (CALA)-catalyzed resolution of some β-aminoesters at room temperature. Preferentially acylated enantiomers are shown: **128–130** [22], **131** [128], **132** [129], **133** [130].

4.2.2.3 Amino Acid Derivatives

Hydrolase-catalyzed enantioselective N-acylation is an important tool for the preparation of enantiopure α- and β-aminoacids. It has been observed that the reactions of many amino acid esters with ester acyl donors catalyzed by CALB is sometimes complicated by interesterification reactions. CALA has, however, emerged as a very chemoselective catalyst in favor of N-acylation of β-aminoesters. Some reviews on CALA and other hydrolases as catalysts for N-acylations of aminoesters are available [109, 126, 127].

Some very efficient resolutions of β-aminoesters using CALA as acylating catalyst are exemplified in Scheme 4.39. The enzyme is immobilized on Celite in the presence of sucrose or on polypropylene (Accurel EP-100).

For the β-aminocarboxamides **134–137**, CALA gave very low enantioselectivity. Because no ester function was present in the substrate, N-acylation proceeded satisfactory with CALB as the catalyst and trifluoroethyl butyrate as the acyl donor in t-butylmethyl ether/t-amylalcohol (TBME)/(TAME) mixtures at 45 °C (Scheme 4.40) [131].

4.2.3
Reaction of Acid Derivatives

4.2.3.1 Acids

Although quite reliable empirical rules exist for the enantioselectivity of hydrolases for secondary alcohols, such rules are not as reliable for acid derivatives. The rule

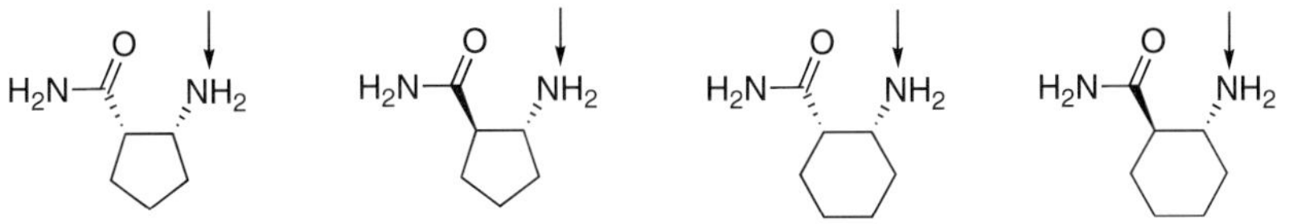

4.2 Enantioselective Reactions in Organic Solvents

134
E > 200
c = 50%, 1 h

135
E > 200
c = 50%, 10 h

136
E = 61
c = 27%, 6 h

137
E > 200
c = 50%, 21 h

Scheme 4.40 CALB-catalyzed acylation of 2-aminocyclopentanecarboxamides. Preferentially acylated enantiomers are shown.

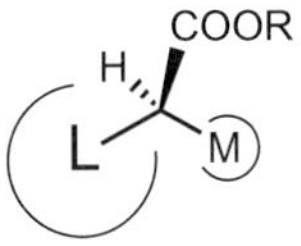

Figure 4.6 Empirical selection rule for 2-substituted acid derivatives.

described in (Figure 4.6) is derived for reactions catalyzed by *Candida rugosa* lipase (CRL) of 2-substituted acid derivatives [132]. Several exceptions to this rule have been demonstrated [132]. Therefore, as soon as acid derivatives are resolved by enzyme catalysis, independent proof of configuration for a previously unknown product is recommended.

Resolution of branched alkanoic acids. Hydrolase-catalyzed esterification of 2-methylalkanoic acids can be fairly efficient, especially for acids with long chains, provided that the conditions are carefully adjusted by immobilization of the enzyme (in some cases), by control of the water activity, and by proper choice of the appropriate alcohol as nucleophile as well as the correct solvent [134]. The alcohol concentration does also influence the E-value [133]. It is important to note that the esterifications are reversible, thus preventing easy access to the remaining substrate in high *ees*. Some representative examples are given in Table 4.4. A procedure based on iterative resolutions can be used to provide both enantiomers of 2-methyloctanoic acid in high *ees* (>99%) and reasonable yields (25% for *S*- and 43% for *R*-acid based on the starting racemic acid) [137].

Methyldecanoic acids with remotely located methyl branches have been subjected to esterifications under similar conditions with 1-hexadecanol in cyclohexane at a_w = 0.8. Here, *R*-selectivity is the result for the 3-, 5-, and 7-methyldecanoic acids (E = 21, 58, and 17, respectively), whereas one obtains *S*-selectivity for the 2-, 4-, 6-, and 8-methyldecanoic acids (E = 91, 68, 3, and 4, respectively) [139]. Several other alkanoic and alkenoic acids can also be resolved by CRL-catalyzed esterification under similar conditions and with similar efficiency (Scheme 4.41) [136]. It is worth noting that either an *E*-double bond in the 6-position or a methyl group in the 6-position correctly placed in space enhances the 2*S*- over the 2*R*-selectivity [136].

Table 4.4 Resolution of 2-methylalkanoic acids by esterification in an organic solvent. In cases where immobilized CRL was used, it was immobilized on Accurel EP100 (350-1000 μm) [136].

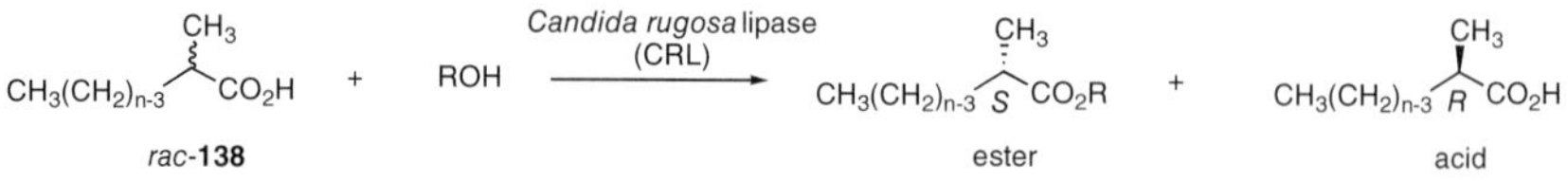

$$CH_3(CH_2)_{n-3}\underset{rac\text{-}138}{\overset{CH_3}{\underset{|}{C}}}CO_2H + ROH \xrightarrow{\text{Candida rugosa lipase (CRL)}} CH_3(CH_2)_{n-3}\underset{ester}{\overset{CH_3}{\underset{S}{C}}}CO_2R + CH_3(CH_2)_{n-3}\underset{acid}{\overset{CH_3}{\underset{R}{C}}}CO_2H$$

138

n	Acid	Alcohol (ROH)	E	Solvent	Water activity	Immob.	Ref.
4	2-Methylbutanoic	1-Octadecanol	5	Heptane	No control	No	[135]
5	2-Methylpentanoic	1-Octanol	70	Heptane	No control	No	[135]
6	2-Methylhexanoic	1-Octadecanol	30	Heptane	No control	No	[135]
7	2-Methylheptanoic	1-Hexadecanol	23	Cyclohexane	$a_w = 0{,}8$	Yes	[136]
8	2-Methyloctanoic	1-Eicosanol	115	Cyclohexane	$a_w = 0{,}8$	Yes	[137]
9	2-Methylnonanoic	1-Hexadecanol	85	Cyclohexane	$a_w = 0{,}8$	Yes	[138]
10	2-Methyldecanoic	1-Octadecanol	115	Cyclohexane	$a_w = 0{,}8$	Yes	[137]

139 E = 24

140 E = 17

141 E = 51

142 E = 23

143 "E" = 53 (6R)

144 "E" = 15 (6S)

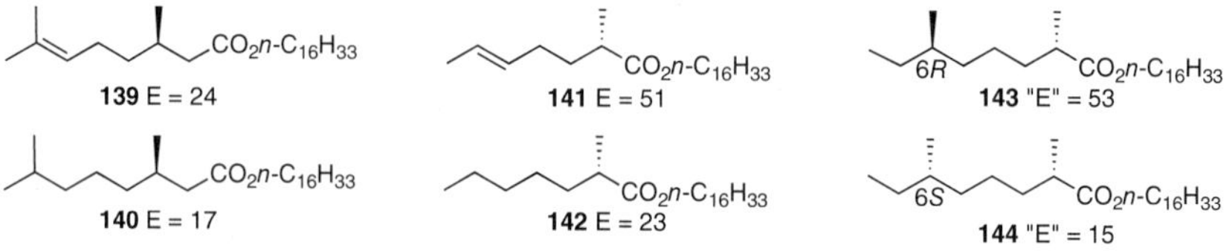

Scheme 4.41 Resolution of a mono- and some disubstituted acids: Major enantiomers of the products **139–144** are shown. Major enantiomers of the products are shown. For **143** and **144** the starting materials were pure (6R)- or (6S)-isomers, respectively, with 50/50 R/S at position 2.

4.2.3.1.1 Resolutions of 2-Arylpropanoic Acids

The 2-arylpropanoic acids (profens) are important as non-steroidal anti-inflammatory drugs (NSAIDs). The (S)-enantiomers are mainly responsible for the desired pharmacological activity. Therefore many strategies have been developed for the enantioselective preparation of the (S)-profens. Among them are esterification reactions in organic media catalyzed by various hydrolases. Hence, racemic naproxen **145** can be resolved in isooctane at controlled water activity ($a_w = 0{,}8$) by esterification with a low concentration of n-butanol with *Candida rugosa* lipase with E > 1000 (Scheme 4.42) [140]. Reaction of *rac*-ibuprofen (**146**) catalyzed by Lipozyme IM20 (*Rhizomucor miehei* lipase) with butanol in water-saturated isooctane furnishes E = 113, and the conditions can be used in a packed-bed reactor to produce the (S)-ester and the remaining (R)-acid [141]. (S)-Ketoprofen can be obtained as the remaining substrate from the racemate **147** via CALB-catalyzed reaction with

Scheme 4.42 Hydrolase-catalyzed esterification of profen drugs.

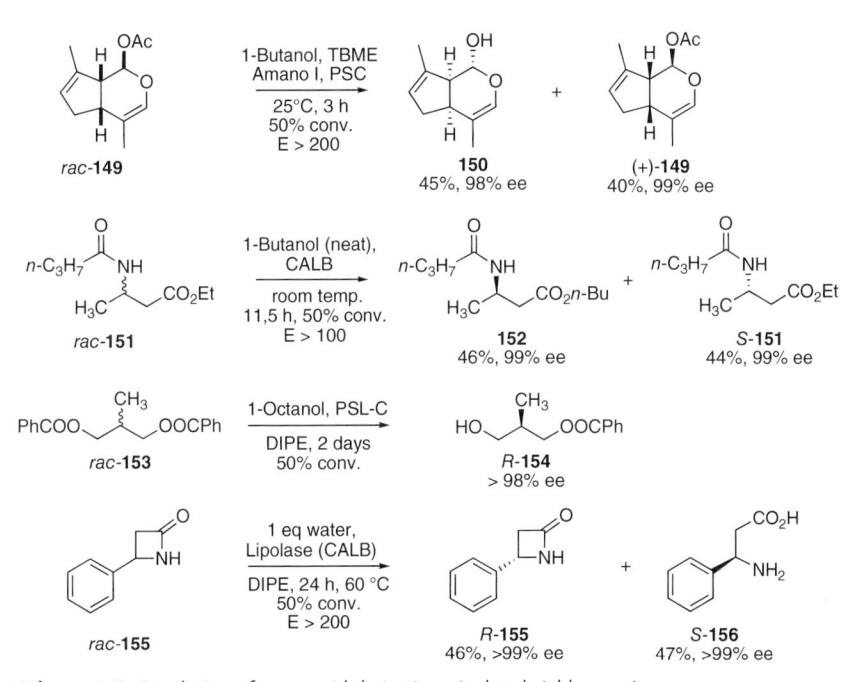

145 Naproxen
S-selective esterification
CRL, isooctane,
n-butanol, a_w = 0,8,
E > 1000

146 Ibuprofen
S-selective esterification
Lipozyme IM20, isooctane,
n-butanol, water sat.,
E = 113

147 Ketoprofen
R-selective esterification
Novozyme 435,
1,2-dichlorethane/hexane (1/4),
ethanol, trace of water,
Remaining substrate:
98% ee (S) at c = 60%.

148 Flurbiprofen
R-selective interesterification
Novozyme 435, acetonitrile
tri-n-propylorthoformate,
trace n-propanol, 45 °C
Remaining substrate:
>98% ee (S) at c = 60%.

Scheme 4.43 Resolution of some acid derivatives. Isolated yields are given.

rac-**149** → (1-Butanol, TBME, Amano I, PSC, 25°C, 3 h, 50% conv., E > 200) → **150** (45%, 98% ee) + (+)-**149** (40%, 99% ee)

rac-**151** → (1-Butanol (neat), CALB, room temp., 11,5 h, 50% conv., E > 100) → **152** (46%, 99% ee) + S-**151** (44%, 99% ee)

rac-**153** → (1-Octanol, PSL-C, DIPE, 2 days, 50% conv.) → R-**154** (> 98% ee)

rac-**155** → (1 eq water, Lipolase (CALB), DIPE, 24 h, 60 °C, 50% conv., E > 200) → R-**155** (46%, >99% ee) + S-**156** (47%, >99% ee)

ethanol in 1,2-dichloroethane/hexane (1/4, v/v) in the presence of a small amount of water (0.15%), and the reaction proceeds with E = 15 [142]. Similarly, (S)-flurbiprofen can be obtained as the remaining substrate from the racemate **148** by reaction to 60% conversion catalyzed by CALB with tri-n-propylorthoformate as an essentially irreversible alcohol donor in acetonitrile at 45 °C [143].

4.2.3.2 Other Acid Derivatives

Chiral racemic esters can of course be resolved by hydrolase-catalyzed, hydrolytic reactions in aqueous media, which are not treated here. Alcoholysis provides an alternative. Some examples are shown in Scheme 4.43. Thus, the racemic gastro-lactyl acetate rac-**149** can be efficiently resolved to give (+)-**149** and the lactol **150** by butanolysis catalyzed by Amano I, PS-C [144]. The enantiopure lactol product can epimerize at the hemiacetal hydroxyl, and the product shown is the major diastereomer **150**. CALB-catalyzed butanolysis of, e.g., the β-aminoacid derivative *rac*-**151**, is highly enantioselective and provides a separable mixture of the enantiopure remaining ethyl ester *S*-**151** along with the enantiopure butyl ester *R*-**152** [145]. Desymmetrization by PSL-C-catalyzed octanolysis of the dibenzoate **153** provides the enantiopure monobenzoate *R*-**154** [145]. Although amides are normally not easily hydrolyzed or alcoholyzed by lipases, β-lactams do react, probably because of the strained nature of the four-membered β-lactam ring. Thus CALB-catalyzed (Novozym 435) ring opening of *rac*-**155** by alcoholysis with racemic 2-octanol in DIPE at 60 °C furnished enantiopure recovered β-lactam *R*-**155**, whereas the alcoholysis product was obtained only in a low yield, albeit in a high *ee* [147]. The recovery of the product **156** was substantially increased by exchanging the nucleophile 2-octanol for an equimolar amount of water in DIPE, as shown in (Scheme 4.43) [148].

4.3
Summary and Outlook

Reactions catalyzed by hydrolases in organic solvents have over the past two decades emerged as one of the most important methods for the preparation of enantiomerically highly enriched compounds, ranging from small-scale organic synthesis to industrial-scale preparation of important synthetic intermediates for the pharmaceutical and agrochemical industries. The potential of hydrolases is by no means exhausted. Medium engineering and new immobilization techniques will yield new advances. The developments allowing one to modify the structure of a hydrolase by either site-specific or random mutagenesis provide the tools for expanding the potential uses of hydrolase catalysis and allow tailoring of the selectivity of an enzyme for a specific substrate.

References

1 (a) G. Carrea, S. Riva, *Angew. Chem. Int. Ed.*, 2000, **39**, 2226–2254. (b) B. G. Davis, V. Boyer, *Nat. Prod. Rep.* 2001, **18**, 618–40. (c) K. Drauz, H. Waldmann, Enzyme Catalysis in Organic Synthesis: A Comprehensive Handbook, Wiley-VCH, Weinheim, 2002. (d) K. Faber, Biotransformations in Organic Chemistry, 5th Edn., Springer Verlag, Heidelberg, 2004. (e) A. Ghanem, H. Y. Aboul-Enein, *Chirality*, 2005, **17**, 1–15 (f) U. T. Bornscheuer, R. J. Kazlauskas, Hydrolases in Organic Synthesis: Regio- and Stereoselective Biotransformations, Wiley-VCH, Weinheim, 2005. (g) V. Gotor-Fernández, R. Brieva, V. Gotor, *J. Mol. Catal. B: Enzymatic*, 2006, **40**, 111–120.

2 Enantiomeric purity is expressed as enantiomeric excess (% *ee*) defined as % *ee* = 100 (*R* − *S*)/(*R* + *S*) for *R* > *S*, where *R* and *S* are the masses of the *R*- and *S*-enantiomers, respectively.

3 C.-S. Chen, Y. Fujimoto, G. Girdaukas, C. J. Sih, *J. Am. Chem. Soc.*, 1982, **104**, 7294–7299.

4 C.-S. Chen, S.-H. Wu, G. Girdaukas, C. J. Sih, *J. Am. Chem. Soc.*, 1987, **109**, 2812–2817.

5 E. Egholm Jacobsen, E. van Hellemond, A. R. Moen, L. Camino Vazquez Prado, T. Anthonsen, *Tetrahedron Lett.*, 2003, **44**, 8453–8455.

6 E. García-Urdiales, I. Alfonso, V. Gotor, *Chem. Rev.*, 2005, **105**, 313–354.

7 U. Hanefeld, *Org. Biomol. Chem.*, 2003, **1**, 2405–2415.

8 K. Faber, S. Riva, *Synthesis*, 1992, 895–910.

9 J. M. Fang, C.-H. Wong, *Synlett*, 1994, 393–402.

10 E. S. Rothman, S. Serota, T. Perlstein, D. Swern, *J. Org. Chem.*, 1962, **27**, 3123–3127.

11 E. S. Rothman, S. Serota, D. Swern, *J. Org. Chem.*, 1966, **31**,629–630.

12 M. Lobell, M. P. Schneider, *Synthesis*, 1994, 375–377.

13 T. Mitsudo, Y. Hori, Y. Yamakawa, Y. Watanabe, *Tetrahedron Lett.*, 1986, **27**, 2125–2126.

14 T. Mitsudo, Y. Hori, Y. Yamakawa, Y. Watanabe, *J. Org. Chem.*, 1987, **52**, 2230–2239.

15 H. Nakagawa, Y. Okimoto, S. Sakaguchi, Y. Ishii, *Tetrahedron Lett.*, 2003, **44**, 103–106.

16 Y. Kita, S. Akai, *The Chemical Record*, 2004, **4**, 363–372, and references cited therein.

17 R. B. Woodward, K. Heusler, J. Gosteli, P. Naegeli, W. Oppolzer, R. Ramage, S. Ranganathan, H. Vorbrüggen, *J. Am. Chem. Soc.*, 1966, **88**, 852–853.

18 T. W. Greene, P. G. M. Wuts, *Protective Groups in Organic synthesis*, John Wiley & Sons, New York, 1999.

19 M. Ono, H. Nakamura, F. Konno, H. Akita, *Tetrahedron: Asymmetry*, 2000, **11**, 2753–2764.

20 T. Miyazawa, S. Nakajo, M. Nishikawa, K. Hamahara, K. Imigawa, E. Ensatsu, R. Yanagihara, T. Yamada, *J. Chem. Soc., Perkin 1*, 2001, 82-86.

21 N. Baba, M. K. Alam, Y. Mori, S. S. Haider, M. Tanaka, S. Nakajima, S. Shimizu, *J. Chem. Soc., Perkin 1*, 2001, 221–223.

22 S. Gedey, A. Liljeblad, L. Lázár, F. Fülöp, L. T. Kanerva, *Tetrahedron: Asymmetry*, 2001, **12**, 105–110.

23 V. K. Aggarwal, D. E. Jones, A. M. Martin-Castro, *Eur. J. Org. Chem.*, 2000, 2939–2945.

24 H. Frykman, N. Öhrner, T. Norin, K. Hult, *Tetrahedron Lett.*, 1993, **34**, 1367–1370.

25 C. Orrenius, N. Öhrner, D. Rottici, A. Matsson, K. Hult, T. Norin, *Tetrahedron: Asymmetry*, 1995, **6**, 1217–1220.

26 H. K. Weber, J. Zuegg, K. Faber, J. Pleiss, *J. Mol. Catal. B: Enzymatic*, 1997, **3**, 131–138.

27 S. Joly, M. S. Nair, *J. Mol. Catal. B: Enzymatic*, 2003, **22**, 151–160.

28 B. Berger, K. Faber, *J. Chem. Soc. Chem. Commun.*, 1991, 1198–1200.

29 H. K. Weber, K. Faber, *Methods Enzymol.*, 1997, **286**, 509–518.

30 H.-E. Högberg, M. Lindmark, D. Isaksson, K. Sjödin, M. C. R. Franssen, H. Jongejan, J. P. B. A. Wijnberg, A. de Groot, *Tetrahedron Lett.*, 2000, **41**, 3193–3196.

31 D. Isaksson, M. Lindmark-Henriksson, T. Manoranjan, K. Sjödin, H.-E. Högberg, *J. Mol. Catal. B: Enzymatic*, 2004, **31**, 31–37.

32 H. Weber, L. Brecker, D. de Souza, H. Griengl, D. W. Ribbons, H. K. Weber, *J. Mol. Catal. B: Enzymatic*, 2002, **19–20**, 149–157.

33 A. Ghanem, H. Y. Aboul-Enein, *Tetrahedron: Asymmetry*, 2004, **15**, 3331–3351.

34 A. Ghanem, V. Schurig, *Tetrahedron: Asymmetry*, 2003, **14**, 57–62.

35 A. O. Magnusson, M. Takwa, A. Hamberg, K. Hult, *Angew. Chem. Int. Ed.*, 2005, **44**, 4582–4585.

36 M. T. Reetz, S. Wilensek, D. Zha, K.-E. Jaeger, *Angew. Chem., Int. Ed.* 2001, **40**, 3589–3591.

37 D. Zha, S. Wilensek, M. Hermes, K. E. Jaeger, M. T. Reetz, *Chem. Commun.* 2001, 2664–2665.

38 Y. Koga, K. Kato, H. Nakano, T. Yamane, *J. Mol. Biol.*, 2003, **331**, 585–592.

39 A. O. Magnusson, J. C. Rotticci-Mulder, A. Santagostino, K. Hult, *ChemBioChem*, 2005, **6**, 1051–1056.

40 M. T. Reetz, J. D. Carballeira, J. Peyrelans, H. Höbenreich, A. Maichele, A. Vogel, *Chem. Eur. J.*, 2006, **12**, 6031–6038.
41 W. V. Tuomi, R. J. Kazlauskas, *J. Org. Chem.*, 1999, **64**, 2638–2647.
42 S. Barth, F. Effenberger, *Tetrahedron: Asymmetry*, 1993, **4**, 823–833.
43 O. Nordin, E. Hedenström, H.-E. Högberg, *Tetrahedron: Asymmetry*, 1994, **5**, 785–788.
44 O. Nordin, B.-V. Nguyen, C. Vörde, E. Hedenström, H.-E. Högberg, *J. Chem. Soc., Perkin Trans. 1*, 2000, 367–376.
45 P. Ferraboschi, S. Casati, A. Manzocchi, E. Santaniello, *Tetrahedron: Asymmetry*, 1995, **6**, 1521–1524.
46 T. Miyazawa, T. Yukawa, T. Koshiba, H. Sakamoto, S. Ueji, R. Yanagihara, T. Yamada, *Tetrahedron: Asymmetry*, 2001, **12**, 1595–1602.
47 L. F. Solares, R. Brieva, M. Quirós, I. Llorente, M. Bayod, V. Gotor, *Tetrahedron: Asymmetry*, 2004, **15**, 341–345.
48 M. Jungen, H.-J. Gais, *Tetrahedron: Asymmetry*, 1999, **10**, 3747–3758.
49 A. Zada, E. Dunkelblum, *Tetrahedron: Asymmetry*, 2006, **17**, 230–233.
50 T. Olsen, F. Kerton, R. Marriot, G. Grogan, *Enz. Microbial Technol.*, 2006, **39**, 621–625.
51 T. Sakai, Y. Liu, H. Ohta, T. Korenaga, T. Ema, *J. Org. Chem.*, 2005, **70**, 1369–1375.
52 G. Laval, G. Audran, S. Sanchez, H. Monti, *Tetrahedron: Asymmetry*, 1999, **10**, 1927–1933.
53 S. Horiuchi, H. Takikawa, K. Mori, *Bioorg. Med. Chem.*, 1999, **7**, 723–726.
54 K. Tanaka, H. Osuga, H. Suzuki, Y. Shogase, Y. Kitahara, *J. Chem. Soc. Perkin Trans.1*, 1998, 935–940.
55 D. Wiktelius, E. K. Larsson, K. Luthman, *Tetrahedron: Asymmetry*, 2006, **17**, 2088–2100.
56 R. Chênevert, M. Simard, J. Bergeron, M. Dasser, *Tetrahedron: Asymmetry*, 2004, **15**, 1889–1892.
57 F.-R. Alexandre, F. Huet, *Tetrahedron: Asymmetry*, 1998, **9**, 2301–2310.
58 R. Chênevert, D. Duguay, F. Touraille, D. Caron, *Tetrahedron: Asymmetry*, 2004, **15**, 863–866.
59 R. Chênevert, G. Courchesne, N. Pelchat, *Bioorg. Med. Chem.*, 2006, **14**, 5389–5396.
60 P. Kiełbasiński, R. Żurawiński, M. Albrycht, M. Mikołajczyk, *Tetrahedron: Asymmetry*, 2003, **14**, 3379–3384.
61 D. Wiktelius, M. J. Johansson, K. Luthman, N. Kann, *Org. Lett.*, 2005, **7**, 4991–4994.
62 R. Chênevert, G. Courchesne, D. Caron, *Tetrahedron: Asymmetry*, 2003, **14**, 2567–2571.
63 R. Chênevert, Y. S. Rose, *J. Org. Chem.*, 2000, **65**, 1707–1709.
64 R. Chênevert, G. M. Ziarani, M. P. Morin, M. Dasser, *Tetrahedron: Asymmetry*, 1999, **10**, 3117–3122.
65 T. Oishi, M. Maruyama, M. Shoji, K. Maeda, N. Kumahara, S. Tanaka, M. Hirama, *Tetrahedron*, 1999, **55**, 7471–7498.
66 T. Lee, R. Sakowicz, V. Martichonok, J. K. Hogan, M. Gold, J. B. Jones, *Acta Chem. Scand.*, 1996, **50**, 697–706.
67 P. A. Fitzpatrick, A. M. Klibanov, *J. Am. Chem. Soc.* 1991, **113**, 3166–3171.
68 T. Ema, R. Okada, M. Fukumoto, M. Jittani, M. Ishida, K. Furuie, K. Yamaguchi, T. Sakai, M. Utaka, *Tetrahedron Lett.*, 1999, **40**, 4367–4370.
69 K. Lemke, M. Lemke, F. Theil, *J. Org. Chem.*, 1997, **62**, 6268–6273.
70 K. Naemura, R. Fukuda, M. Konishi, K. Hirose, Y. Tobe, *J. Chem. Soc., Perkin Trans. 1*, 1994, 1253–1256.
71 K. Naemura, R. Fukuda, M. Murata, M. Konishi, K. Hirose, Y. Tobe, *Tetrahedron: Asymmetry*, 1995, **6**, 2385–2394.
72 R. J. Kazlauskas, A. N. E. Weissfloch, A. T. Rappaport, L. A. Cuccia, *J. Org. Chem.*, 1991, **56**, 2656–2665.
73 T. Matsuda, K. Tsuji, T. Kamitanaka, T. Harada, K. Nakamura, T. Ikariya, *Chem. Lett.*, 2005, **34**, 1102–1103.
74 L. Borén, B. Martín-Matute, Y. Xu, A. Córdova, J.-E. Bäckvall, *Chem. Eur. J.*, 2006, **12**, 225–232.
75 J. Uenishi, T. Hiraoka, S. Hata, K. Nishiwaki, O. Yonemitsu, *J. Org. Chem.*, 1998, **63**, 2481–2487.
76 E. Sundby, M. M. Andersen, B. H. Hoff, T. Anthonsen, *ARKIVOC*, 2001, 76–84.

77 J. Kaminska, I. Górnicka, M. Sikora, J. Góra, *Tetrahedron: Asymmetry*, 1996, **7**, 907–910.
78 A. Bierstedt, J. Stöltin, R. Fröhlich, P. Metz, *Tetrahedron: Asymmetry*, 2001, **12**, 3399–3407.
79 N. Bouzemi, L. Aribi-Zouioueche, J.-C. Fiaud, *Tetrahedron: Asymmetry*, 2006, **17**, 797–800.
80 D. Rottici, F. Haeffner, C. Orrenius, T. Norin, K. Hult, *J. Mol. Catal. B: Enzymatic*, 1998, **5**, 267–272.
81 J. Ottosson, K. Hult, *J. Mol. Catal. B: Enzymatic*, 2001, **11**, 1025–1028.
82 M. Lindström, E. Hedenström, S. Bouilly, K. Velonia, I. Smonou, *Tetrahedron: Asymmetry*, 2005, **16**, 1355–1360.
83 E. Hedenström, H. Edlund, S. Lund, M. Abersten, D. Persson, *J. Chem. Soc., Perkin Trans. 1*, 2002, 1810–1817.
84 R. Chênevert, S. Gravil, J. Bolte, *Tetrahedron: Asymmetry*, 2005, **16**, 2081–2086.
85 B. S. Morgan, D. Hoenner, P. Evans, S. M. Roberts, *Tetrahedron: Asymmetry*, 2004, **15**, 2807–2809.
86 D. Isaksson, K. Sjödin, H.-E. Högberg, *Tetrahedron: Asymmetry*, 2006, **17**, 275–280.
87 A. Abate, E. Brenna, G. Fronza, C. Fuganti, F. G. Gatti, S. Maroncelli, *Chemistry & Biodiversity*, 2006, **3**, 677–694.
88 T. O. Vieira, H. M. C. Ferraz, L. H. Andrade, A. L. M. Porto, *Tetrahedron: Asymmetry*, 2006, **17**, 1990–1994.
89 M. C. R. Franssen, H. Jongejan, H. Kooijman, A. L. Spek, R. P. L. Bell, J. P. B. A. Wijnberg, A. de Groot, *Tetrahedron: Asymmetry*, 1999, **10**, 2729–2738.
90 A. J. Pihko, K. Lundell, L. Kanerva, A. M. P. Koskinen, *Tetrahedron: Asymmetry*, 2004, **15**, 1637–1643.
91 M. Larsson, J. Andersson, R. Liu, H.-E. Högberg, *Tetrahedron: Asymmetry*, 2004, **15**, 2907–2915.
92 E. Vänttinen, L. T. Kanerva, *Tetrahedron: Asymmetry*, 1995, **6**, 1779–1786.
93 H.-L. Liu, T. Anthonsen, *Chirality*, 2002, **14**, 25–27.
94 A. Wallner, H. Mang, S. M. Glueck, A. Steinreiber, S. F. Mayer, K. Faber, *Tetrahedron: Asymmetry*, 2003, **14**, 2427–2432.
95 S. Akai, T. Naka, S. Omura, K. Tanimoto, M. Imanishi, Y. Takebe, M. Matsugi, Y. Kita, *Chem. Eur. J.*, 2002, **8**, 4255–4264.
96 S. Akai, K. Tanimoto, Y. Kita, *Angew. Chem. Int. Ed.*, 2004, **43**, 1407–1410.
97 S. Akai, K. Tanimoto, Y. Kanao, S. Omura, Y. Kita, *Chem. Commun.*, 2005, 2369–2371.
98 E. Sundby, L. Perk, T. Anthonsen, A. J. Aasen, T. V. Hansen, *Tetrahedron*, 2004, **60**, 521–524.
99 M. Fujii, M. Fukumura, Y. Hori, Y. Hirai, H. Akita, K. Nakamura, K. Toriizuka, Y. Ida, *Tetrahedron: Asymmetry*, 2006, **17**, 2292–2298.
100 N. Toyooka, A. Nishino, T. Momose, *Tetrahedron*, 1997, **53**, 6313–6326.
101 R. Chênevert, G. Courchesne, F. Jacques, *Tetrahedron: Asymmetry*, 2004, **15**, 3587–3590.
102 B. Olofsson, K. Bogár, A.-B. L. Fransson, J.-E. Bäckvall, *J. Org. Chem.*, 2006, **71**, 8256–8260.
103 K. Toyama, S. Iguchi, H. Sakazaki, T. Oishi, M. Hirama, *Bull. Chem. Soc. Jpn.*, 2001, **74**, 997–1008.
104 E. Henke, J. Pleiss, U. T. Bornscheuer, *Angew. Chem. Int. Ed.*, 2002, **41**, 3211–3213.
105 S. H. Krishna, M. Persson, U. T. Bornscheuer, *Tetrahedron: Asymmetry*, 2002, **13**, 2693–2696.
106 M. Juárez-Hernandez, D. V. Johnson, H. L. Holland, J. McNulty, A. Capretta, *Tetrahedron: Asymmetry*, 2003, **14**, 289–291.
107 C. Sanfilippo, G. Nicolosi, G. Delogu, D. Fabbri, M. A. Dettori, *Tetrahedron: Asymmetry*, 2003, **14**, 3267–3270.
108 N. Aoyagi, N. Ogawa, T. Izumi, *Tetrahedron Lett.*, 2006, **47**, 4797–4801.
109 F. van Rantwijk, R. A. Sheldon, *Tetrahedron*, 2004, **60**, 501–519.
110 S. Puertas, R. Brieva, F. Rebolledo, V. Gotor, *Tetrahedron*, 1993, **49**, 4007–4014.
111 M. T. Reetz, C. Dreisbach, *Chimia*, 1994, **48**, 570.
112 B. A. Davis, D. A. Durden, *Synth. Commun.*, 2001, **31**, 569–578.
113 C. D. Reeve, WO 9931264, 1999; *Chem. Abstr.*, 1999, **131**, 43668q.

114 (a) H. Kitaguchi, P. A. Fitzpatrick, J. E. Huber, A. M. Klibanov, *J. Am. Chem. Soc.*, 1989, **111**, 3094–3095. (b) H. Kitaguchi, P. A. Fitzpatrick, J. E. Huber, A. M. Klibanov, *Ann. N.Y. Acad. Sci.*, 1990, **613**, 656–658.

115 E. García-Urdiales, F. Rebolledo, V. Gotor, *Tetrahedron: Asymmetry*, 2000, **11**, 1459–1463.

116 C. Nübling, K. Ditrich, C. Dully, USP 6465223, 2002; *Chem. Abstr.*, 2000, **132**, 165214e.

117 S. Takayama, S. T. Lee, S.-C. Hung, C.-H. Wong, *Chem. Commun.*, 1999, 127–128.

118 K. Ditrich, F. Balkenhohl, W. Ladner, US 5905167, 1999; *Chem. Abstr.*, 1997, **126**, 277259f.

119 U. Stelzer, C. Dreisbach, US 6387692, 2002; *Chem. Abstr.*, 1997, **127**, 108759i.

120 L. E. Iglesias, V. M. Sánchez, F. Rebolledo, V. Gotor, *Tetrahedron: Asymmetry*, 1997, **8**, 2675–2677.

121 K. A. Skupinska, E. J. McEachern, I. R. Baird, R. T. Skerlj, G. J. Bridger, *J. Org. Chem.*, 2003, **68**, 3546–3551.

122 J. González-Sabín, V. Gotor, F. Rebolledo, *Tetrahedron: Asymmetry*, 2002, **13**, 1315–1320.

123 J. González-Sabín, V. Gotor, F. Rebolledo, *Tetrahedron: Asymmetry*, 2004, **15**, 481–488.

124 A. Luna, I. Alfonso, V. Gotor, *Org. Lett.*, 2002, **4**, 3627-3629.

125 N. Aoyagi, T. Izumi, *Tetrahedron Lett.*, 2002, **43**, 5529–5531.

126 A. Liljeblad, L. T. Kanerva, *Tetrahedron*, 2006, **62**, 5831–5854.

127 P. Domínguez de María, C. Carboni-Oerlemans, B. Tuin, G. Bargeman, A. van der Meer, R. van Gemert, *J. Mol. Cat. B: Enzymatic*, 2005, **37**, 36–46.

128 M. Solymár, F. Fülöp, L. T. Kanerva, *Tetrahedron: Asymmetry*, 2002, **13**, 2383–2388.

129 X.-G. Li, L. T. Kanerva, *Tetrahedron: Asymmetry*, 2005, **16**, 1709–1714.

130 Z. C. Gyarmati, A. Liljeblad, M. Rintola, G. Bernáth, L. T. Kanerva, *Tetrahedron: Asymmetry*, 2003, **14**, 3805–3814.

131 M. Fitz, K. Lundell, F. Fülöp, L. T. Kanerva, *Tetrahedron: Asymmetry*, 2006, **17**, 1129–1134.

132 S. N. Ahmed, R.J. Kazlauskas, A. H. Morinville, P. Grochulski, J. D., Schrag, M. Cygler, *Biocatalysis*, 1994, **9**, 209–225.

133 P. Berglund, M. Holmquist, K. Hult, H.-E. Högberg, *Biotechnology Letters*, 1995, **17**, 55–60.

134 P. Berglund, C. Vörde, H.-E. Högberg, *Biocatalysis*, 1994, **9**, 123–130.

135 K.-H. Engel, *Tetrahedron: Asymmetry*, 1991, **2**, 165–168.

136 B.-V. Nguyen, E. Hedenström, *Tetrahedron: Asymmetry*, 1999, **10**, 1821–1826.

137 H. Edlund, P. Berglund, M. Jensen, E. Hedenström, H.-E. Högberg, *Acta Chem. Scand.*, 1996, **50**, 666–671.

138 P. Berglund, E. Hedenström, in E. N. Vulfson, P. J. Halling, H. L. Holland, Eds, "*Methods in Biotechnology vol 15. Enzymes in Nonaqueous solvents: Methods, Protocols*" (Humana Press, Totowa, NJ), 2001, 307–317.

139 E. Hedenström, B.-V. Nguyen, L. A. Silks, III, *Tetrahedron: Asymmetry*, 2002, **13**, 835–844.

140 J.-Y. Wu, S.-W. Liu, *Enzyme Microbial Techn.*, 2000, **26**, 124–130.

141 A. Sánchez, F. Valero, J. Lafuente, C. Solà, *Enzyme Microbial Techn.*, 2000, **27**, 157–166.

142 H. J. Park, W. J. Choi, E. C. Huh, E. Y. Lee, C.Y. Choi, *J. Biosc. Bioeng.*, 1999, **87**, 545–547.

143 R. Morrone, M. Piattelli, G. Nicolosi, *Eur. J. Org. Chem.*, 2001, 1441–1443.

144 E. M. Santangelo, D. Rottici, I. Liblikas, T. Norin, C. R. Unelius, *J. Org. Chem.*, 2001, **66**, 5384–5387.

145 E. Santaniello, P. Ciuffreda, S. Casati, L. Alessandrini, A. Repetto, *J. Mol. Catal. B: Enzymatic*, 2006, **40**, 81–85.

146 S. Gedey, A. Liljeblad, L. Lázár, F. Fülöp, L. T. Kanerva, *Can. J. Chem.*, 2002, **80**, 565–570.

147 S. Park, E. Forró, H. Grewal, F. Fülöp, R. J. Kazlauskas, *Adv. Synth. Catal.*, 2003, **345**, 986–995.

148 E. Forró, T. Paál, G. Tasnádi, F. Fülöp, *Adv. Synth. Catal.*, 2006, **348**, 917–923.

5
Chemoenzymatic Deracemization Processes
Belén Martín-Matute and Jan-E. Bäckvall

5.1
Introduction

5.1.1
Asymmetric Synthesis Methods

The development of efficient methods for the synthesis of enantiomerically pure chiral molecules is of tremendous importance. For example, in the year 2004 it was reported that in nine of the top ten drugs produced, the active ingredients were enantiomerically pure chiral molecules [1]. Pharmaceuticals containing only one enantiomer, the active drug, offer the possibility of being administered in a lower dosage, therefore minimizing side effects. New methods to control the stereochemistry in a molecule are of great importance, not only for the synthesis of pharmaceuticals, but also for the synthesis of flavor and aroma chemicals, agricultural chemicals, and specialty materials [2]. In principle, asymmetric catalysis is the best and most refined strategy. This approach uses a catalytic amount of one enantiomerically pure molecule (chiral catalyst) which is able to transform many prochiral molecules (starting material) into enantiomerically pure products (chiral multiplication). The chiral catalyst can be an enzyme, a simple organic molecule, or a metal complex bearing chiral ligands. Another approach employed in asymmetric synthesis is the use of natural chiral starting materials in which the required stereocenters are incorporated or can be obtained by modification of the existing stereocenters during the synthesis. Nature provides us with many valuable chiral starting materials, such as amino acids, alkaloids, and carbohydrates. However, an adequate starting material that can be efficiently transformed in only a few steps to the desired product must be available. Finally, kinetic resolution (KR) of a racemic mixture is the third approach that leads us to enantiomerically pure compounds. A KR is the total or partial separation of two enantiomers from a racemic mixture [3]. A KR is based on the different reaction rates of the enantiomers with a chiral molecule (a reagent, a catalyst, etc.). In the ideal case, the difference is large, and one of the enantiomers reacts very fast to give the product, whereas the other does not react at all or very slowly and can be recovered

$$S \xrightarrow[fast]{cat^*} p \quad 50\%$$

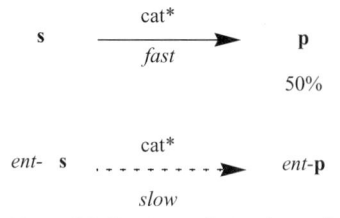

$$ent\text{-}S \xdashrightarrow[slow]{cat^*} ent\text{-}p$$

Figure 5.1 Kinetic resolution (s = substrate, p = product, cat* = homochiral catalyst).

enantiomerically pure when the transformation is completed (Figure 5.1). However, even in this ideal situation, KRs suffer from major drawbacks: a maximum yield of 50%, and the need for a method of separating the product from the remaining substrate. Furthermore, most of the KRs, including those catalyzed by enzymes, do not show an ideal behavior: at 50% conversion, the reaction does not stop, but only slows down. The concentration of each enantiomer is not constant during the reaction. Therefore, the reaction rate of both enantiomers changes with the conversion. To obtain the product with high enantiomeric excess (ee), it is necessary to stop the reaction before reaching 50% conversion. However, despite this drawback and despite the growing number of available methods for the preparation of enantiopure compounds via asymmetric catalysis, kinetic resolution is still the most employed method in industry [4], and in most cases biocatalysts (enzymes) are used.

During the past few years great efforts have been made to overcome the 50% threshold of enzyme-catalyzed KRs. Among the methods developed, deracemization processes have attracted considerable attention. Deracemizations are processes during which a racemate is converted into a non-racemic product in 100% theoretical yield without intermediate separation of materials [5]. This chapter aims to provide a summary of chemoenzymatic dynamic kinetic resolutions (DKRs) and chemoenzymatic cyclic deracemizations.

It should be mentioned that the great majority of dynamic kinetic resolutions reported so far are carried out in organic solvents, whereas all cyclic deracemizations are conducted in aqueous media. Therefore, formally, this latter methodology would not fit the scope of this book, which is focused on the synthetic uses of enzymes in non-aqueous media. However, to fully present and discuss the applications and potentials of chemoenzymatic deracemization processes for the synthesis of enantiopure compounds, chemoenzymatic cyclic de-racemizations will also be briefly treated in this chapter, as well as a small number of other examples of enzymatic DKR performed in water.

5.1.2
Dynamic Kinetic Resolution (DKR)

The opposite of a resolution, that is the racemization of a chiral compound, can sometimes be highly desirable and be applicable in enantioselective synthesis. By

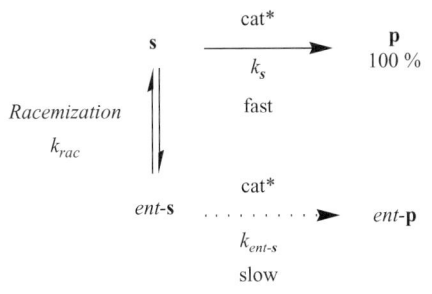

```
                    cat*
        s    ─────────────►   p
              k_s             100 %
Racemization  fast
   k_rac
                    cat*
        ent-s ·········►  ent-p
              k_ent-s
               slow
```

Figure 5.2 Dynamic kinetic resolution.

combining a racemization with an enzyme-catalyzed KR, a highly efficient asymmetric transformation to only one enantiomer can be obtained. Such dynamic kinetic resolutions (DKRs), with a theoretical yield of 100%, represent a powerful approach to the preparation of enantiomerically pure molecules (Figure 5.2). Racemization can be performed by a chemocatalyst or a biocatalyst, or it can occur spontaneously.

Some kinetic implications in KRs and DKRs must be considered first. In a KR, the ee of substrate and product changes with the conversion [6]. Thus, to compare different KRs we should stop them at the same extent of conversion. The "Enantiomeric Ratio" ($E = k_s/k_{ent-s}$) [7] measures the ability of an enzyme to distinguish between two enantiomers [8]. This enantioselectivity of an enzyme-catalyzed kinetic resolution can easily be calculated by applying the equations developed by Sih [9]. E remains constant throughout the reaction, and allows two KRs to be easily compared [10]. In DKRs where the substrate stays racemic throughout the reaction process, the optical purity should depend only on the Enantiomeric Ratio (E) [$ee = (E - 1)/(E + 1)$], and should be independent of the extent of conversion. The ee of the product formed under racemizing conditions should equal the initial ee under non-racemizing conditions. Thus, it is very important to maintain an efficient racemization during the DKR until the starting material is consumed to avoid depletion of the product ee.

For an efficient enzymatic DKR the following requirements must be fulfilled: (a) the KR must be very selective ($E > 20$); (b) the racemization must be fast (at least 10 times faster than the enzyme-catalyzed transformation of the slow-reacting enantiomer, $k_{rac} \geq 10\ k_{ent-s}$); (c) the racemization catalyst must not react with the product of the reaction, and (d) the KR and the racemization must be compatible under the same reaction conditions. Most often, the optimal activity of the two catalysts in the same pot is the major problem.

5.1.3
Cyclic Deracemizations

This approach transforms a racemic compound into the enantiomerically pure form of the same compound. Cyclic deracemizations have been achieved for

5 Chemoenzymatic Deracemization Processes

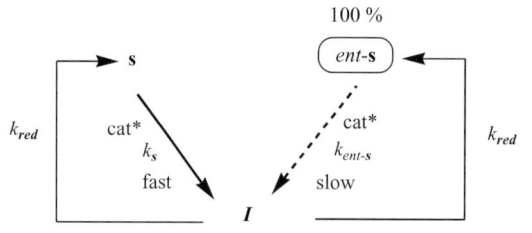

Figure 5.3 Cyclic deracemization.

sec-alcohols and for amines linked to a sec-carbon. This technique combines an enantioselective oxidation of one of the enantiomers of the starting material (s) yielding a prochiral intermediate (I) with a non-selective reduction to obtain again the starting material. When this oxidation-reduction sequence is repeated several times, a chiral inversion of the faster-reacting enantiomer, yielding its enantiomer as the final product in 100% yield, occurs (Figure 5.3). To the best of our knowledge, this oxidation-reduction process has not yet been achieved with the use of two chemocatalysts because of problems with compatibility. However, the high specificity of biocatalysts has allowed several successful cyclic deracemizations to be performed by combination with chemical reagents. As in the case of DKRs, the feasibility of cyclic deracemization relies on the compatibility of the chiral biocatalyst employed for the oxidation and the reagent used for the non-selective reduction in the same pot.

Some kinetics will also be considered in this case. A detailed study of the kinetics of this kind of deracemization has been reported by Faber et al. [11]. Here, we present some important parameters that must be taken into account. The maximum ee is obtained when the equilibrium has been reached. If the activity of both catalysts remains constant, and in the absence of any side reaction, ee_{max} can be calculated using the following expression: $ee_{max} = (E - 1)/(E + 1)$. Note that this equation is the same as that employed for DKRs. At the equilibrium, the concentrations of both enantiomers, s and ent-s, and that of the intermediate, I, depend not only on the selectivity of the oxidation (k_s and k_{ent-s}), but also on the reverse reaction (k_{red}). Faber has defined the "percent of equilibrium" (Poe) as the term to describe the progress of the reaction. This parameter describes how far the deracemization has proceeded toward the equilibrium. Since the equilibrium is reached when ee equals ee_{max}, the progress of the reaction can be calculated by using the following equation: $Poe = (ee) \cdot [(E + 1)/(E - 1)]$. Finally, the number of cycles (oxidation-redution sequence) required to reach the equilibrium depends on the selectivity of the asymmetric oxidation. Thus, for highly selective processes only a few cycles are required. Let us consider the following example. If the E value of the biocatalytic oxidation is very high ($E > 200$), in the first cycle only one enantiomer (s) will be oxidized, yielding 50% of I and leaving ent-s (50%) untouched. The first cycle is completed when I (50%) is reduced to rac-s. Thus, after the first cycle, the reaction mixture contains 25% of s and 75% of ent-s. Repeating the cycle, it is very easy to conclude that after only four rounds the *ee* of the reaction mixture is above 90%. However, when the selectivity of the oxidation is not very high, not only will the ee_{max} be below 100%, but also a higher number of cycles to reach ee_{max} will be needed. This is a disadvantage, since the probability of formation of by-products can also increase.

5.2
Dynamic Kinetic Resolution

DKRs presented in this chapter are catalyzed by lipases or proteases, which belong to the group of hydrolases [2e, 12]. Hydrolases are excellent catalysts, and a large number of them are commercially available. They do not require a cofactor and they efficiently catalyze the hydrolysis of non-natural esters. During the past two decades their applications in organic synthesis have grown exponentially. In the middle of the 80s it was discovered that enzymes can be employed in organic media since they keep their native structures in anhydrous organic solvents [13]. Before that time, hydrolases were only employed in aqueous media, which was a drawback in synthesis because of the insolubility of organic compounds in water and the tendency of many substrates to decompose in this medium. Since then, many new enzyme-catalyzed transformations have been discovered, such as transesterifications, aminolysis, and thiotransesterifications, and undoubtedly one of the most important contributions that has been developed is the combination of enzyme-catalyzed KRs with racemizations (DKRs), which, in many instances, can only work in organic solvents.

In this chapter, DKRs are categorized, according to the racemization method employed, as being base-, acid-, aldehyde-, enzyme-, or metal-catalyzed. Also, radical-induced racemizations, and racemizations that take place through continuous cleavage/formation of the substrate or through S_N2 displacement, are among other methods that are also discussed. In most cases, the racemization method of choice depends on the structure of the substrate. In all cases, the KR is catalyzed by an enzyme.

5.2.1
Base-Catalyzed Racemization

In this section, dynamic kinetic resolution of substrates having a proton with low p*K*a is discussed. Racemization occurs by performing the DKR in the presence of a weak base. Enzyme- and base-catalyzed DKRs are categorized, according to the nature of the substrates, as being thioesters, α-activated esters, oxazolones, hydantoins or acyloins.

5.2.1.1 **DKR of Thioesters**
In contrast to oxoesters, the α-protons of thioesters are sufficiently acidic to permit continuous racemization of the substrate by base-catalyzed deprotonation at the

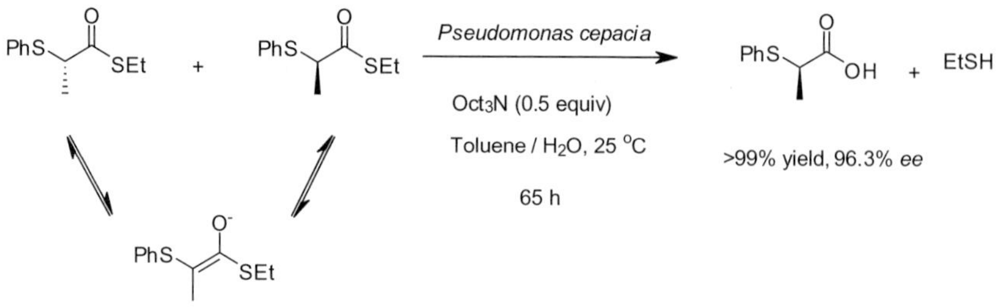

Scheme 5.1 DKR of thioesters using a base for racemization.

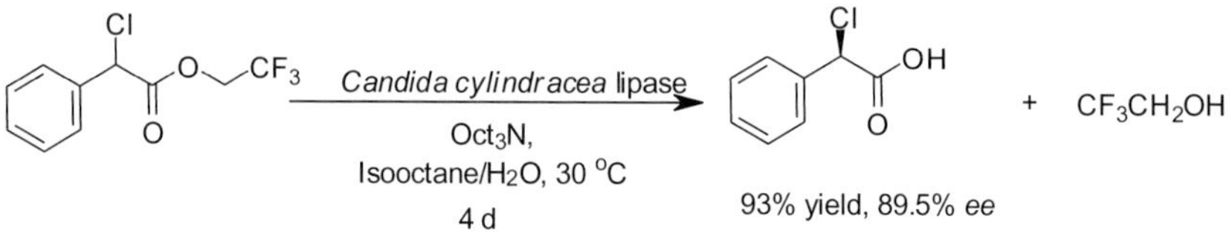

Scheme 5.2 DKR of activated esters using a base for racemization.

α-carbon. Drueckhammer et al. first demonstrated the feasibility of this approach by performing DKR of a propionate thioester bearing a phenylthiogroup, which also contributes to the acidity of the α-proton (Scheme 5.1) [14a]. The enzymatic hydrolysis of the thioester was coupled with a racemization catalyzed by trioctylamine. Because of the insolubility of the substrate and base in water, they employed a biphasic system (toluene/H_2O). Using *Pseudomonas cepacia* (Amano PS-30) as the enzyme and a catalytic amount of trioctylamine, they obtained a quantitative yield of the corresponding carboxylic acid in 96.3% ee. In this DKR, a maximum ee of 96.8% is expected, since the E value for the enzymatic resolution is 62.4. Drueckhammer et al. also reported DKR of thioesters not having activating groups at the α position [14b].

5.2.1.2 DKR of Activated Esters

DKR of esters bearing an electron-withdrawing group at the α-carbon can be performed easily under mild reaction conditions because of the low pK_a of the α-proton. Tsai et al. have reported an efficient DKR of *rac*-2,2,2-trifluoroethyl α-chorophenyl acetate in water-saturated isooctane [15]. They used lipase MY from *Candida rugosa* for the KR and trioctylamine as the base for racemization. (*R*)-α-chlorophenylacetic acid was obtained in 93% yield and 89.5% ee (Scheme 5.2).

5.2.1.3 DKR of Oxazolones

A very attractive and efficient method for the synthesis of L-aminoacids via DKR has been reported by Turner et al. [16a, b]. They employed enzyme-catalyzed ring opening of 5(4*H*)-oxazolones in combination with a catalytic amount of Et_3N. The relatively low pK_a of the C-4 proton (8.9) of oxazolones makes racemization take place easily. Hydrolysis of the ester obtained through DKR, followed by debenzo-

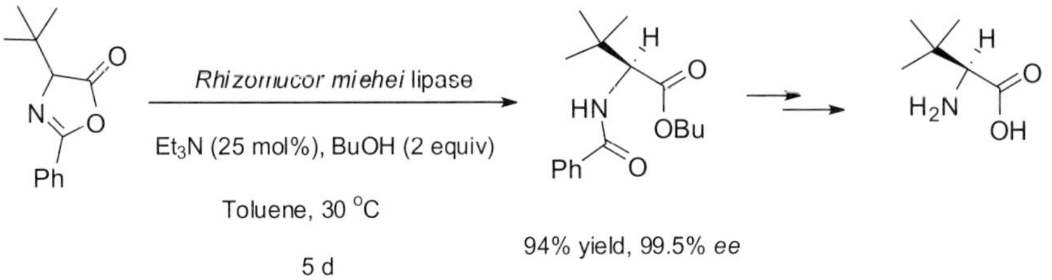

Scheme 5.3 DKR of oxazolones using a base for racemizing.

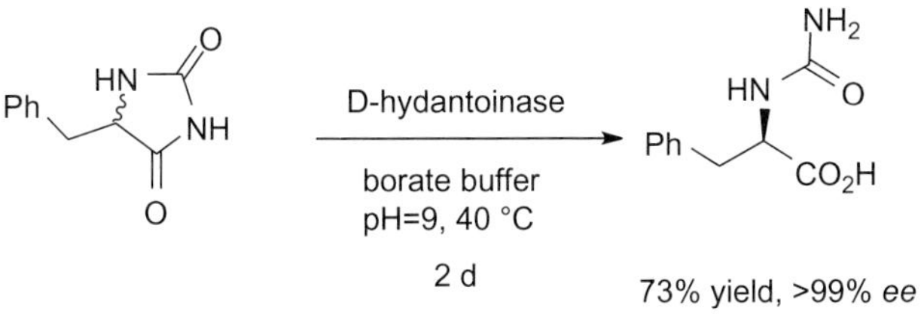

Scheme 5.4 DKR of hydantoins under weak basic conditions.

ylation, yields L-aminoacids in excellent *ee* (99.5%) (Scheme 5.3). In their initial studies, they employed *Rhizomucor miehei* lipase (Lipozyme) as the biocatalyst [16a]. More recently, they have obtained excellent results employing *Candida antarctica* lipase B (CALB) [16b]. This method has also been employed by Bevinakatti [16c, d] and Sih [16e, f].

5.2.1.4 DKR of Hydantoins

Another approach to the synthesis of enantiopure amino acids or amino alcohols is the enantioselective enzyme-catalyzed hydrolysis of hydantoins. As in the case discussed above, hydantoins are very easily racemized in weak alkaline solutions via keto enol tautomerism. Sugai et al. have reported the DKR of the hydantoin prepared from DL-phenylalanine. DKR took place smoothly by the use of D-hydantoinase at pH 9 employing a borate buffer (Scheme 5.4) [17].

5.2.1.5 DKR of Acyloins

Ogasawara et al. took advantage of the easy racemization of acyloins in the presence of a weak base for the DKR of *endo*-3-hydroxytricyclo[4.2.1.0^{2,5}]non-7-en-4-one (Scheme 5.5) [18]. Acylation of the hydroxyl group was catalyzed by a lipase, and racemization took place via a transient *meso*-enediol.

5.2.2
Acid-Catalyzed Racemization

In contrast to enzyme- and base-catalyzed DKRs, there are only a few reports of enzyme- and acid-catalyzed DKRs. A plausible explanation is that deactivation of

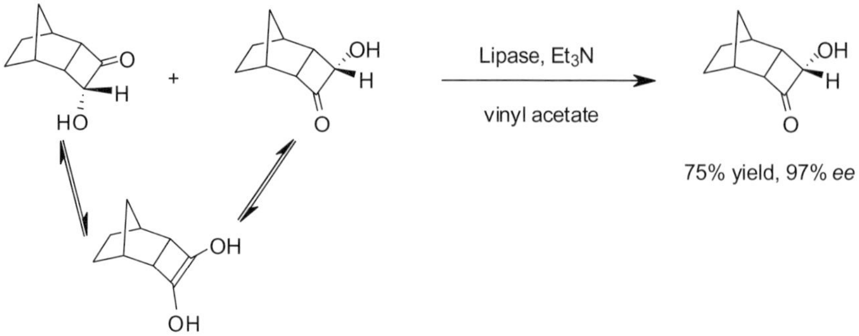

Scheme 5.5 DKR of acyloins through *meso*-enediol intermediates.

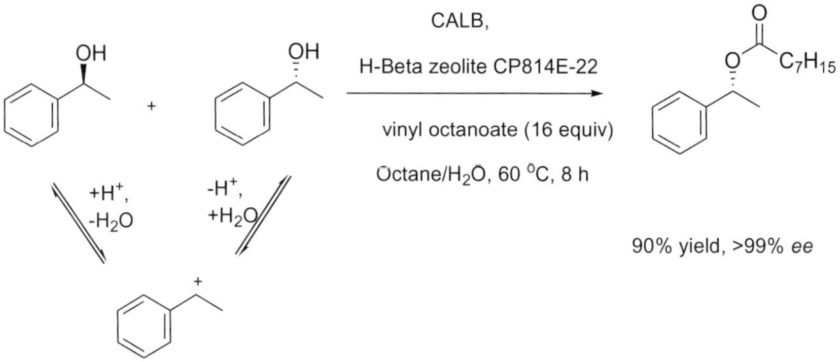

Scheme 5.6 DKR of *sec*-alcohols catalyzed by acid zeolites and a lipase.

the enzyme can occur under acidic conditions. Also, decomposition of the substrate has been observed.

Jacobs et al. employed an acidic zeolite catalyst for the racemization of *sec*-alcohols, which occurs through the formation of carbocations (Scheme 5.6) [19]. The KR was catalyzed by CALB in the presence of vinyl octanoate as acyl donor. DKR took place successfully in a biphasic system (octane/H$_2$O, 1:1) at 60 °C.

Another example of enzyme- and acid-catalyzed DKR has been reported by Bornscheurer [20]. Acyloins were racemized by using an acidic resin through the formation of enol intermediates. The enzymatic resolution was catalyzed by CALB. Since deactivation of this enzyme occurred in the presence of the acidic resin, they designed a simple reactor setup with two glass vials connected via a pump to achieve a spatial separation between the acidic resin and the enzyme (Scheme 5.7).

5.2.3
Racemization through Continuous Reversible Formation–Cleavage of the Substrate

Racemization of some substrates can take place through reversible formation of the substrate via an addition/elimination process. The racemization can be

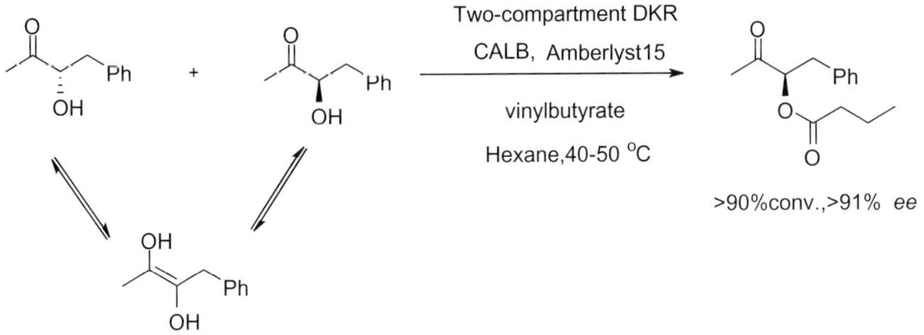

Scheme 5.7 DKR of sec-alcohols catalyzed by an acidic resin and a lipase.

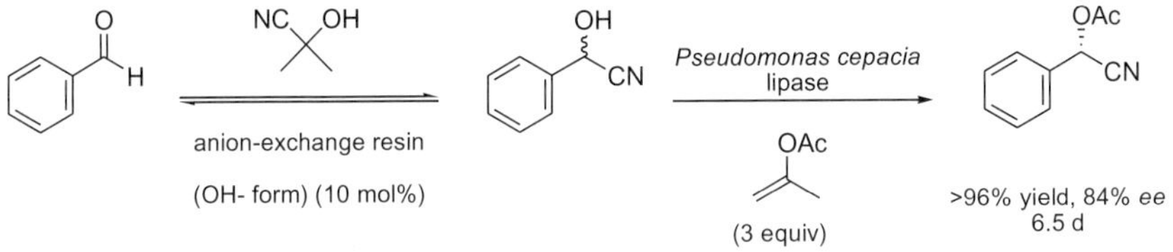

Scheme 5.8 DKR of cyanohydrins.

acid- or base-catalyzed. In this section we discuss DKR of cyanohydrins and hemithioacetals.

5.2.3.1 DKR of Cyanohydrins

Several reports on DKR of cyanohydrins have been developed using this methodology. The unstable nature of cyanohydrins allows continuous racemization through reversible elimination/addition of HCN under basic conditions. The lipase-catalyzed KR in the presence an acyl donor yields cyanohydrin acetates, which are not racemized under the reaction conditions.

Oda et al. reported in 1992 a one-pot synthesis of optically active cyanohydrin acetates from aldehydes, which were converted to the corresponding racemic cyanohydrins through transhydrocyanation with acetone cyanohydrin, catalyzed by a strongly basic anion-exchange resin [21]. The racemic cyanohydrins were acetylated by a lipase from *Pseudomonas cepacia* (Amano), with isopropenyl acetate as the acyl donor. The reversible nature of the base-catalyzed transhydrocyanation enabled continuous racemization of the unreacted cyanohydrins, thereby effecting the total conversion (Scheme 5.8).

Kanerva et al. have also reported DKR of cyanohydrins [22]. In particular, they obtained very good results with *Candida antarctica* lipase A (CAL-A) as the catalyst for the KR of a variety of substrates for which other enzymes such as CALB or PS-C do not give good results (Scheme 5.9) [22a].

The group of Burk has used a variety of nitrilases for the DKR of cyanohydrins [23]. Nitrilases catalyze the hydrolytic conversion of cyanohydrins directly to the corresponding carboxylic acid. Racemization was performed under basic

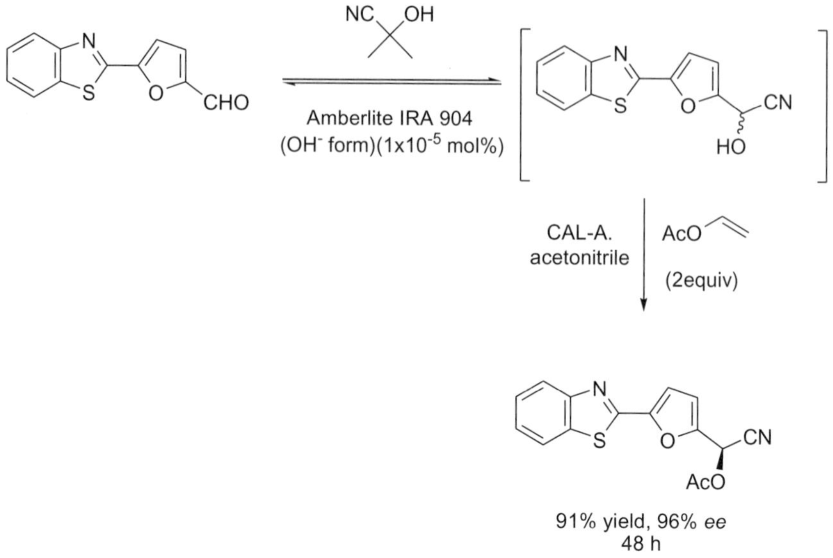

Scheme 5.9 Lipase-catalyzed DKR of cyanohydrins.

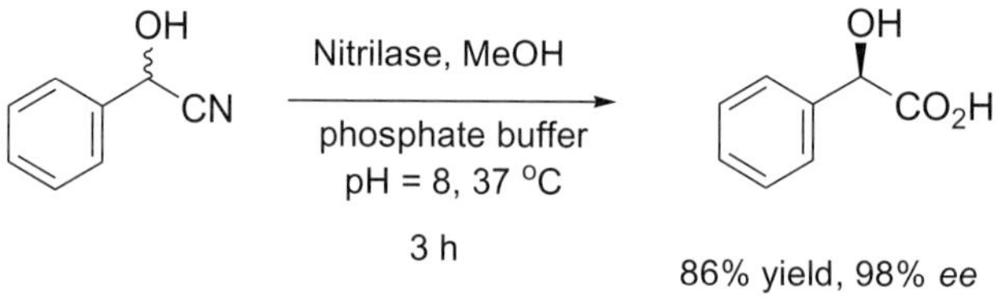

Scheme 5.10 Nitrilase-catalyzed DKR of cyanohydrins.

conditions (phosphate buffer, pH 8) through reversible loss of HCN. (R)-Mandelic acid was formed after 3 h in high yield (86% yield) and high enantioselectivity (98% ee) (Scheme 5.10).

5.2.3.2 DKR of Hemithioacetals
As in the case discussed above, hemithioacetals can be racemized by elimination/addition of a small molecule (a thiol in this case) under weak acidic conditions. Rayner et al. reported the first example of DKR of this kind of substrate yielding homochiral α-acetoxysulfides (Scheme 5.11) [24].

5.2.4
Racemization Catalyzed by Aldehydes

In 1983 Yamada et al. developed an efficient method for the racemization of amino acids using a catalytic amount of an aliphatic or an aromatic aldehyde [25]. This method has been used in the DKR of amino acids. Scheme 5.12 shows the mecha-

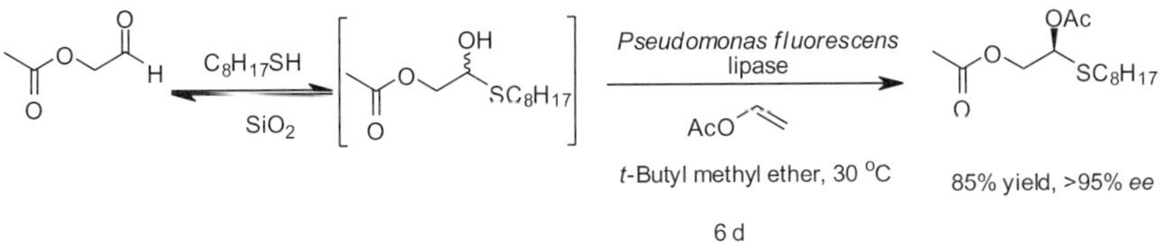

Scheme 5.11 DKR of hemithioacetals.

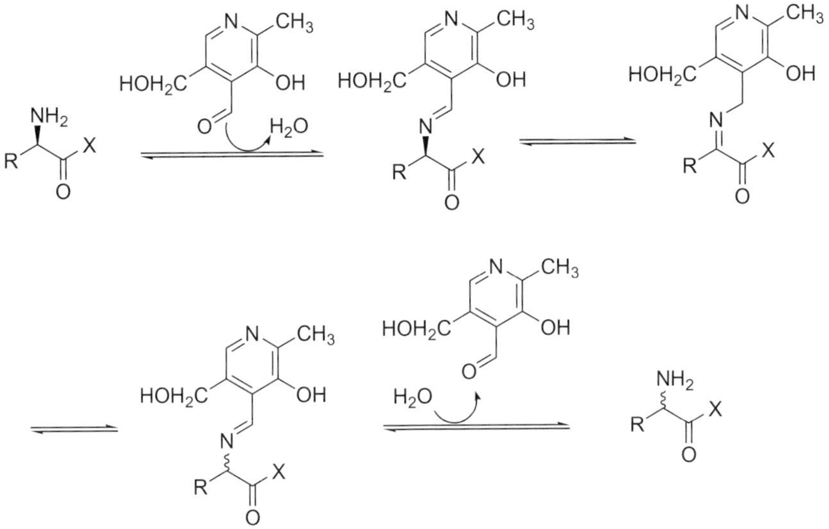

Scheme 5.12 Racemization of amino acids through formation of Schiff-base intermediates.

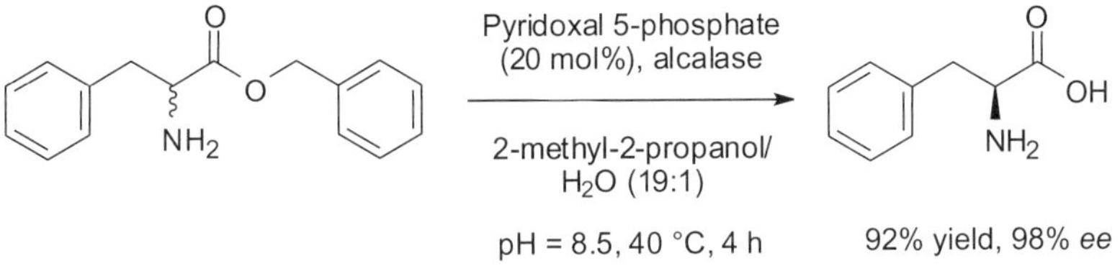

Scheme 5.13 DKR of amino acids catalyzed by pyridoxal 5-phosphate.

nism of the racemization of a carboxylic acid derivative catalyzed by pyridoxal. Racemization takes place through the formation of Schiff-base intermediates.

Wang et al. reported in 1994 the DKR of amino acid derivatives by using pyridoxal 5-phosphate as the racemization catalyst [26]. The enzyme employed to catalyze the KR was an alcalase, substilisin Carlsberg being the major enzyme component. To avoid racemization of the final product, they employed a mixture of 2-methyl-2-propanol/H_2O (19:1). Under these conditions, the product precipitated out during the course of hydrolysis (Scheme 5.13). A very similar DKR process was reported two years later by Parmar [27].

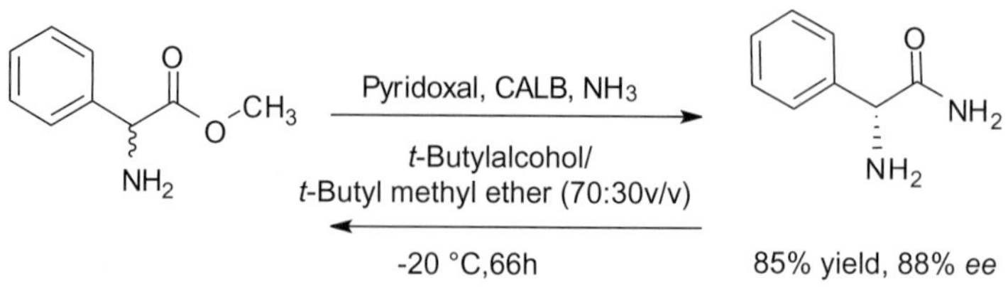

Scheme 5.14 DKR of amino acid derivatives.

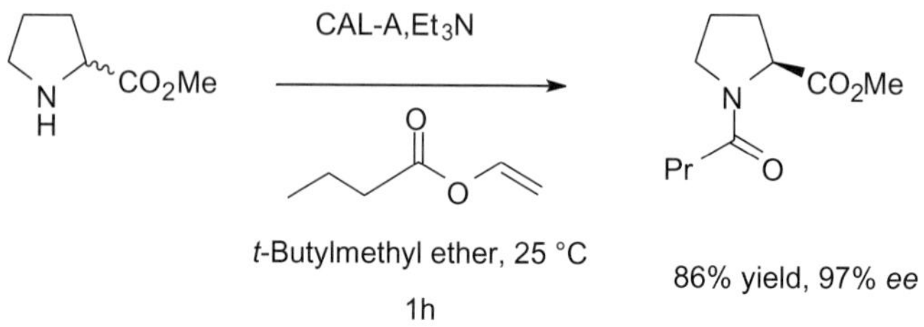

Reagents: Pyridoxal, CALB, NH$_3$, t-Butylalcohol/t-Butyl methyl ether (70:30 v/v), −20 °C, 66 h; 85% yield, 88% ee.

Scheme 5.15 DKR of α-amino esters catalyzed by CAL-A and acetaldehyde released in situ.

Reagents: CAL-A, Et$_3$N, t-Butylmethyl ether, 25 °C, 1 h; 86% yield, 97% ee.

Sheldon et al. have reported the DKR of phenylglycine esters via lipase-catalyzed ammonolysis [28]. Racemization was carried out by an aldehyde, such as salicylaldehyde or pyridoxal, under basic conditions. The major problem they found was racemization by these aldehydes of the final products. However, when performing the DKR at low temperature (−20 °C) the substrate was racemized much faster than the product, and DKR was feasible, yielding the product in good yield and high ee (Scheme 5.14).

A very elegant approach has been developed by Kanerva et al. DKR of N-heterocyclic α-amino esters is achieved using CAL-A [29]. Racemization occurs when acetaldehyde is released in situ from the acyl donor. In this case, aldehyde-catalyzed racemization of the product cannot occur (Scheme 5.15). This is one of the few reported examples of DKR of secondary amines (see also Reference [69]).

5.2.5
Enzyme-Catalyzed Racemization

The group of Asano et al. has developed an approach for the synthesis of D-amino acids through DKR using a two-enzyme system [30, 31]. They had previously reported the discovery of new D-stereospecific hydrolases that can be applied to KR of racemic amino acid amides to yield D-amino acids. Combination of a D-stereospecific hydrolase with an amino acid amide racemase allows DKR of L-amino acid amides, yielding enantiomerically pure D-amino acids in excellent yields (Scheme 5.16).

The group of Faber has reported a DKR of mandelic acid by using a lipase-catalyzed O-acylation followed by a racemization catalyzed by mandelate racemase. However, these two transformations do not take place simultaneously in the same

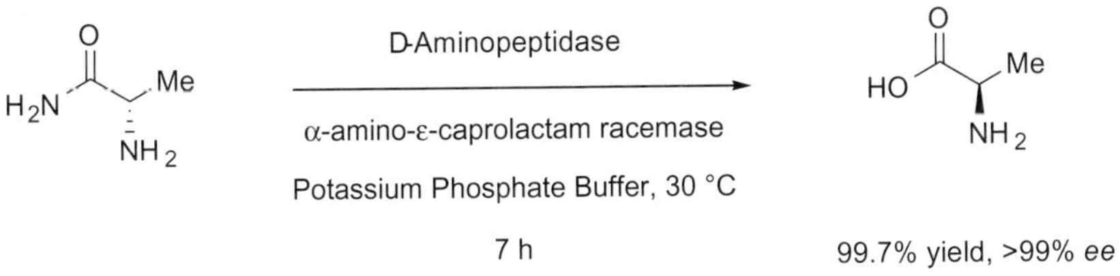

Scheme 5.16 DKR using a two-enzyme system.

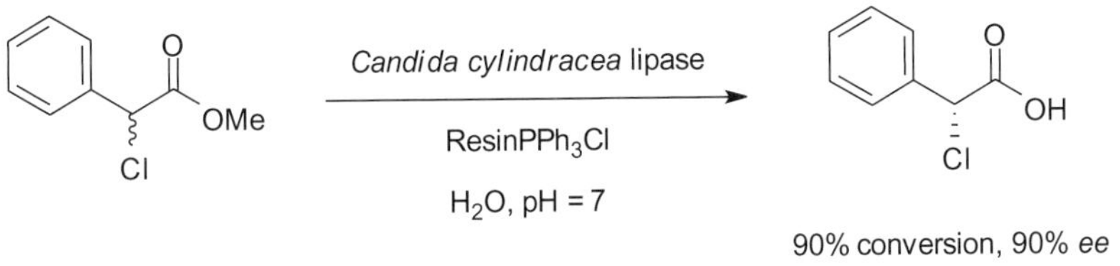

Scheme 5.17 Enzyme- and chloride-catalyzed DKR.

pot. When the sequence was repeated four times, (S)-O-acetylmandelic acid was obtained in 80% isolated yield and >98% ee [32].

5.2.6
Racemization through S_N2 Displacement

The group of Williams has reported a DKR of halo α-bromo [33a] and α-chloro esters [33b]. In the latter case, the KR is catalyzed by commercially available cross-linked enzyme crystals derived from *Candida cylindracea* lipase. The racemization takes place through halide S_N2 displacement. The DKR is possible because the racemization of the substrates is faster than that of the products (carboxylates). For the ester, the empty $\pi^*(C=O)$ orbital is able to stabilize the S_N2 transition state by accepting electron density. However, the carboxylate is more electron-rich and therefore less able to facilitate an S_N2 reaction. Racemization is performed by the use of a chloride source. The best results where obtained by using a resin-bound phosphonium choride (Scheme 5.17).

5.2.7
Radical Racemization

Racemization of amines can be performed via reversible hydrogen abstraction at the chiral center in the position α to the nitrogen atom by using alkyl sulfanyl radicals. In this radical racemization, the prochiral intermediate is an α-amino radical. Racemization of the product (amide) does not occur because its α-C—H bond dissociation enthalpy is much stronger than that of the corresponding amine starting material. DKR of amines was performed using CALB and AIBN (azobisisobutyronitrile, a radical initiator) in the presence of a specifically designed thiol (Scheme 5.18) [34].

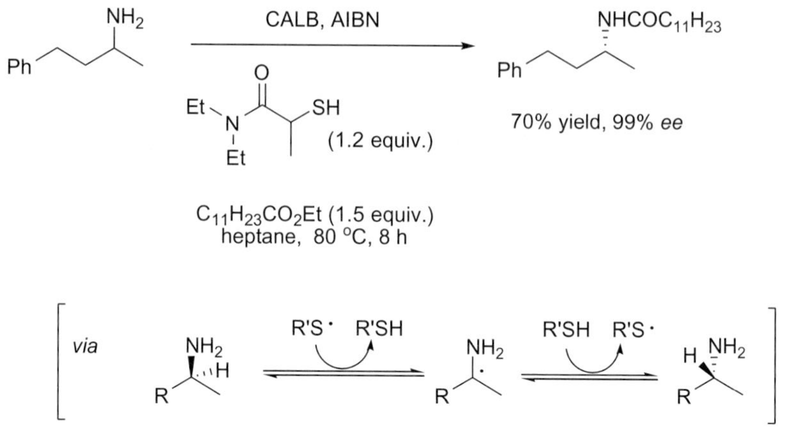

Scheme 5.18 DKR of amines via radical racemization.

5.2.8
Metal-Catalyzed Racemization

A range of metal-catalyzed racemization reactions are known in the literature, but it is not always straightforward to combine these with an enzymatic resolution. For example, several complexes of rhodium, iridium, and ruthenium, among others, are known to catalyze rapid racemization of a variety of alcohol substrates [35], but only a few have proved compatible with an enzymatic reaction. Most often, the compatibility of the two catalysts under the same reaction conditions is the major problem: the metal may interfere with the enzyme to give poor resolution, or the enzyme may slow down or inhibit the racemization by the metal catalyst. Furthermore, the by-products produced in the metal- and/or enzyme-catalyzed reaction, or the surfactants employed to support the enzyme, can also interfere with the racemization or with the enzymatic resolution. Another problem that has to be overcome is to find the appropriate solvent for optimal activity of both catalysts, the enzyme and the metal complex. Lipases show excellent activity and selectivity in water, but most organometallic complexes do not work in water. Lipases also work very well in apolar organic solvents, such as hexane. However, most organometallic complexes are not soluble in this solvent, resulting in very low racemization reactions. Also, racemization is usually faster at high temperatures, at which most enzymes lose their catalytic activity. Despite this, a deep understanding concerning the compatibility of the two catalysts has been achieved, and excellent results have been obtained.

In 1997, Stürmer highlighted the importance of the combination of enzymes and transition metals in one pot [36]. Since then, this concept has aroused much interest within the scientific community. In all the DKRs presented in this section, the enzyme catalyzes a transesterification process. Thus, enzyme- and metal-catalyzed DKRs are categorized according to the nature of the substrates as being allylic substrates, secondary alcohols, or primary amines. In the first case,

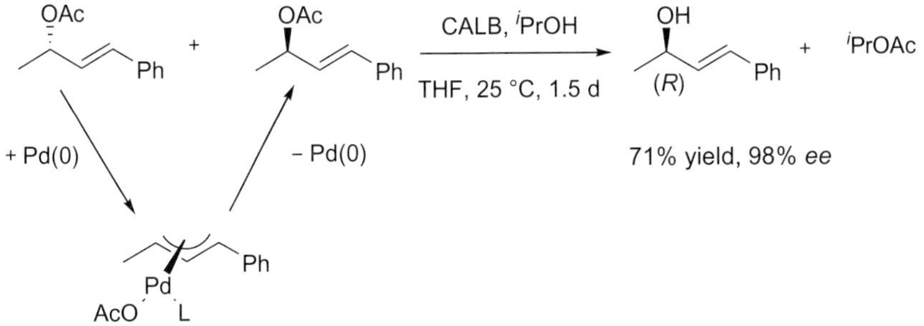

Scheme 5.19 DKR of allylic acetates catalyzed by a lipase and Pd(0).

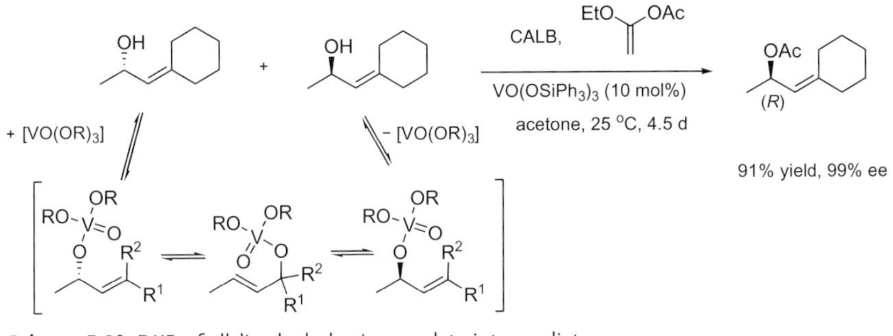

Scheme 5.20 DKR of allylic alcohols via vanadate intermediates.

racemization occurs through 1,3-shift of an acetate or a hydroxyl group, and in the last two cases through hydrogen transfer processes.

5.2.8.1 DKR of Allylic Acetates and Allylic Alcohols

The first example of chemoenzymatic DKR of allylic alcohol derivatives was reported by Williams et al. [37]. Cyclic allylic acetates were deracemized by combining a lipase-catalyzed hydrolysis with a racemization via transposition of the acetate group, catalyzed by a Pd(II) complex. Despite a limitation of the process, i.e. long reaction times (19 days), this work was a significant step forward in the combination of enzymes and metals in one pot. Some years later, Kim et al. considerably improved the DKR of allylic acetates using a Pd(0) complex for the racemization, which occurs through π-allyl(palladium) intermediates. The transesterification is catalyzed by a lipase (*Candida antarctica* lipase B, CALB) using isopropanol as acyl acceptor (Scheme 5.19) [38].

A novel approach was developed very recently by Kita et al. [39]. DKR of allylic alcohols was performed by combining a lipase-catalyzed acylation with a racemization through the formation of allyl vanadate intermediates. Excellent yields and enantioselectivities were obtained. An example is shown in Scheme 5.20. With this approach, choice of substrates is limited, in that the allylic alcohol must be equally disubstituted in the allylic position ($R^1 = R^2$), since C—C single bond rotation is required in the tertiary alkoxy intermediate. Alternatively, R^1 or R^2 can be H if the two allylic alcohols formed by migration of the hydroxyl group are enantiomers (e.g., cyclic allylic alcohols).

DKR of allylic alcohols can be also performed using ruthenium complexes using racemization that occurs through hydrogen transfer reactions (see below) [40].

5.2.8.2 DKR of sec-Alcohols

The racemization mechanism of sec-alcohols has been widely studied [40, 41]. Metal complexes of the main groups of the periodic table react through a direct transfer of hydrogen (concerted process), e.g., aluminum complexes in the Meerwein-Ponndorf-Verley/Oppenauer reaction. However, racemization catalyzed by transition metal complexes occurs via a hydrogen transfer process through metal hydrides or metal dihydrides as intermediates (Scheme 5.21) [42].

The groups of Williams [43] and Bäckvall [44] first reported the combination of enzymes and transition metals for DKR of sec-alcohols.

Williams employed complexes of Al, Rh, or Ir in combination with PFL (*Pseudomonas fluorescens* lipase) for the DKR of 1-phenylethanol. The best results were obtained using $Rh_2(OAc)_4$ as the catalyst for the racemization which gave 60% conversion of the alcohol to 1-phenylethyl acetate in 98% *ee* (Scheme 5.22) [43].

The group of Bäckvall employed the Shvo's ruthenium complex (1) [45] for the racemization. This complex is activated by heat. For the KR, *p*-chlorophenyl acetate

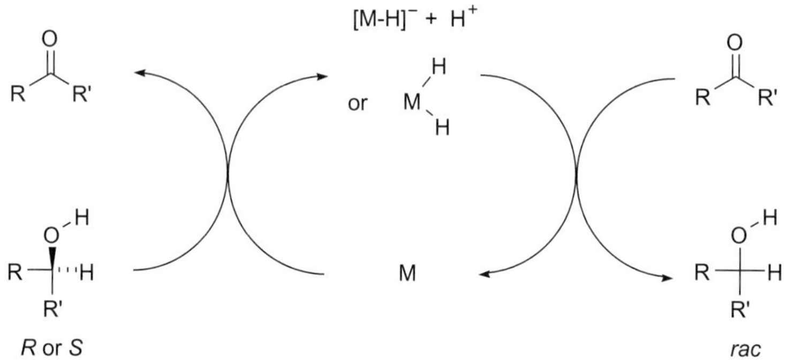

Scheme 5.21 Simplified mechanism of the racemization of sec-alcohols catalyzed by transition metal complexes.

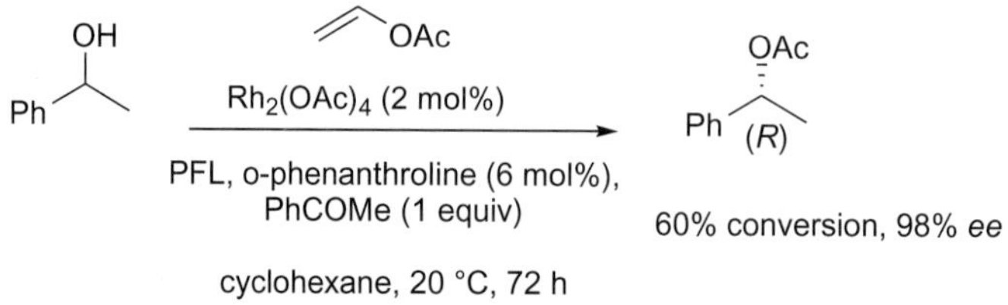

Scheme 5.22 DKR of sec-alcohols catalyzed by PFL and Rh complexes.

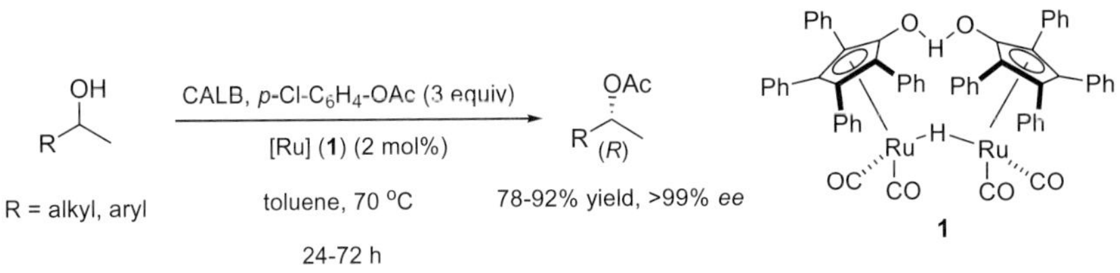

Scheme 5.23 DKR of sec-alcohols using CALB and Shvo's complex (1).

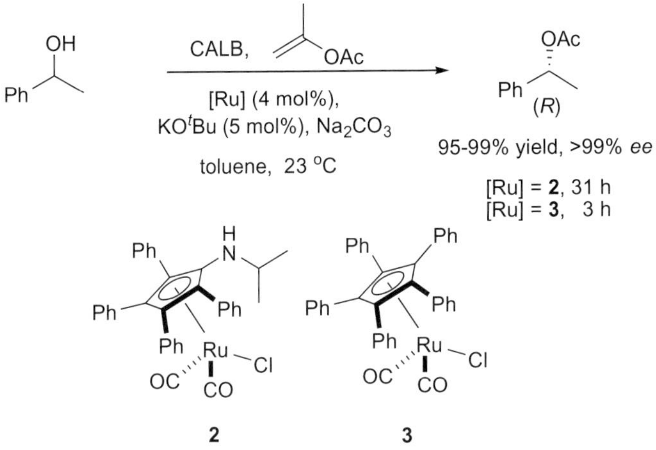

Scgene 5.24 DKR of sec-alcohols catalyzed by a lipase and Ru complexes at ambient temperature.

was used as the acyl donor in combination with thermostable enzymes, such as *Candida antarctica* lipase B (CALB) or *Pseudomonas cepacia* lipase (PS-C) [44] (Scheme 5.23). This was the first practical chemoenzymatic DKR affording acetylated sec-alcohols in high yields and excellent enantioselectivities. This method was subsequently applied to a variety of different substrates [46, 47, 48, 49, 50, 51], and a modified version is employed by the Dutch company DSM for the large-scale production of (R)-phenylethanol [52].

Kim and Park subsequently reported that ruthenium pre-catalyst 2 racemizes alcohols within 30 min at room temperature [53]. However, when combined with an enzyme (lipase) in DKR at room temperature, very long reaction times (1.3 to 7 days) were required, in spite of the fact that the enzymatic KR takes only a few hours (Scheme 5.24). Despite these compatibility problems, their results constituted an important improvement, since chemoenzymatic DKR could now be performed at ambient temperature to give high yields, which enables non-thermostable enzymes to be used. More recently, we communicated a highly efficient metal- and enzyme-catalyzed DKR of alcohols at room temperature (Scheme 5.24) [40, 54]. This is the fastest DKR of alcohols hitherto reported by the combination of transition metal and enzyme catalysts. Racemization was effected by a new class of very potent hydrogen transfer catalysts (**3**). Catalysts **2** and **3** have been employed in the DKR of a broad variety of secondary alcohols. We have studied the racemization and found that it proceeds via a new mechanism where the intermediate ketone formed in the process stays coordinated to ruthenium [40, 41g].

Important improvements in the dynamic kinetic asymmetric transformation (DYKAT) of diols have been obtained recently by using the highly efficient catalyst **3** to achieve very fast epimerization [55]. In the DYKAT of 2,5-hexanediol catalyzed by **3**, the faster epimerization and the low reaction temperature reduce the concentration of intermediates that can evolve by anomalous enzyme-catalyzed S-acylations, such as alcohols bearing a carbonyl group or an (R)-acetoxy group at the δ position that give rise to the unwanted formation of *meso* diacetates. In the DYKAT of 2,4-pentanediol catalyzed by **3**, the fast epimerization outruns unwanted acyl migrations that yield *meso* diacetates. Thus, by using complex **3**, (R,R)-2,5-hexanediol diacetate and (R,R)-2,4-pentanediol diacetate can now be obtained in quantitative yield with excellent *ee* (>99%) and high (R,R)/*meso* ratio (≥94:6).

Kim and Park et al. have reported a polymer-supported derivative (**4**), which, together with the enzyme, can be recycled and reused [56].

4

Sheldon et al. have combined a KR catalyzed by CALB with a racemization catalyzed by a Ru(II) complex in combination with TEMPO (2,2,6,6-tetramethylpiperidine 1-oxyl free radical) [57]. They proposed that racemization involves initial ruthenium-catalyzed oxidation of the alcohol to the corresponding ketone, with TEMPO acting as a stoichiometric oxidant. The ketone is then reduced to the racemic alcohol by ruthenium hydrides which are proposed to be formed under the reaction conditions. Under these conditions, they obtained a 76% yield of enantiopure 1-phenylethanol acetate at 70 °C after 48 h.

Naturally occurring lipases are (R)-selective for alcohols according to Kazlauskas' rule [58, 59]. Thus, DKR of alcohols employing lipases can only be used to transform the racemic alcohol into the (R)-acetate. Serine proteases, a sub-class of hydrolases, are known to catalyze transesterifications similar to those catalyzed by lipases, but, interestingly, often with reversed enantioselectivity. Proteases are less thermostable enzymes, and for this reason only metal complexes that racemize secondary alcohols at ambient temperature can be employed for efficient (S)-selective DKR of *sec*-alcohols. Ruthenium complexes **2** and **3** have been combined with subtilisin Carlsberg, affording a method for the synthesis of

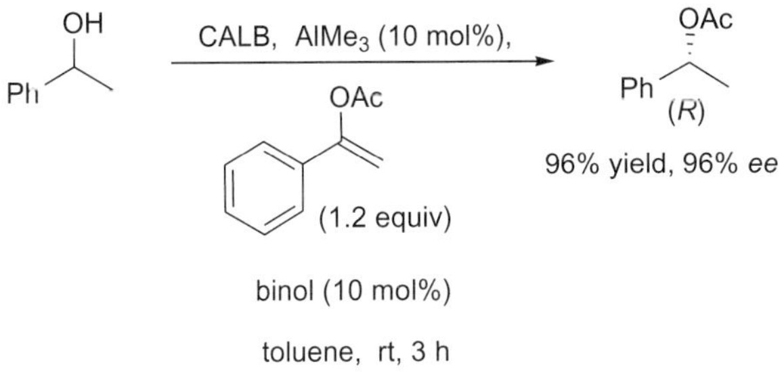

Scheme 5.25 DKR of sec-alcohols using a lipase and inexpensive Al complexes.

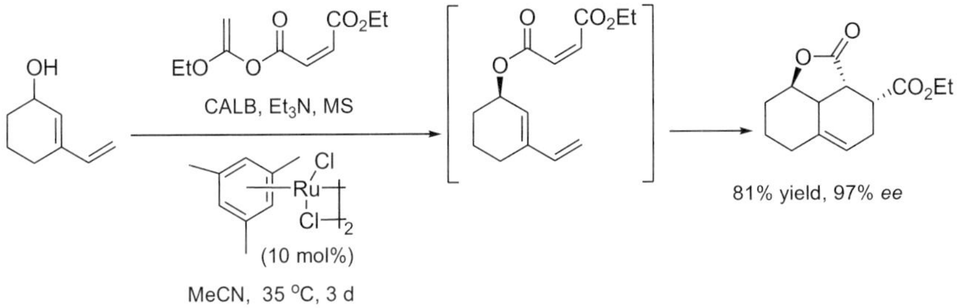

Scheme 5.26 Tandem DKR-intramolecular Diels-Alder reaction.

acetylated (S)-sec-alcohols [60]. With complex 2, the DKR reaction took 3–4 days [60a], whereas with catalyst 3, DKR was obtained, with reaction times down to 18 h, in both cases at room temperature [60b].

Very recently the Meerwein-Ponndorf-Verley-Oppenauer (MPVO) reaction has been exploited for the racemization of alcohols using inexpensive aluminum-based catalysts. Combination of these complexes with a lipase (CALB) results in an efficient DKR of sec-alcohols at ambient temperature. To increase the reactivity of the aluminum complexes, a bidentate ligand, such as binol, is required. Also, specific acyl donors need to be used for each substrate (Scheme 5.25) [61].

Kita et al. have made use of the structure of the acyl donor to develop a domino transformation: a DKR followed by an intramolecular Diels-Alder reaction [62]. They have found reaction conditions for the three different transformations to effectively take place in one pot: a KR, a racemization, and a Diels-Alder reaction (Scheme 5.26).

The groups of Meijer and Heise have developed iterative tandem catalysis with the aim of generating chiral oligomers [63]. The polymerization of 6-methyl-ε-caprolactone was achieved by combining enzymatic ring-opening with ruthenium-catalyzed racemization in one pot yielding enantioenriched oligomers (Scheme 5.27).

More recently, the group of Heise has shown that DKR can be combined with enzymatic polymerization for the synthesis of chiral polyesters from racemic secondary diols in one pot [64] (Scheme 5.28).

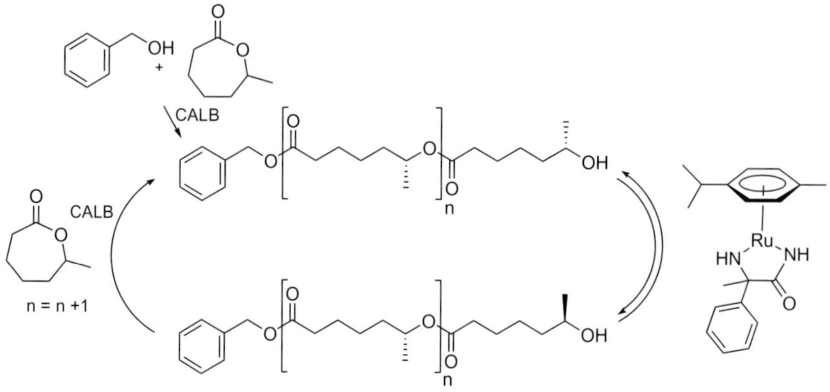

Scheme 5.27 Synthesis of chiral oligomers by DKR.

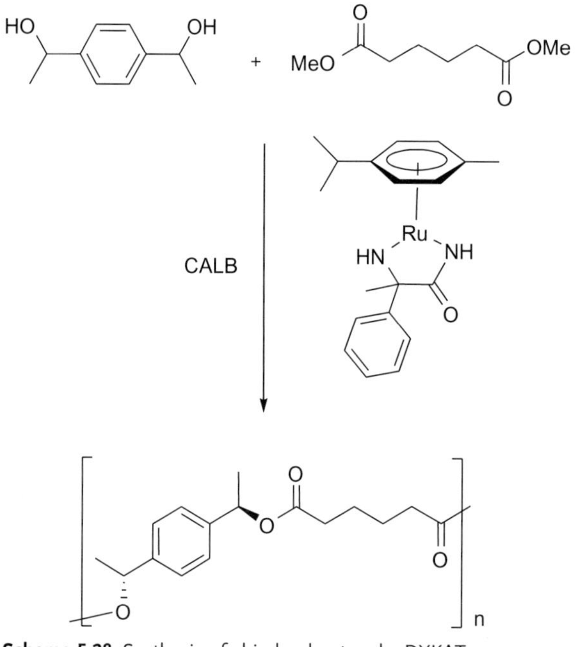

Scheme 5.28 Synthesis of chiral polyesters by DYKAT.

5.2.8.3 DKR of Amines

Racemization of amines is difficult to achieve and usually requires harsh reaction conditions. Reetz et al. developed the first example of DKR of amines using palladium on carbon for the racemization and CALB for the enzymatic resolution [65]. This combination required long reaction times (8 days) to obtain 64% yield in the DKR of 1-phenylethylamine. More recently, Bäckvall et al. synthesized a novel Shvo-type ruthenium complex (**5**), which, in combination with CALB, made it

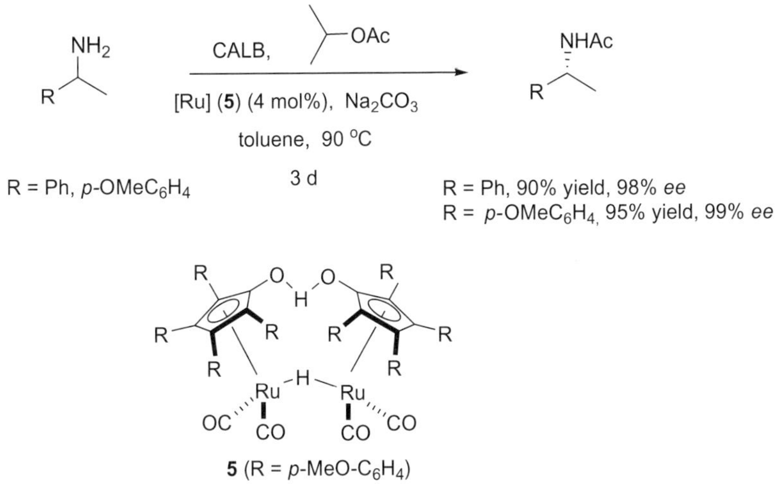

Scheme 5.29 Ru- and lipase-catalyzed DKR of primary amines.

possible to perform DKR of a variety of primary amines with excellent yields and enantioselectivities (Scheme 5.29) [66].

Jacobs and De Vos have found that the efficiency of the Pd-catalyzed racemization of amines can be improved by using Pd immobilized on supports such as $BaSO_4$, $CaCO_3$, or $BaCO_3$. The racemization was combined with a KR catalyzed by CALB affording enantiopure acetylated benzylamines in high yields [67]. The group of Kim and Park has also reported DKR of amines by combining a Pd nanocatalyst with CALB [68]. The group of Page has reported very recently the DKR of sec-amines using a low catalyst loading of an Ir complex for the racemization and *Candida rugosa* lipase for the KR [69].

5.2.9
Other Racemization Methods

In this section we discuss some DKRs in which racemization occurs spontaneously during the enzymatic resolution and without further addition of any reagent.

5.2.9.1 DKR of 5-Hydroxy-2-(5H)-Furanones
During the KR of this kind of substrate, racemization occurs spontaneously as a consequence of the labile stereogenic center at C-5 [70]. The interconversion occurs by mutarotation, allowing a DKR. The KR is catalyzed by a lipase in the presence of vinyl acetate (Scheme 5.30).

5.2.9.2 DKR of Hemiaminals
Spontaneous racemization also occurs during the enzyme-catalyzed acetylation of hemiaminals [71]. It is thought that the racemization occurs by ring opening at

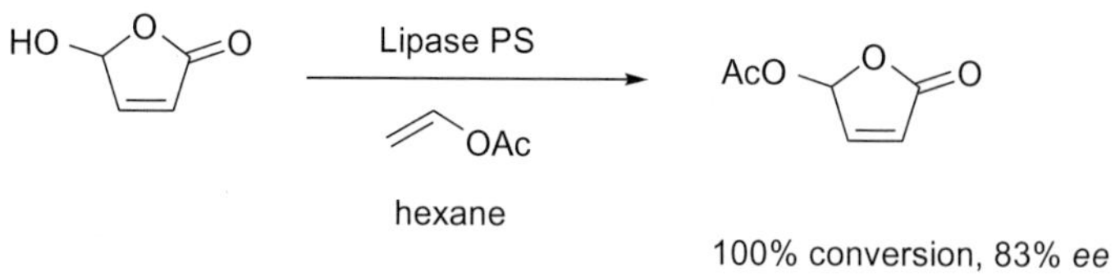

Scheme 5.30 DKR in which racemization occurs spontaneously.

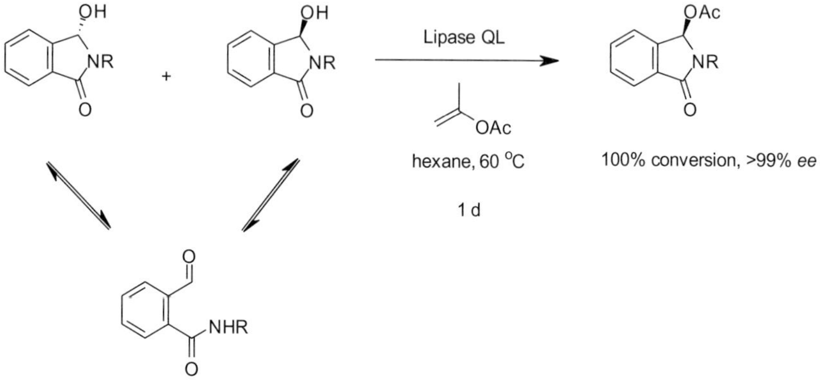

Scheme 5.31 DKR in which racemization occurs spontaneously.

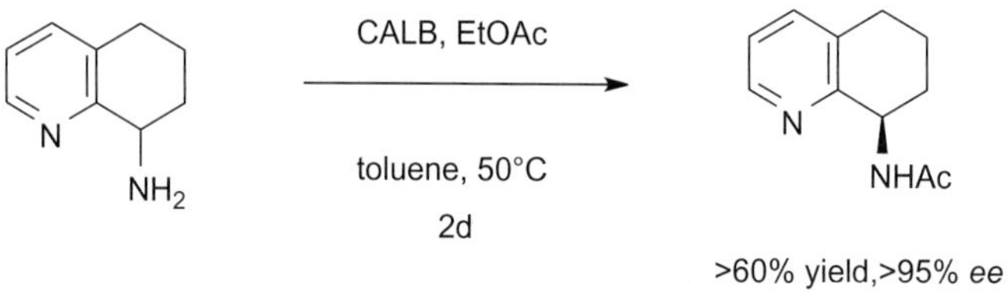

Scheme 5.32 DKR in which racemization occurs spontaneously.

the C-3 position under the reaction conditions (Scheme 5.31). The KR is catalyzed by lipase QL (*Alcaligenes species*) in the presence of isopropenyl acetate.

5.2.9.3 DKR of 8-Amino-5,6,7,8-Tetrahydroquinoline

Very recently Crawford has reported an interesting KR catalyzed by CALB in which the yield was >60% and the product was obtained in very high enantiopurity (Scheme 5.32) [72]. Furthermore, the remaining starting material was racemic.

A mechanistic study performed by the authors showed that CALB and EtOAc are required for racemization to take place. They claim that enzyme-catalyzed oxidation of the amine substrate takes place, and a ketone intermediate is produced. The ketone reacts with the substrate to form an enamine. This process would result in racemization of the substrate (Scheme 5.33).

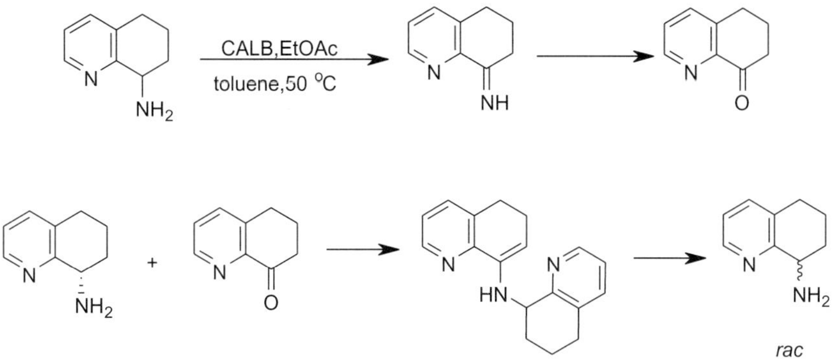

Scheme 5.33 Mechanism of the racemization suggested in ref [72].

To confirm this racemization mechanism, Crawford et al. added 5 mol% of the ketone to the reaction mixture and obtained the product in 78% yield in >98% ee. This DKR is therefore catalyzed by a carbonyl compound and can be compared to those shown in Section 5.2.4.

5.3
Cyclic Deracemizations

As in the case of DKRs, cyclic deracemizations overcome the 50% yield drawback of kinetic resolutions. Although these two processes are similar in the sense that they convert a racemate into one enantiomer with high efficiency, there is an important difference between them. Whereas DKR gives a 100% yield of an enantiopure product that is a functionalized starting material (for example, an enantiopure acetylated alcohol), a cyclic deracemization converts a racemic starting material to its enantiomerically pure form. Thus, an advantage of cyclic deracemization over DKR is that the final product does not need to be deprotected (e.g., deacetylated). Nevertheless, we have already shown some examples where a structurally defined acyl donor has been employed in DKRs in order to keep it in the molecule for further transformations (see above). Another difference between these two deracemization processes is that whereas most DKRs reported are performed in organic solvents, all cyclic deracemizations reported to date are performed in aqueous media, where many organic molecules are insoluble or decompose. Thus, whereas for DKRs great efforts have been made to obtain high activity of the biocatalyst in organic solvents, in the case of cyclic deracemizations, more attention has been paid to the stability in water of the chemical reagents. Finally, it is important to note that cyclic deracemization has been employed for stereoinversions as well, e.g., for the production of an enantiomerically pure amino acid with the opposite configuration to that of the naturally occurring enantiomer.

In Section 5.3.1, chemoenzymatic cyclic deracemizations or stereoinversions are categorized according to the type of substrate, namely amino acids, hydroxy acids, or amines. In the Section 5.3.2, deracemizations where only biocatalysts are employed are discussed.

5.3.1
Chemoenzymatic Cyclic Deracemizations

5.3.1.1 Cyclic Deracemizations of Amino Acids

The first example of this kind of deracemization was reported as early as 1971 by Hafner and Wellner [73]. They transformed D-leucine and D-alanine into the corresponding L-amino acids by using porcine kidney D-amino acid oxidase and NaBH$_4$. Despite the low yields obtained, this example constitutes the first report of a cyclic deracemization by combining a selective enzyme-catalyzed oxidation with a non-selective chemical reduction. Some years later, the group of Soda employed this technique for deracemizing cyclic and acyclic amino acids [74]. For example, D-proline is transformed into the L-enantiomer by using a D-amino acid oxidase in combination with NaBH$_4$. Although a great excess of the reducing agent was employed, the stereoinversion successfully occurred through formation of the achiral pyrroline-2-carboxylate (Scheme 5.34).

More recently, this process was improved by Turner et al. [75]. The use of NaCNBH$_3$, which is a milder and water-stable reducing agent, instead of NaBH$_4$ allowed the use of only 3 equiv. of the reducing agent instead of the 500 equiv. required in the original report (Scheme 5.35) [74].

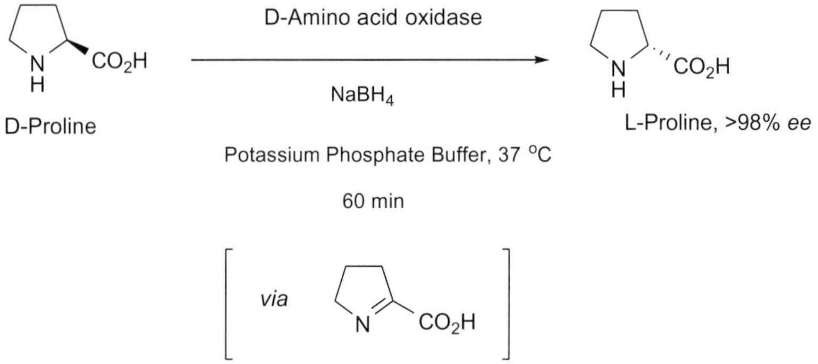

Scheme 5.34 Chemo-enzymatic production of L-proline from D-proline.

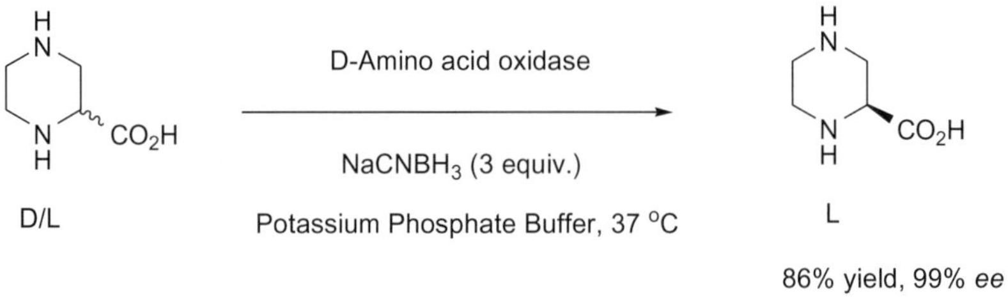

Scheme 5.35 Chemo-enzymatic deracemization of cyclic amino acids.

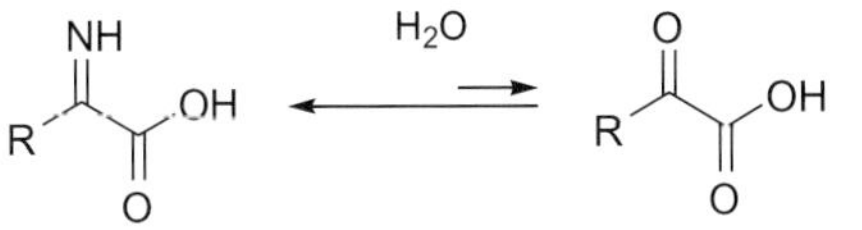

Scheme 5.36 Undesired hydrolysis of imino acid intermediates.

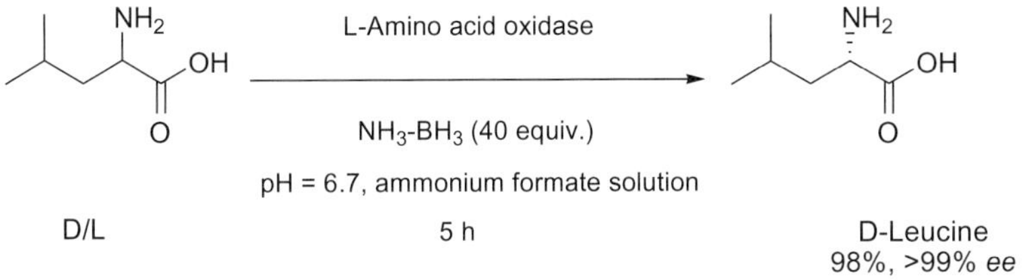

Scheme 5.37 Deracemization of acyclic amino acids.

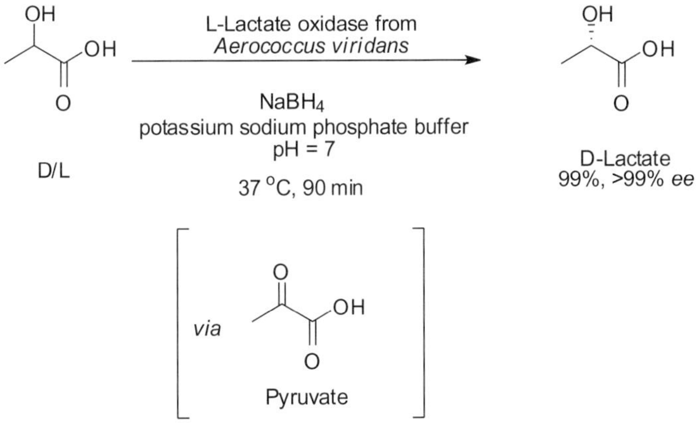

Scheme 5.38 Deracemization of hydroxy acids.

Deracemization of acyclic amino acids is more difficult than that of cyclic ones because of the higher instability of the imino acid intermediates toward hydrolysis (Scheme 5.36).

Nevertheless, it can be achieved by the use of a variety of reducing agents (amine-boranes, NaBH$_4$, HCOOH/Pd/C) [76, 77] (Scheme 5.37). In particular, amine-borane gives excellent results and although some excess of this reagent need to be used, the number of equivalents is much smaller than that required when NaBH$_4$ is the reducing agent.

5.3.1.2 Chemoenzymatic Cyclic Deracemizations of Hydroxy Acids

Similarly to the case of amino acids, hydroxy acids can also be deracemized by combining an enantioselective oxidation with a non-enantioselective reduction with sodium borohydride. For example, the group of Soda has reported the transformation of DL-lactate into D-lactate in >99% ee (Scheme 5.38) [78].

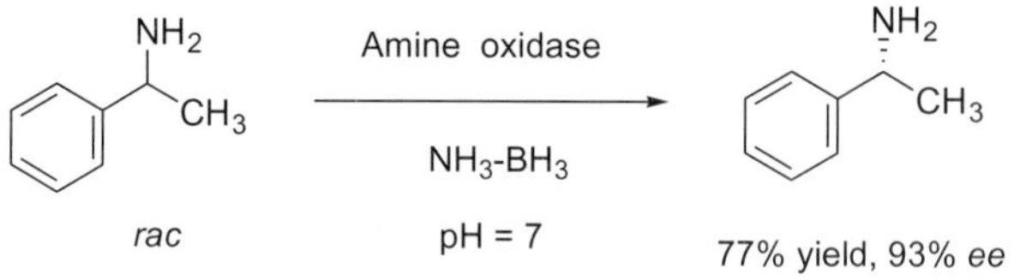

Scheme 5.39 Deracemization of amines.

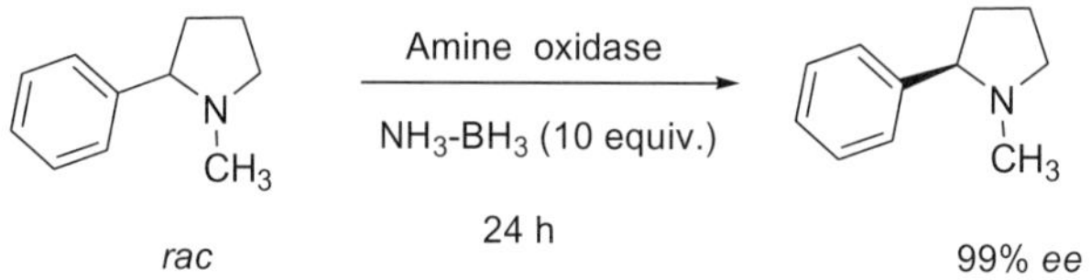

Scheme 5.40 Deracemization of tertiary amines.

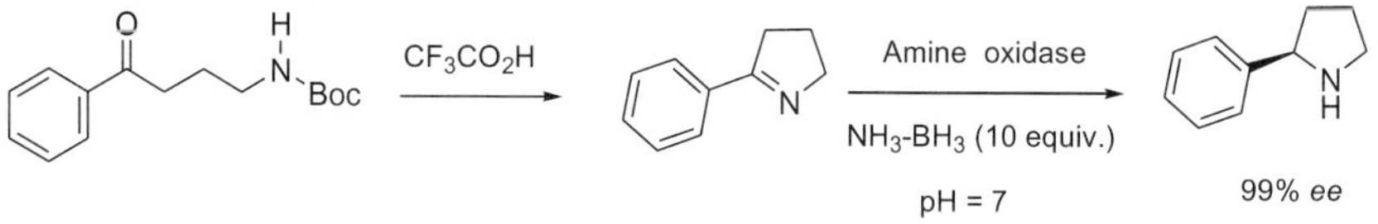

Scheme 5.41 Transformation of γ-amino ketones into enantiomerically pure secondary amines.

5.3.1.3 Chemoenzymatic Cyclic Deracemizations of Amines

The group of Turner has reported the deracemization of amines [79]. The wild type of Type II monoamine oxidase from *Aspergillus niger* possesses very low but measurable activity toward the oxidation of L-α-methylbenzylamine. The oxidation of the D enantiomer is even slower. *In vitro* evolution led to the identification of a mutant with enhanced enantioselectivity, showing high E values (>100) for a variety of primary and secondary amines. An example is shown in Scheme 5.39.

An important breakthrough was made very recently in this area. A chemoenzymatic method developed by Turner has allowed the cyclic deracemization of tertiary amines [80]. Enantiopure tertiary amines cannot be obtained via DKR. One of the variants obtained by directed evolution of the monoamine oxidase from *Aspergillus niger* showed high activity and enantioselectivity toward cyclic tertiary amines (Scheme 5.40).

Turner has applied this deracemization process to a very interesting tandem transformation where γ-amino ketones are transformed into enantiopure secondary amines via intramolecular reductive amination followed by deracemization (Scheme 5.41) [80].

5.3.2
Biocatalyzed Deracemizations

This type of transformation occurs through enantiospecific stereochemical inversion of one of the enantiomers of a racemate into the opposite one. Although in this chapter we have focused on chemoenzymatic methods, biocatalyzed deracemizations deserve some attention, since there are many reports of enantiomerically pure products being obtained in very high yields. The stereoinversion of the reacting enantiomer occurs via an achiral intermediate, as in the case of cyclic deracemizations. However, in contrast to cyclic deracemizations, two consecutive stereoselective transformations occur: enantioselective oxidation followed by enantioselective reduction. Therefore, the kinetics of biocatalyzed deracemizations is different from that of DKRs or cyclic deracemizations. At least one of the two reactions has to be irreversible. Some biocatalytic stereoinversions are achieved by the use of microbial cells containing two enzymes with the desired activity. Thus, one enzyme (dehydrogenase) oxidizes one enantiomer (s) to the achiral intermediate (*I*), whereas the other enzyme reduces this intermediate, yielding the opposite enantiomer (*ent*-s). The other enantiomer (*ent*-s) remains untouched throughout the reaction. Thus, 100% yield of only one enantiomer (*ent*-s) can be obtained. The net redox balance is zero, since the redox equivalents are recycled by the cell system. Other biocatalytic stereoinversions occur using two microorganisms, such as two dehydrogenase enzymes with complementary specificity, that perform successive oxidation-reduction reactions. Examples of deracemizations of alcohols have been reported using two microorganisms or a single microorganism. For example, *sec*-alcohols can be deracemized by combining an oxidation by *B. stereothermophilus* and another microorganism, *Y. lipolytica*, which can perform the reduction of the ketone intermediate (Scheme 5.42) [81].

Nakamura et al. have reported the deracemization of *sec*-alcohols using one microorganism, a strain of a fungus, *Geotrichum candidum* IFO 5767, and obtained *sec*-alcohols in high yield and enantiomeric excess [82].

Biocatalytic deracemizations of amino acids and hydroxy acids have also been reported [83].

The final example constitutes one of the few deracemizations of carboxylic acids [84]. We would like to highlight it because it shares some features with DKRs and

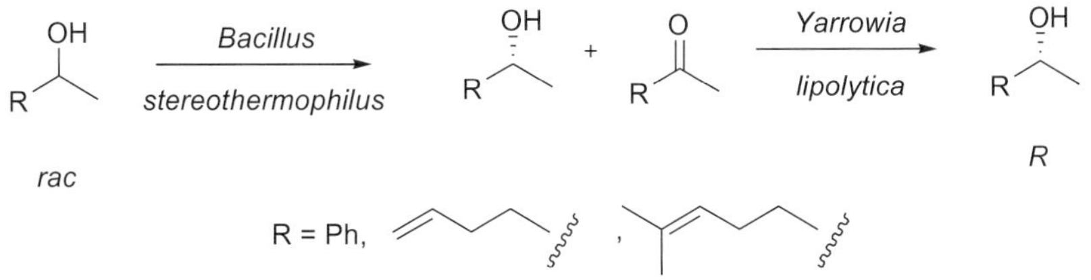

Scheme 5.42 Stereoinversion of *sec*-alcohols using two microorganisms.

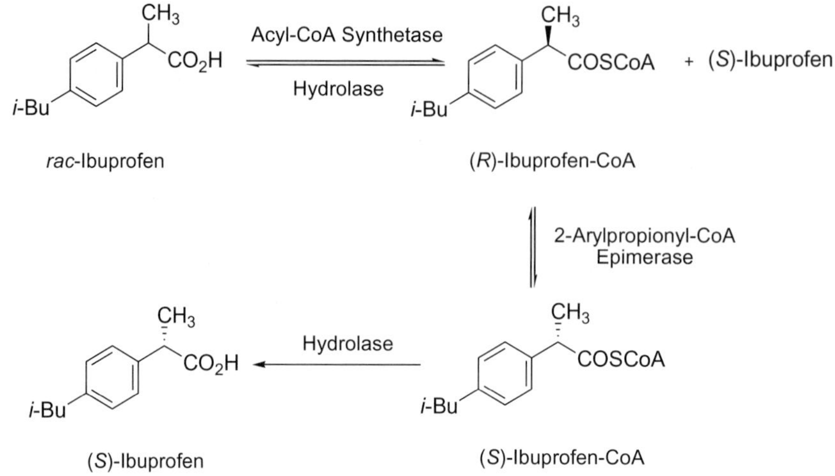

Scheme 5.43 Stereoinversion of an α-substituted carboxylic acid by combining an epimerization with an enantioselective thioester formation.

with biocatalytic steroinversions, in that it involves a racemization and that the same starting material is obtained enantiopure or enantioenriched by the use of biocatalysts, respectively (Scheme 5.43).

5.4
Concluding Remarks

Biocatalysts offer an excellent alternative for the synthesis of chiral molecules. When enzyme-catalyzed kinetic resolutions are combined with chemical methods, deracemizations affording 100% yield of enantiomerically pure products are achieved. The efficiency of chemoenzymatic deracemization processes has improved dramatically during the past few years. DKRs can be performed by combining enzymatic resolutions with a variety of racemization methods, such as metal-catalyzed racemizations, base- or acid-catalyzed racemizations, aldehyde- or enzyme-catalyzed racemizations, or even spontaneous racemizations that take place under the reaction conditions. The use of enzymes in organic solvents has contributed to the development of DKRs in which racemization is catalyzed by a metal complex. Also, several highly efficient cyclic deracemizations have been developed recently. Nowadays, enzyme-catalyzed DKRs and deracemizations can be applied to a broad range of substrates, such as secondary alcohols, primary amines, secondary and tertiary amines, amino acids, and hydroxy acids, to give excellent yields and enantioselectivites. A variety of applications can also be found in the literature, such as the synthesis of chiral polymers (impossible to accomplish without the presence of a racemization catalyst) or the development of highly atom-economical processes where the structure of the acyl donor is employed for further transformations. In all cases, excellent yields and enantioselectivities are obtained, making enzyme-catalyzed DKRs and cyclic deracemizations useful and efficient strategies for the synthesis of enantiopure compounds. Furthermore, both processes – DKR and cyclic deracemization – are useful techniques that nowadays are employed in the industry. The next few years will also witness many new contributions to this area. New racemization catalysts, which are more efficient, more environmentally friendly, and also cheaper, are being developed. Also, uses of other enzymes will broaden the scope of this transformation. In particular, the new methods available today such as directed evolution, coupled with advances in high-throughput screening technologies will provide new enzymes with the desired properties to perform new transformations.

References

1 A. M. Rouhi, *Chem Eng News* 2004, **82**, 47–62.
2 a) J. Halpern, B. M. Trost, *Proc. Natl. Acad. Sci. U.S.A.* 2004, **101**, 5347; b) Catalytic Asymmetric Synthesis, 2nd ed.; I. Ojima, Ed.; Wiley-VCH: New York, 2000; c) Comprehensive Asymmetric Catalysis; E. N. Jacobsen, A. Pfaltz, H. Yamamoto, Eds.; Springer: Berlin, 1999; d) R. Noyori in Asymmetric Catalysis in Organic Synthesis; John Wiley & Sons: New York, 1994; e) Enzyme Catalysis in Organic Synthesis: A Comprehensive Handbook, 2nd ed.; K. Drauz, H. Waldmann Eds.; Wiley-VCH: Weinheim, 2002; Vols. I–III; f) K. Faber, Biotransformations in Organic Chemistry, 4th ed; Springer: Berlin, 2000; g) P. I. Dalko, L. Moisan, L. *Angew. Chem. Int. Ed.* 2004, **43**, 5138–5175.
3 E. L. Eliel, S. H. Wilen, L. N. Mander, *Stereochemistry of Organic Compounds*; Wiley & Sons: New York, 1994.
4 M. Breuer, K. Ditrich, T. Habicher, B. Hauer, M. Kesseler, R. Stürmer, T. Zelinski, *Angew. Chem. Int. Ed.* 2004, **43**, 788–824.
5 K. Faber, *Chem. Eur. J.* 2001, **7**, 5004–5010.
6 H. B. Kagan, J. C. Fiaud, *Top. Stereochem.* 1988, **18**, 249–330.
7 For biocatalyzed reactions, the "binding" of the substrate enantiomers (which can be neglected with chemical catalysts) usually plays an important role in the chiral selection process, and E-values of enzyme-catalyzed reactions are therefore defined through Michaelis-Menten kinetics: $E = (k_{cat}/K_M)_R/(k_{cat}/K_M)_S$.
8 The selectivity factor used in non-enzymatic kinetic resolutions is known as s. See: a) Reference [6]; b) V. S. Martin, S. S. Woodard, T. Katsuki, Y. Yamada, M. Ikeda, K. B. Sharpless, *J. Am. Chem. Soc.* 1981, **103**, 6237–6240.
9 a) C. J. Sih, S.-H. Wu, *Top. Stereochem.* 1989, **19**, 63–125; b) C.-S. Chen, S.-H. Wu, G. Girdaukas, C. J. Sih, *J. Am. Chem. Soc.* 1987, **109**, 2812–2817; c) C.-S. Chen, Y. Fujimoto, G. Girdaukas, C. J. Sih, *J. Am. Chem. Soc.* 1982, **104**, 7294–7299.
10 Faber has developed a simple program (Selectivity) that allows one to calculate the E value and to draw plots of the variation of ee_s and ee_p as a function of time for an irreversible kinetic resolution. The program can be obtained in: http://borgc185.kfunigraz.ac.at/
11 W. Kroutil, K. Faber, *Terahedron: Asymmetry*, 1998, **9**, 2901–2913.
12 U. T. Bornscheuer, R. J. Kazlauskas, *Hydrolases in Organic Synthesis*; Wiley-VCH: Weinheim, 1999.
13 a) A. M. Klibanov, *Nature* 2001, **409**, 241–246; b) P. J. Halling, *Curr. Opin. Chem. Biol.* 2000, **4**, 74–80; c) A. Zaks, A. M. Klibanov, *Proc. Natl. Acad. Sci. U.S.A.* 1985, **82**, 3192–3196. d) G. Kirchner, M. P.

Scollar, A. M. Klibanov, *J. Am. Chem. Soc.* 1985, **107**, 7072–7076.

14 a) D. S. Tan, M. M. Günter, D. G. Drueckhammer, *J. Am. Chem. Soc.* 1995, **117**, 9093–9094; b) P.-J. Um, D. G. Drueckhammer, *J. Am. Chem. Soc.* 1998, **120**, 5605–5610.

15 W.-Y. Wen, I-S. Ng, S.-W. Tsai, *J. Chem. Technol. Biotechnol.* 2006, **81**, 1715–1721.

16 a) N. J. Turner, J. R. Winterman, R. McCague, J. S. Parrat, S. J. C. Taylor, *Tetrahedron Lett.* 1995, **36**, 1113–1116; b) S. A. Brown, M.-C. Parker, N. J. Turner, *Tetrahedron: Asymmetry* 2000, **11**, 1687–1690; see also: c) H. S. Bevinakatti, R. V. Newadkar, A. A. Banerji, *J. Chem. Soc., Chem. Commun.* 1990, 1091–1092; d) H. S. Bevinakatti, A. A. Banerji, R. V. Newadkar, A. Mokashi, *Tetrahedron: Asymmetry* 1992, **3**, 1505–1508; e) R.-L. Gu, I. S. Lee, C. J. Sih, *Tetrahedron Lett.* 1992, **33**, 1953–1956; f) J. Crich, R. Brieva, P. Marquart, R.-L. Gu, S. Flemming, C. J. Sih. *J. Org. Chem.* 1993, **58**, 3252–3258.

17 a) M. Suzuki, T. Yamazaki, H. Ohta, K. Shima, K. Ohi, S. Nishiyama, T. Sugai, *Synlett*, 2000, 189–192; b) See also: H.-H. Lo, C.-H. Kao, D.-S. Lee, T.-K. Yang, W.-H. Hsu, *Chirality*, 2003, **15**, 699–702.

18 T. Taniguchi, K. Ogasawara, *Chem. Commun.* 1997, 1399–1400.

19 a) S. Wuyts, D. De Temmerman, D. De Vos, P. A. Jacobs, *Chem. Commun.* 2003, 1928–1929; b) S. Wuyts, K. De Temmerman, D. E. De Vos, P. A. Jacobs, *Chem. Eur. J.* 2005, **11**, 386–397.

20 P. Ödman, L. A. Wessjohann, U. T. Bornscheuer, *J. Org. Chem.* 2005, **70**, 9551–9555.

21 M. Inagaki, J. Hiratake, T. Nishioka, J. Oda, *J. Org. Chem.* 1992, **57**, 5643–5649.

22 a) C. Paizs, M. Toşa, C. Majdik, P. Tähinen, F.-D. Irimie, L. T. Kanerva, *Tetrahedron: Asymmetry*, 2003, **14**, 619–627; b) C. Paizs, P. Tähinen, K. Lundell, L. Poppe, F.-D. Irimie, L. T. Kanerva, *Tetrahedron: Asymmetry*, 2003, **14**, 1895–1904; c) C. Paizs, P. Tähinen, M. Toşa, C. Majdik, F.-D. Irimie, L. T. Kanerva, *Tetrahedron*, 2004, **60**, 10533–10540; d) L. Veum, L. T. Kanerva, P. J. Halling, T. Maschmeyer, U. Hanefeld, *Ad. Synth. Catal.* 2005, **347**, 1015–1021.

23 G. Desantis, Z. Zhu, W. A. Greenberg, K. Wong, J. Chaplin, S. R. Hanson, B. Farwell, L. W. Nicholson, C. L. Rand, D. P. Weiner, D. E. Robertson, M. J. Burk, *J. Am. Chem. Soc.* 2002, **124**, 9024–9025.

24 S. Brand, M. F. Jones, C. M. Rayner, *Tetrahedron Lett.* 1995, **36**, 8493–8496.

25 S. Yamada, C. Hongo, R. Yoshioka, I. Chibata, *J. Org. Chem.* 1982, **48**, 843–846.

26 S.-T. Chen, W.-H. Huang, D.-T. Wang, *J. Org. Chem.* 1994, **59**, 7580–7581.

27 V. S. Parmar, A. Singh, K. S. Bisht, N. Kumar, Y. N. Belokon, K. A. Kochetkov, N. S. Iknooikov, S. A. Orlova, V. I. Tararov, T. F. Saveleva, *J. Org. Chem.* 1996, **61**, 1223–1227.

28 a) M. A. P. J. Hacking, M. A. Wegman, J. Rops, F. van Rantwijk, R. A. Sheldon, *J. Mol. Catal. B: Enzym.* 1998, **5**, 155–157; b) M. A. Wegman, M. A. P. J. Hacking, J. Rops, P. Pereira, F. Van Rantwijk, R. A. Sheldon, *Tetrahedron: Asymmetry*, 1999, **10**, 1739–1750.

29 A. Liljeblad, A. Kiviniemi, L. T. Kanerva, *Tetrahedron* 2004, **60**, 671–677.

30 a) Y. Asano, S. Yamaguchi, *J. Am. Chem. Soc.* 2005, **127**, 7696–7697; b) See also: O. May, S. Verseck, A. Bommarius, D. Drauz, *Org. Process Res. Dev.* 2002, **6**, 452–457.

31 Review on enzymatic racemization and its application to synthetic biotransformations: B. Schnell, K. Faber, W. Kroutil, *Adv. Synth. Catal.* 2003, **345**, 653–666.

32 U. T. Strauss, K. Faber, *Tetrahedron: Asymmetry*, 1999, **10**, 4079–4081.

33 a) M. M. Jones, J. M. J. Williams, *Chem. Commun.* 1998, 2519–2520; b) L. Haughton, J. M. J. Williams, *Synthesis*, 2001, 943–946.

34 S. Gastaldi, S. Escoubet, N. Vanthuyne, G. Gil, M. P. Bertrand, *Org. Lett.* 2007, **9**, 837–839.

35 J. H. Koh, H. M. Jeong, J. Park, *Tetrahedron Lett.* 1998, **39**, 5545–5548.

36 R. Stürmer, *Angew. Chem. Int. Ed.* 1997, **36**, 1173–1174.

37 J. V. Allen, J. M. J. Williams, *Tetrahedron Lett.* 1996, **37**, 1859–1862.

38 K. L. Choi, J. H. Suh, D. Lee, I. T. Lim, J. Y. Jung, M.-J. Kim, *J. Org. Chem.* 1999, **64**, 8423–8424.

39 S. Akai, K. Tanimoto, Y. Kanao, M. Egi, T. Yamamoto, Y. Kita, *Angew. Chem. Int. Ed.* 2006, **45**, 2592–2595.

40 B. Martín-Matute, M. Edin, K. Bogár, F. B. Kaynak, J.-E. Bäckvall, *J. Am. Chem. Soc.* 2005, **127**, 8817–8825.

41 a) S. E. Clapham, A. Hadzovic, R. H. Morris, *Coord. Chem. Rev.* 2004, **248**, 2201–2237; b) S. Gladiali, E. Alberico, in Transition Metals for Organic Synthesis, 2nd ed; M. Beller, C. Bolm, Eds.; Wiley-VCH: Weinheim, 2004; Vol. 2, pp 145–166; c) J.-E. Bäckvall, *J. Organomet. Chem.* 2002, **652**, 105–111; d) F. F. Huerta, A. B. E. Minidis, J. E. Bäckvall, *Chem. Soc. Rev.* 2001, **30**, 321–331; e) M. Wills, M. Palmer, A. Smith, J. Kenny, T. Walsgrove, *Molecules* 2000, **5**, 4–18; f) M. Palmer, M. Wills, *Tetrahedron: Asymmetry* 1999, **10**, 2045–2061; g) B. Martín-Matute, J. B. Åberg, M. Edin, J. E. Bäckvall, *Chem. Eur. J.* 2007 DOI: 10.1002/chem.200700373.

42 a) O. Pàmies, J.-E. Bäckvall, *Chem. Eur. J.* 2001, **7**, 5052–5058; b) J. S. M. Samec. J.-E. Bäckvall, P. G. Andersson, P. Brandt, *Chem. Soc. Rev.* 2006, **35**, 237–248.

43 P. M. Dinh, J. A. Howarth, A. R. Hudnott, J. M. J. Williams, W. Harris *Tetrahedron Lett.* 1996, **37**, 7623–7626.

44 a) A. L. E. Larsson, B. A. Persson, J.-E. Bäckvall, *Angew. Chem. Int. Ed. Engl.* 1997, **36**, 1211–1212; b) B. A. Persson, A. L. E. Larsson, M. Le Ray, J.-E. Bäckvall, *J. Am. Chem. Soc.* 1999, **121**, 1645–1650.

45 N. Menashe, Y. Shvo, *Organometallics* 1991, **10**, 3885–3891.

46 Hydroxyacid derivatives: a) F. F. Huerta, J.-E. Bäckvall, *Org. Lett.* 2001, **3**, 1209–1212; b) A.-B. Runmo, O. Pàmies, K. Faber, J.-E. Bäckvall, *Tetrahedron Lett.* 2002, **43**, 2983–2986; c) F.-F. Huerta, Y. R. S. Laxmi, J.-E. Bäckvall, *Org. Lett.* 2000, **2**, 1037–1040; d) O. Pàmies, J.-E. Bäckvall, *J. Org. Chem.* 2002, **67**, 1261–1265.

47 Hydroxynitriles: a) O. Pàmies, J.-E. Bäckvall, *Adv. Synth. Catal.* 2001, **343**, 726–731; b) O. Pàmies, J.-E. Bäckvall, *Adv. Synth. Catal.* 2002, **344**, 947–952.

48 Azidoalcohols: O. Pàmies, J.-E. Bäckvall, *J. Org. Chem.* 2001, **66**, 4022–4025.

49 Haloalcohols: O. Pàmies, J.-E. Bäckvall, *J. Org. Chem.* 2002, **67**, 9006–9010.

50 Hydroxyphosphonates: O. Pàmies, J.-E. Bäckvall, *J. Org. Chem.* 2003, **68**, 4815–4818.

51 Diols: a) B. A. Persson, F. F. Huerta, J.-E. Bäckvall, *J. Org. Chem.* 1999, **64**, 5237–5240; b) M. Edin, J. Steinreiber, J.-E. Bäckvall, *Proc. Natl. Acad. Sci. U.S.A.* 2004, **101**, 5761–5766; c) B. Martín-Matute, J.-E. Bäckvall, *J. Org. Chem.* 2004, **69**, 9191–9195; d) A. B. L. Fransson, Y. Xu, K. Leijondahl, J. E. Bäckvall, *J. Org. Chem.* 2006, **71**, 6309–6316.

52 G. K. M. Verzijl, J. G. De Vries, Q. B. Broxterman, PCT. Int. Appl. 2003: WO 0190396 Al 20011129.

53 a) J. E. Choi, Y. H. Kim, S. H. Nam, S. T. Shin, M. J. Kim, J. Park, *Angew. Chem. Int. Ed.* 2002, **41**, 2373–2376; b) J. H. Choi, Y. K. Choi, Y. H. Kim, E. S. Park, E. J. Kim, M.-J. Kim, J. Park, *J. Org. Chem.* 2004, **69**, 1972–1977.

54 B. Martín-Matute, M. Edin, K. Bogár, J.-E. Bäckvall, *Angew. Chem. Int. Ed.* 2004, **43**, 6535–6539.

55 B. Martín-Matute, M. Edin, J.-E. Bäckvall, *Chem. Eur. J.* 2006, **12**, 6053–6061.

56 N. Kim, S.-B. Ko, M. S. Kwon, M.-J. Kim, J. Park, *Org. Lett.* 2005, **7**, 4523–4526.

57 A. Dijksman, J. M. Elzinga, Y.-X. Li, I. W. C. E. Arends, R. A. Sheldon, *Tetrahedron: Asymmetry*, 2002, **13**, 879–884.

58 R. J. Kazlauskas, A. N. E. Weissfloch, A. T. Rappaport, L. A. Cuccia, *J. Org. Chem.* 1991, **56**, 2656–2665.

59 (R)- and (S)-selective are used for typical sec-alcohols where the large group has the higher priority in the sequential rule for determining the configuration (according to the Cahn-Ingold-Prelog system).

60 S-selective DKR: a) M. J. Kim, Y. Chung, Y. Choi, H. Lee, D. Kim, J. Park, *J. Am. Chem. Soc.* 2003, **125**, 11494–11495; b) L. Borén, B. Martín-Matute, Y. Xu, A. Córdova, J.-E. Bäckvall, *Chem. Eur. J.* 2006, **12**, 225–232.

61 A. Berkessel, M. L. Sebastian-Ibarz, T. N. Müller, *Angew. Chem. Int. Ed.* 2006, **45**, 6567–6570.

62 S. Akai, K. Tanimoto, Y. Kita, *Angew. Chem. Int. Ed.* 2004, **43**, 1407–1410.
63 B. A. C. van As, J. van Buijtenen, A. Heise, Q. B. Broxterman, G. K. M. Verzjil, A. R. A. Palmans, E. W. Meijer, *J. Am. Chem. Soc.* 2005, **127**, 9964–9965.
64 I. Hilker, G. Rabani, G. K. M. Verzijl, A. R. A. Palmans, A. Heise, *Angew. Chem. Int. Ed.* 2006, **45**, 2130–2132.
65 M. T. Reetz, K. Schimossek, *Chimia*, 1996, **50**, 668–669.
66 J. Paetzold, J.-E. Bäckvall, *J. Am. Chem. Soc.* 2005, **127**, 17620–17621.
67 A. Parvulescu, P. Jacobs, D. De Vos, *Chem. Eur. J.* 2007, **13**, 2034–2043.
68 M. J. Kim, W.-H. Kim, K. Han, Y. K. Choi, J. Park, *Org. Lett.* 2007, **9**, 1157–1159.
69 M. Stirling, J. Blacker, M. I. Page, *Tetrahedron Lett.* 2007, **48**, 1247–1250.
70 J. W. J. F. Thuring, A. J. H. Klunder, G. H. L. Nefkens, M. A. Wegman, B. Zwanenburg, *Tetrahedron Lett.* 1996, **37**, 4759–4760.
71 M. Sharfuddin, A. Narumi, Y. Iwai, K. Miyazawa, S. Yamada, T. Kakuchi, H. Kaga, *Tetrahedron: Asymmetry*, 2003, **14**, 1581–1885.
72 J. B. Crawford, R. T. Skerlj, G. J. Bidger, *J. Org. Chem.* 2007, **72**, 669–671.
73 E. W. Hafner, D. Wellner, *Proc. Nat. Acd. Sci.* 1971, **68**, 987–981.
74 J. W. Huh, K. Yokoigawa, N. Esaki, K. Soda, *J. Ferment. Bioeng.* 1992, **74**, 189–190.
75 T. M. Beard, N. J. Turner, *Chem. Commun.* 2002, 246–247.
76 F.-R. Alexandre, D. P. Pantaleone, P. P. Taylor, I. G. Fotheringham, D. J. Ager, N. J. Turner, *Tetrahedron Lett.* 2002, **43**, 707–710.
77 G. J. Roff, R. C. Lloyd, N. J. Turner, *J. Am. Chem. Soc.* 2004, **126**, 4098–4099.
78 T. Oikawa, S. Mukoyama, K. Soda, *Biotech. Bioengineering* 2001, **73**, 80–82.
79 a) M. Alexeeva, A. Enright, M. J. Dawson, M. Mahmoudian, J. J. Turner, *Angew. Chem. Int. Ed.* 2002, **41**, 3177–3180; b) R. Carr, M. Alexeeva, A. Enright, T. S. C. Eve, M. J. Dawson, N. J. Turner, *Angew. Chem. Int. Ed.* 2003, **42**, 4807–4810.
80 C. J. Dunsmore, R. Carr, T. Fleming, N. J. Turner, *J. Am. Chem. Soc.* 2006, **128**, 2224–2225.
81 G. Fantin, M. Gogagnolo, P P. Giovannini, A. Medici, P. Pedrini, *Tetrahedron: Asymmetry*, 1995, **6**, 3047–3053.
82 K. Nakamura, Y. Inoue, T. Matsuda, A. Ohno, *Tetrahedron. Lett.* 1995, **36**, 6263.
83 For some reviews, see: a) A. J. Carnell, *Adv. Biochem. Eng. Biotechnol.* 1999, **63**, 57–72; b) C. C. Gruber, I. Lavandera, K. Faber, W. Kroutil, *Adv. Synth. Catal.* 2006, **348**, 1789–1805.
84 D.-I. Kato, S. Mitsuda, H. Ohta, *J. Org. Chem.* 2003, **68**, 7234–7242, and references therein.

6
Exploiting Enzyme Chemoselectivity and Regioselectivity
Sergio Riva

6.1
Introduction

Chemoselectivity and *regioselectivity* are enzymatic properties of significant synthetic interest [1]. Their common feature is the fact that the enzymatic preference toward one of the several functional groups present on a substrate molecule is dictated by its accessibility to the protein active site (steric effect) and not necessarily by its chemical reactivity.

Specifically, in non-aqueous enzymology the synthetic exploitation of chemoselectivity and regioselectivity is a well-defined area both in terms of the enzymes used (proteases and lipases) and the substrates (mainly peptides, polyols, sugars, and natural glycosides). The birth date of these investigations is also well defined, taking us back to the late eighties and to the work of Klibanov and coworkers. A seminal report was a paper published at the beginning of 1988, in which the performance of the protease subtilisin on a series of polyfunctionalized compounds was reported [2]. Figure 6.1. shows some of the representative molecules that were acylated by subtilisin suspended in anhydrous DMF (the disaccharide sucrose, **1**, the glycoside salicin, **2**, and the nucleoside adenosine, **3**), the arrows indicating the preferential esterification site. The first two compounds offered clear examples of regioselective acylation. In the case of sucrose (**1**), the esterification of the C-1' fructose hydroxyl (the less chemically reactive of the three primary OHs of the molecule) took place preferentially, whereas salicin (**2**) was acylated only at its glucopyranosidic primary OH, despite the fact that its aglycon was carrying a primary alcohol with similar reactivity. Acylation of adenosine (**3**) was also directed toward the sugar OHs of the molecule (in a regioselective mode), but in this case the biocatalyzed transformation was also chemoselective, as the more reactive amino group of the adenosine moiety was not acylated at all.

These findings, together with other previously collected information on the regioselectivity of hydrolases toward alcohols, monosaccharides, and polyhydroxylated steroids dissolved in organic solvents [3–6], have since been extended and exploited by several research groups as a protective step in a synthetic sequence [7] or as an efficient way to obtain specific compounds with potential applications

Organic Synthesis with Enzymes in Non-Aqueous Media. Edited by Giacomo Carrea and Sergio Riva
Copyright © 2008 WILEY-VCH Verlag GmbH & Co. KGaA, Weinheim
ISBN: 978-3-527-31846-9

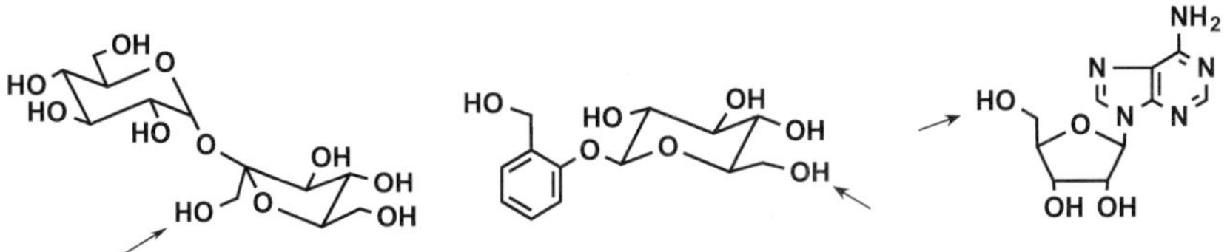

Figure 6.1 Chemo- and/or regioselective acylation of sucrose (1), salicin (2), and adenosine (3) catalyzed by subtilisin.

as pharmaceuticals, biodegradable surfactants, food additives, and monomers suitable for radical polymerization. This chapter briefly discusses the state of the art–updated at the end of 2006–in this interesting area of applied biocatalysis.

6.2
Chemoselectivity of Hydrolases

Chemoselectivity is the preferential reaction of a chemical reagent with one functional group in the presence of other similar functional groups [1].

The above-described acylation of the sugar moiety of the nucleotide adenosine (3) [2] has been followed by a series of papers reporting on the chemoselective enzymatic modification of natural compounds carrying both hydroxyl and amino groups. In addition to the extensive work on nucleosides developed by Gotor and coworkers [8], the biocatalyzed esterification of the hydroxylated alkaloids castanospermine (4) and 1-deoxynojirimicin (5) should be mentioned. Both compounds were selectively acylated at their C-6 and/or C-2 OH by the protease subtilisin, despite the presence of a potentially more reactive amino functions [9].

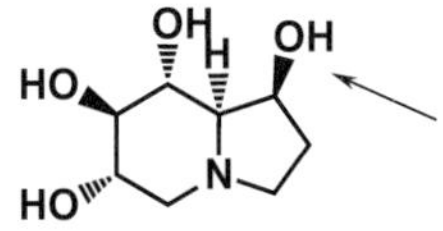

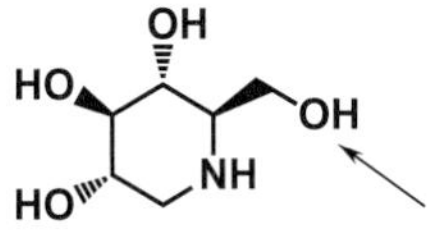

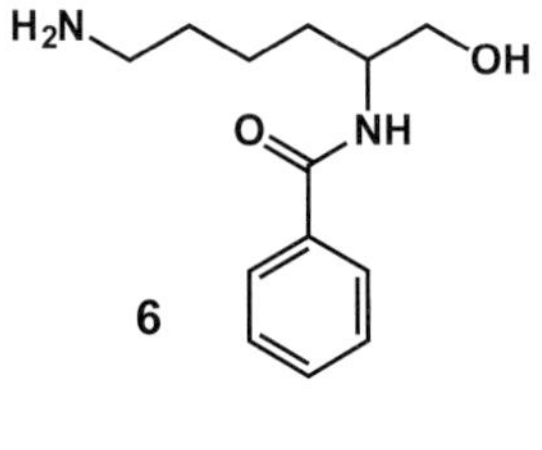

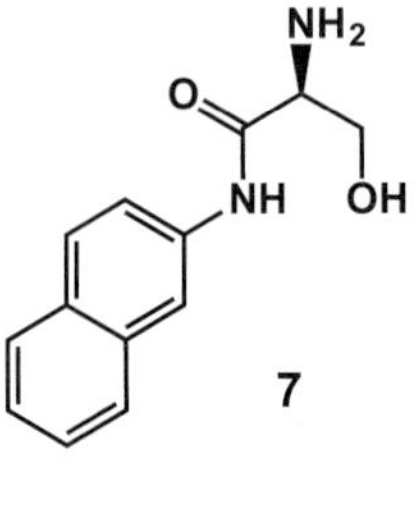

The chemoselective monoacylation of simpler aminoalcohols was also described. In the early days, Klibanov and coworkers reported on the unexpected selectivity of *Aspergillus niger* lipase toward the O-acylation of 6-amino-1-hexanol [10]. The overwhelming preference for the esterification of the hydroxyl group allowed the authors to prepare several O-acylated amino alcohols in good yield without requiring protecting groups. This study was then extended to small peptides [11] and exploited for the chemo-enzymatic construction of a four-component Ugi combinatorial library [12]. In the meantime, it was reported that the ratio of O- to N-acylation was markedly dependent on the solvent [13]. For example, this ratio varied from 1.1 in *tert*-butyl alcohol to 21 in 1,2-dichloroethane in the lipase-catalyzed acylation of N-α-benzoyl-L-lynisol (**6**) with trifluoroethyl butyrate [13a] and from 10.5 to 0.05 when moving from pyridine to *t*-amyl alcohol in the biocatalyzed acylation of L-serine-β-napthylamide (**7**) [13b].

Other authors have described the lipase-catalyzed chemoselective acylation of alcohols in the presence of phenolic moities [14], the protease-catalyzed acylation of the 17-amino moiety of an estradiol derivative [15], the chemoselectivity in the aminolysis reaction of methyl acrylate (amide formation vs the favored Michael addition) catalyzed by *Candida antarctica* B lipase (Novozym 435) [16], and the lipase preference for the O-esterification in the presence of thiol moieties, as, for instance, in 2-mercaptoethanol and dithiotreitol [17]. This last finding was recently exploited for the synthesis of thiol end-functionalized polyesters by enzymatic polymerization of ε-caprolactone initiated by 2-mercaptoethanol (Figure 6.2) [18].

Finally, a recent example on the biocatalyzed synthesis of new cytotoxic derivatives of the anthracycline doxurobicin (**8**) deserves to be mentioned. In the presence of *Mucor javanicus* lipase or subtilisin Carlsberg ion–paired with Aerosol OT, the primary C-14 OH was chemoselectively acylated to give the corresponding derivatives, such as the valerate **8a** [19]. However, optimization of this enzymatic synthesis did not provide a synthetic approach suitable for the preparative synthesis of gram quantities of Valrubicin (*N*-trifluoroacetyl doxurobicin-14-valerate, **8b**) [20]. The problem was solved using a more traditional chemo-enzymatic approach in which the sugar amino group was preliminarily protected by reacting **8** with

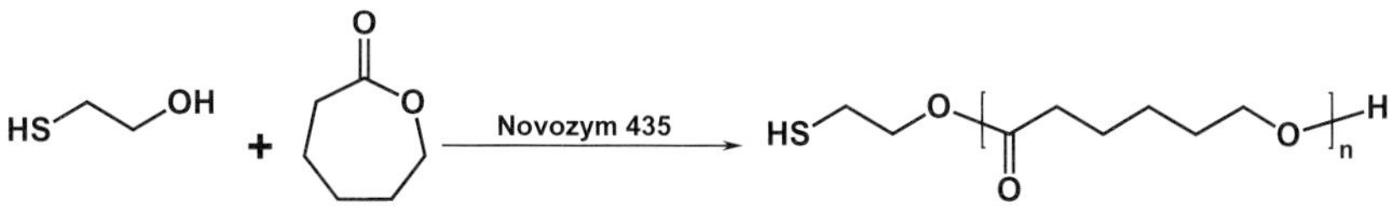

Figure 6.2 Enzymatic ring-opening polymerization of ε-caprolactone initiated by 2-mercaptoethanol.

trifluoroacetic anhydride to give **8c**, followed by *Candida antarctica* lipase-catalyzed acylation to give **8b** [20].

8 : $R_1 = R_2 = H$
8a : $R_1 = CO(CH_2)_3CH_3$; $R_2 = H$
8b : $R_1 = CO(CH_2)_3CH_3$; $R_2 = COCF_3$
8c : $R_1 = H$; $R_2 = COCF_3$

6.3
Regioselectivity of Hydrolases

According to the IUPAC definition "a regioselective reaction is one in which one direction of bond making or breaking occurs preferentially over all the other possible directions" [1]. Since the first report on the regioselective acylation of the primary OHs of simple aliphatic glycols by the action of porcine pancreatic lipase, described by Klibanov in 1985 [3], the ability of hydrolases to perform this kind of transformation in organic solvents has been extensively exploited in the modification of other diols and of polyfunctionalized compounds, such as carbohydrates, steroids, and alkaloids. As several comprehensive reviews have been published on this topic also in recent years [21], this chapter discusses some significant examples only, chosen to exemplify the synthetic versatility of these biotransformations

6.3.1
Regioselective Acylation of Polyols

Regioselective (and in some cases also enantioselective) mono-acylation of diols (acyclic and cyclic) has been investigated by several groups [22], with compounds as simple as 1,4-aliphatic diols [23] or (Z,E)-hexa-2,4-diene-1,6-diol [24]. The synthetic potential of these biocatalyzed transformations can better be appreciated in the modification of more complex natural compounds, as has been exemplified, for instance, by the vitamin-B_6 derivative piridoxime [25], quinic and shikimic acids [26], various sesquiterpenes [27], or the immunosuppressant rapamycin (**9**). The latter compound has been monoacylated at its C-42 OH by two lipases to give a series of aliphatic esters (**9a**) in 95–99 % yields [28]. Interestingly, using the appropriate acylating agent, specifically functionalized C-42 esters have been isolated, like the hemiadipate **9b** and the so-called temsirolimus (CCI-779, **9c**) drug.

9 : R = H
9a : R = CO(CH$_2$)$_n$CH$_3$, n = 0, 1, 8
9b : R = CO(CH$_2$)$_4$COOH
9c : R = COC(CH$_3$)(CH$_2$OH)$_2$

The biocatalyzed regioselective acylation of polyhydroxylated steroids has been studied in detail [29]. A first report described the complementary regioselectivity of two hydrolases, the protease subtilisin and the lipase from *Chromobacterium viscosum*, toward different hydroxylated positions of the steroid skeleton [6]. As shown in Figure 6.3, the model steroid 5α-androstane-3β,17β-diol (**10**) was selectively monoacylated at one of its two OHs by changing enzyme (this "complementary" regioselectivity is quite a common outcome when different hydrolases are screened, see below). Later on, three other lipases were found to be able to catalyze the esterification of hydroxysteroids, specifically the enzymes isolated from *Candida cylindracea* (subsequently identified as *Candida rugosa* lipase) [30], *Candida antarctica* (lipase B, Novozym 435) [31], and *Pseudomonas cepacia* (lipase PS) [32], and their selectivity was evaluated. These reports indicated that no enzymes are able to acylate hydroxyl groups located on the inner B and C steroid rings, whereas esterification of OHs on the A ring (mainly at position C-3, but in some cases also at C-2 and C-4) [30–32] and the D ring (at C-17) [6] as well as on the side chain [6, 31] can be obtained by a proper choice of the hydrolase.

Natural polyols have been used as substrates for the so-called "combinatorial biocatalysis", a proposed approach to drug discovery [33]. For instance, complementary enzymatic regioselectivity was applied to produce a combinatorial library of 167 distinct selectively acylated derivatives of the flavonoid bergenin (**11**) on a robotic workstation [34]. Another lead compound, the antitumoral paclitaxel (**12**, a molecule with very low water solubility) has been similarly derivatized, initially exploiting the selectivity of the protease thermolysin for its side-chain C-2′ OH.

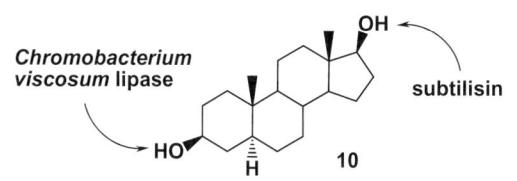

Figure 6.3 Complementary biocatalyzed regioselective acylation of the steroid model 5α-androstan-3β,17β-diol (**10**).

Further enzymatic modifications allowed the isolation of the glucosylated derivatives **12a** (60 times more soluble than **12**) and of the acid derivative **12b**, which was three orders of magnitude more soluble [35].

11

12 : R = H
12a : R = -OC(CH$_2$)$_4$COO-[sugar]
12b : R = CO(CH$_2$)$_4$COOH

The above-described combinatorial biocatalytic approach shows the versatility of hydrolases toward the acylating agents, which are not limited to simple aliphatic acids like acetate and butanoate. As discussed below, there are examples related (a) to the preparation of specific esters of natural glycosides (malonates, cinnamates, coumarates, feroulates, ... [36]) and of sugars esterified with amino acids [2, 37], (b) to the use of acyl moieties that can also be removed by non-hydrolytic reactions (like the levulinates) [38], and (c) to the use of bifunctionalized esters (typically bicarboxylates, like succinic, glutaric, and adipic acid derivatives) that can be further elaborated at the other "free" carboxylic moieties [36, 39]. However, generally speaking, it is not true that it is possible to acylate any substrate with any kind of ester. As has been pointed out in some reports, it is likely that a reciprocal steric hindrance occurs between large acyl groups and large nucleophiles, the latter being excluded by the catalytic site and thus prevented from attacking the acyl-enzyme intermediate.

6.3.2
Regioselective Acylation of Carbohydrates

Selective esterification of sugars still represents a significant challenge for organic chemists because of the multiple hydroxyl groups (with very similar reactivity) of these compounds. Therefore it is not surprising that the first reports of Klibanov 's group on the use of lipases and proteases for the regioselective esterification of primary or secondary hydroxyls of mono-, di- and oligosaccharides, published in the eighties [2, 4, 5], was the start of a rich scientific literature [40]. Specifically, lipases and proteases have been exploited for the following purposes:

1. *To catalyze a protective step in a chemo-enzymatic synthesis of sugar derivatives or of more complex oligosaccharides.*

For instance, subtilisin-catalyzed esterification of lactosides has been used for a chemo-enzymatic approach to 6′-deoxy-6′-fluoro- and 6-deoxy-6-fluoro-lactosides [41]. In other reports, the lipase-catalyzed acylation of benzylidene derivatives of sugars, useful intermediates in the synthesis of oligosaccharides, has been described [42]. For example, the esterification of 4,6-O-benzylidene-α-D-glucopyranoside (**13**) with vinyl acetate by action of *Pseudomonas cepacia* lipase gave quantitatively the 2-O-acetate **13a**. [42b] As a third case, it deserves to be mentioned the extensive work of Russo and coworkers for the chemo-enzymatic synthesis of milk oligosaccharides [43].

Finally, Kren and coworkers have proposed the use of selectively acylated glycosides to be used as acceptors in a bi-enzymatic approach to the synthesis of di- and trisaccharides exploiting lipases/proteases and glycosidases (for instance, 6-O-acetyl-N-acetyl-D-glucosamine, **14**, to give the corresponding chitobiose derivative **15**) [44].

2. *For the large-scale production of new bio-surfactants [45] and, more recently, of sugar-containing self-assembled organogels with nanostructured morphologies [46].*

The latter materials were based on diesters of trehalose (**16a**), a symmetrical disaccharide with an α-1,1 glycosidic bond (**16**), synthesized by *Candida antarctica* lipase B suspended in acetone.

16 : R = H
16a : R = CO(CH$_2$)$_n$CH$_3$
(n = 0 , 2 , 8 , 12 , 16)

3. *For the synthesis of sugar-based polymers.*

Examples of hydrolase-catalyzed preparation of reactive sugar derivatives are the large-scale production of sugar acrylates and vinyl fatty acid esters (obtained by mono-acylation with divinyl esters of dicarboxylic acids), both suitable monomers for the preparation of new polymeric hydrogels [47].

The Novozym 435-catalyzed ring-opening polymerization of ε-caprolactone shown in Figure 6.2 was also performed in the presence of methyl or ethyl glucopyranosides. In this way, polyesters with low polydispersity index (PDI) values and initiated by alkyl glucopyranosides (acylated at their primary OH) could be prepared [48].

"Sweet polyesters" containing alternate units of carbohydrates (the "diol" units) and aliphatic acids (the "diacid" units) were prepared by different groups. In early work it was assumed that, in order to perform an efficient enzyme-catalyzed polymerization, it was necessary for both the activation of the dicarboxylic acids by electron-withdrawing groups (halogenated esters) and the use of polar solvents to dissolve the sugars [49]. More recently it was demonstrated that these polymerization reactions can be performed without activation of the diacid or addition of solvent [50]. In this specific report, the monomers (adipic acid, 1,8-octandiol, and a sugar polyol, i.e. erythritol, xylitol, or D-glucitol) were combined to form a monophasic mixture, and then immobilized CalB lipases was added; the water formed during the polymerization reaction was continuously removed under vacuum.

4. *For the modification of natural polysaccharides.*

Regioselective enzymatic acylation of large, insoluble polysaccharides is still a quite difficult task and therefore it is not surprising that only scant data have been reported up to now, most of them describing reaction outcomes which met with limited success. Nevertheless, enzymatic derivatization of polysaccharides has been performed in nonpolar organic solvents using insoluble polysaccharides with soluble [51] or suspended enzymes [52]. Chemically modified celluloses with either enhanced solubility or more readily accessible hydroxyl groups, like cellulose acetate or hydroxypropyl cellulose, were acylated by CalB, as reported by Sereti and coworkers [53]. However, the same authors failed to modify crystalline cellulose under the same reaction conditions.

More recently, the chemo-enzymatic synthesis of inulin-containing hydrogels was reported [54]. The key point was the solubility of inulin [a mixture of oligomers and polymers containing 2–60 (or more) β-2,1 linked D-fructose molecules having a glucose unit as the initial residue] in dimethylformamide (DMF), a fact that allowed its esterification by action of a protease from *Bacillus subtilis*.

It is commonly agreed that enzymes are inactive in nearly anhydrous dimethyl sulfoxide (DMSO) [55] and that such inactivity might be a direct result of protein solubilization in the organic milieu, which causes deleterious changes in the proteins' secondary and tertiary structures [56]. However, different authors have recently reported that some proteases, namely thermolysin (from *Bacillus thermoproteolyticus*) and Proleather (from *Bacillus subtilis*), were still active in pure DMSO, despite the fact that proteins were indeed solubilized. The former enzyme catalyzed the esterification of sucrose [57] and sucrose-containing oligosaccharides [58], while the latter protease catalyzed the regioselective acryloylation of dextran, a glucose-containing polysaccharide [59].

Additionally, some lipases have also been found to show significant residual activity in reaction media composed of mixtures of DMSO and other cosolvents [60]. This property has been exploited by Hult and coworkers in an elegant approach to the modification of cellulose fiber surfaces by the combined use of a lipase (CalB) and a xyloglucan *endo-trans*-glycosidase. CalB suspended in a mixture of DMSO and 2-methyl-2-butanol catalyzed the preliminary regioselective esterification of a mixture of xyloglucan oligosaccharides, using either vinyl stearate or γ-thiobutyrolactone as the acylating agent. In turn, these modified oligosaccharides were covalently linked to cellulose by the action of the second enzyme [61].

In closing this section, it must be pointed out that neither lipases nor proteases have been developed by Nature to work on sugars, their natural substrates being triglycerides and proteins, respectively. Their ability to catalyze the regioselective acylation of carbohydrates was a surprising discovery, almost as unexpected as the discovery of enzymatic activity in pure organic solvents. Using a modern definition, it was a nice example of enzymatic catalytic "promiscuity", specifically of "substrate" promiscuity.

6.3.3
Regioselective Acylation of Natural Glycosides

Glycosides of various classes of natural compounds are widely distributed in nature, and they are often acylated with aromatic and aliphatic acids (mainly ferulic, *p*-coumaric, malonic, and acetic) at some of their specific sugar(s) hydroxyl groups [62]. Many of these molecules are bioactive compounds or possess other interesting properties. In nature, the esterification process is the last step in the biosynthetic pathway and is catalyzed by different acyltransferases. These enzymes show relative flexibility toward the acyl moiety, but strict selectivity for the substrate to be esterified [63]. Moreover, they are not very convenient for *in vitro* laboratory synthesis as they require stoichiometric amounts of the corresponding acyl-coenzyme A. On the other hand, direct selective chemical acylation of glycosides is still a distant target because of the present lack of suitable reagents and protocols of general applicability.

As previously discussed, Kibanov and coworkers showed that subtilisin was able to regioselectively acylate natural glycosides [2]. Four model substrates were considered [salicin (**2**), riboflavin, adenosine (**3**), and uridine, see Figure 6.1], and it was found that not only was subtilisin active on these compounds, but it also showed absolute selectivity for their sugar moieties, even in the presence of reactive functional groups on the aglycons (so-called "site-selectivity") [64]. The same protocol has been applied to several other natural glycosides [21d], and only in a very few cases did esterification take place at a reactive position of the aglycon moieties [65].

6.3.3.1 Terpene Glycosides

Saponins (e.g., asiaticoside, **17**) are widely distributed in plants as well as in some lower marine animals and possess a broad range of biological activities [62a]. Examples of acylation of these complex molecules have been reported by different authors. For instance, the regioselective esterification of **17** by CalB was quite remarkable, and only the 4'''-O-acetate (**17a**) was isolated in 59% yields [66].

17: R = H
17a: R = Ac

Ginsenosides are dammarane-type triterpene saponins isolated from *Panax ginseng* C. A. Meyer, a plant widely used in traditional Chinese medicine. Some of these ginsenosides exist as monoesters of malonic acid, linked at the primary OHs of their sugar moieties. Danieli and coworkers, using vinyl acetate as acyl donor, found that CalB suspended in *tert*-amyl alcohol was able to catalyze the regioselective esterification of ginsenoside Rg_1 (**18**) to give the monoester **18a** with high yields and selectivity [67]. Subsequently, a chemo-enzymatic approach allowed the isolation of the corresponding malonate **18b**. Excellent results were obtained with the more complex ginsenoside Rb_1, containing four glucose units, and with its cognate derivatives ginsenoside Rg_3 [68]. The acylation of other ginsenosides derivatives has been reported more recently [69].

18: R = H
18a: R = Ac
18b: R = $COCH_2COOH$

The regioselective esterification of other groups of saponins, e.g., a series of diosgenins [70], the haemolytic compound digitonin [71], and the so-called cardiac glycosides [72], have also been investigated by different groups.

Acylation of the sweet terpene tri-glucoside stevioside (**19**) and of its derivative steviolbioside (**20**) catalyzed by lipase PS was efficient in terms of isolated yields of the respective monoesters (almost quantitative) and quite surprising in terms of regioselectivity. In fact, while **19** gave the 6″-O-acetyl derivative **19a**, steviolbioside underwent acetylation at the primary hydroxyl group of the inner glucose moiety to give **20a** [73].

19 : R = H
19a : R = Ac

20 : R = H
20a : R = Ac

6.3.3.2 Flavonoid Glycosides

Flavonoid glycosides and their esters are an important group of natural compounds isolated from plants. They are widely used in pharmaceutical, cosmetic, and food preparations, and it is therefore not surprising that the search for new derivatives with improved antioxidant and antimicrobial activity, or – more simply – with more suitable physico-chemical properties for specific applications (i.e., increased stability and/or solubility), has identified enzyme-catalyzed regioselective esterification as a particularly promising approach to the target molecules [21a].

Riva, Danieli, and coworkers were pioneers in the investigation of the regioselectivity of flavonoid ester synthesis [74]. Exploiting the catalytic performances of the protease subtilisin [74a–c] and of the lipase CalB [74d], they reported the regioselective esterification of flavonoid mono- or di-glycosides (i.e., isoquercitrin **20**, quercitin **21**, rutin **22**, and naringin **23**) with simple aliphatic acids to give the esters reported in Figure 6.4. Later, a chemo-enzymatic approach to the preparation of malonic monoesters, i.e. **23c**, was described [74e]. The two-step protocol was based on the regioselective enzymatic introduction (catalyzed by CalB) of a benzylmalonyl group to naringin to give the mixed ester **23b** followed by Pd/C hydrogenolysis of the benzyl moiety to give **23c**. More recently, different groups showed that it is also possible to use the versatile CalB to introduce phenyl propenoic esters, like cinnamate, coumarate, and feruloate, to flavonoid glycosides [75].

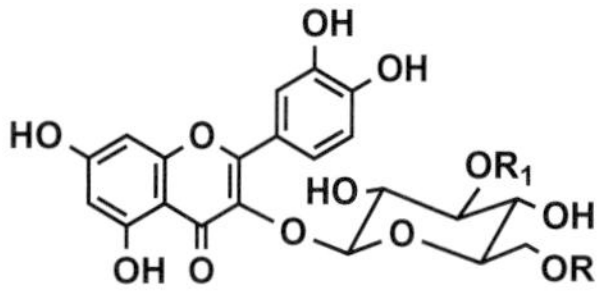

Figure 6.4 Esters of flavonoid mono- or di-glycosides (isoquercitrin **20**, quercitin **21**, rutin **22**, naringin **23**) produced by enzyme-catalyzed regioselective acylation.

20 : R = R$_1$ = H
20a : R = COCH$_2$CH$_2$CH$_3$; R$_1$ = H
20b : R = H ; R$_1$ = COCH$_2$CH$_2$CH$_3$
20c : R = R$_1$ = COCH$_2$CH$_2$CH$_3$

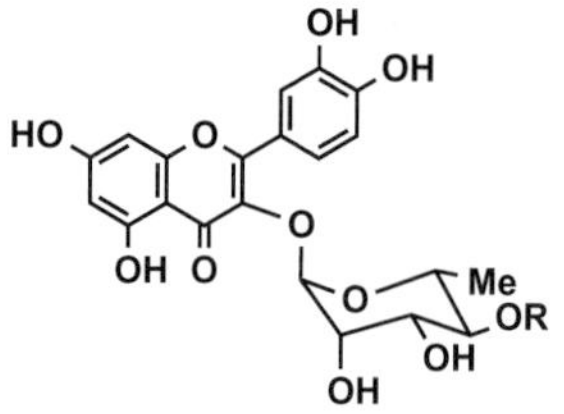

21 : R = H
21a : R = COCH$_3$

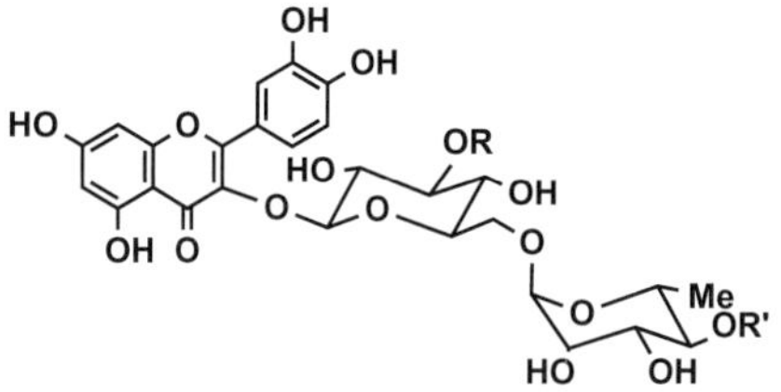

22 : R = R' = H
22a : R = COCH$_2$CH$_2$CH$_3$; R' = H
22b : R = R' = COCH$_3$

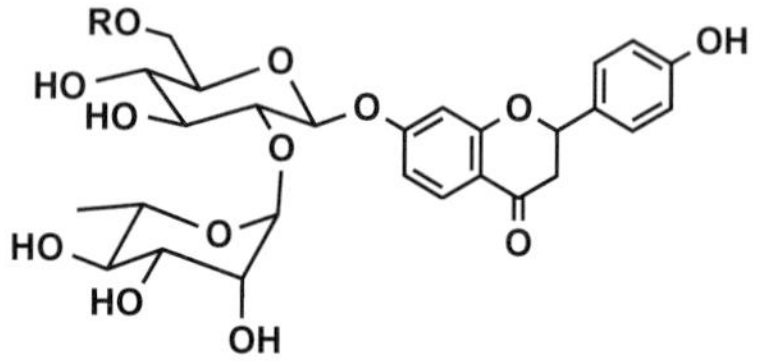

23 : R = H
23a : R = COCH$_2$CH$_2$CH$_3$
23b : R = COCH$_2$COOCH$_2$C$_6$H$_5$
23b : R = COCH$_2$COOH

Because of industrial interest in these derivatives, several papers have appeared in the scientific literature, mainly focused on the optimization of the reaction conditions for the esterification of rutin (**22**) and naringin (**23**) [76]. When the best operating conditions were established, the conversion yield of the substrate could exceed 95%. Several pilot plants are at an advanced stage, and it seems that these acylated flavonoids will soon be marketed [21a].

6.3.3.3 Nucleosides

Following the first examples of the acylation of adenosine and uridine [2], the regioselective lipase-catalyzed esterification of these important molecules has been extensively investigated by Gotor's group [21g]. Attention has been mainly focused on deoxynucleosides – compounds of special interest in medicinal chemistry because of their antiviral and antitumoral properties. Some significant examples are discussed below.

As shown in Figure 6.5, two lipases – CalB and PS – showed a remarkable complementary selectivity toward the hydroxyl groups of different 2'-deoxynucleosides. Specifically, CalB acylated the expected C-5' OH, whereas lipase PS directed its action toward the secondary C-3' OH. In this way, several monoesters were prepared in high yields [77]. Moreover, the same selectivity was

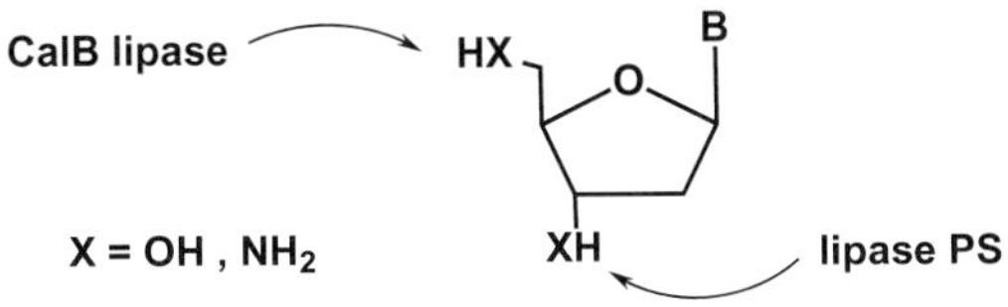

Figure 6.5 Complementary lipases' regioselectivity toward the hydroxyl or amino groups of different 2′-deoxynucleosides. (B, base).

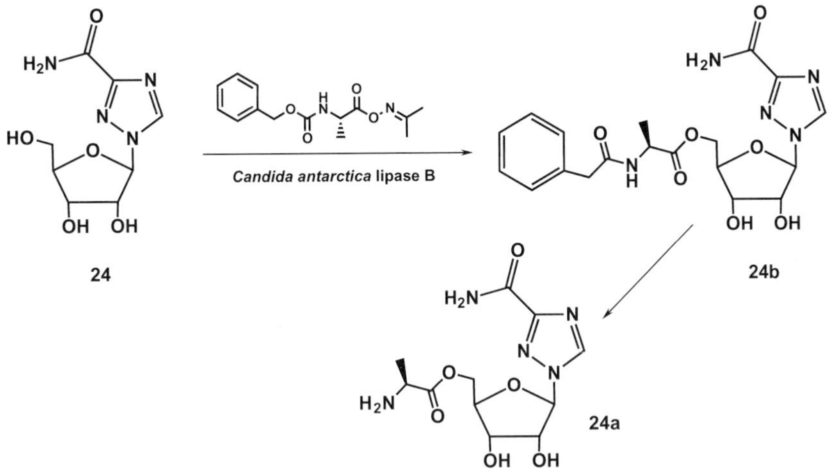

Figure 6.6 Pilot-scale lipase-catalyzed regioselective acylation of ribavirin (**24**).

observed with the corresponding 3′,5′-diamino-2′-desoxynucleosides (Figure 6.5) [78]. These protocols proved to be easily scalable, as was demonstrated by the chemo-enzymatic synthesis of 3′-O-dimethoxytrityl-thymidine on a 25 g scale [79]. Other interesting synthetic exploitations of this unexpected lipase PS selectivity were the chemo-enzymatic synthesis of a polyamine library partly built on an adenosine core [80], the parallel kinetic resolution of racemic mixtures of D/L-nucleosides [81], and the separation of anomeric mixtures of α/β-D-nucleosides [82]. The successful application of the latter protocol was demonstrated (*inter alia*) by a convenient separation of an α/β-mixture of thymidine derivatives from an industrial waste stream [82].

Finally, as a specific example, the regioselective esterification of ribavirin (**24**) deserves to be discussed (Figure 6.6) [83]. This compound is a powerful antiviral agent used to treat hepatitis C. In order to overcome some significant side effects, it has been suggested that its administration in the form of a prodrug might improve its pharmacokinetic profile. Indeed a series of preclinical evaluations showed that the alanine ester **24a** possesses improved bioavailability. In order to satisfy the prodrug requirements to enable it to be used in toxicological studies, researchers at Schering-Plough synthesized the intermediate **24b** in a pilot plant, achieving an overall production of 82 Kg and isolated yields of 82% [83b].

6.3.3.4 Other Glycosides

Despite the simple structure of a glucopyranoside linked to a colchicine or thiocolchicine moiety, subtilisin-catalyzed acylation of colchicoside and thiocolchicoside still represents the only example of regioselective esterification of alkaloid glycosides [84].

Sophorolipids are a group of interesting extracellular glycolipids produced by resting cells of *Candida bombicola*. They consist of a disaccharide core (a sophorose unit) glycosylated with an (ω-1)-hydroxy-fatty acid (i.e., **25**) and are produced as a mixture of variously acylated derivatives. Well-defined sophorolipid analogs have been prepared by the action of CalB and lipase PS, which once again gave the complementary acetylated monoesters **25a** and **25b**, respectively [85].

Finally, the elegant work described by John and coworkers deserves to be mentioned [86]. They reported the efficient CalB-catalyzed acylation of amygdalin (**26**), a byproduct of the fruit juice industry, to give the fatty acid derivatives with the structure **26a**. These compounds proved to have excellent gelation ability, and the authors proposed to exploit this property for the encapsulation of hydrophobic drugs that, in turn, could be released by *in vivo* enzymatic hydrolysis of the gelating agent (an enzyme-triggered drug delivery model).

6.3.4
Regioselective Alcoholyses

All the examples discussed so far have been related to the esterification of polyhydroxylated or polyaminated compounds. However, regioselectivity of hydrolases can also be exploited in the selective hydrolysis, alcoholysis, or aminolysis of

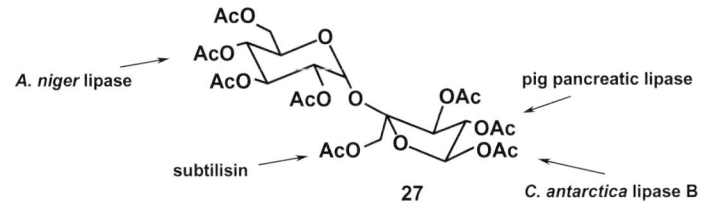

Figure 6.7 Regioselective deacylation of sucrose octaacetate (**27**) catalyzed by different hydrolases.

Figure 6.8 Regioselective deacylation of the polyacetilated flavonoid quercetin (**28a**) by action of different lipases (CalB, lipase B from *Candida antarctica*; Mml, lipase from *Mucor miehei*; t-BME, *tert*-butyl methyl ether)

polyacylated compounds. Hydrolyses performed in water are beyond the scope of this book, but there are examples in which these reactions have been performed in organic solvents at controlled water activity (in this case water is still the nucleophile).

As shown in Figure 6.7, different hydrolases have allowed the regioselective deacylation of sucrose octaacetate (**27**) at different positions [87].

Regioselective alcoholysis of polyacylated steroids catalyzed by *Candida rugosa* lipase suspended in acetonitrile or di-isopropyl ether in the presence of octanol has also been reported [88]. Similarly, the alcoholysis of peracetylated chalcones, acetophenones, and benzopyranones has been reported [89].

Nicolosi and coworkers have intensively investigated the exploitation of lipases for the selective deprotection of bioactive compounds [90]. For instance, as shown in Figure 6.8, the alternative use of the lipases from *Candida antarctica* (CalB) and *Mucor miehei* (Mml) enabled the preparation of different derivatives of the flavonoid quercetin (**28**) [91]. Similar results, this time exploiting the lipase from *Pseudomonas cepacia*, were obtained with the polyacetylated catechin **29** [92].

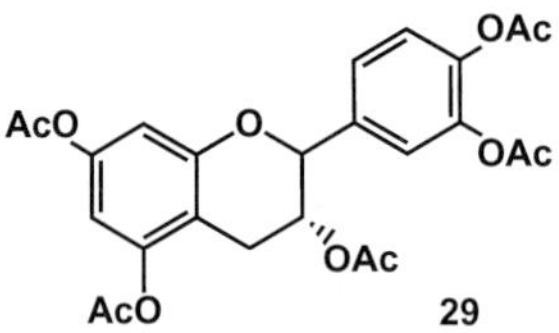

The methodology has also been applied to the preparation of chiral conduritols (5-cyclohexen-1,2,3,4-tetrols, i.e. (−)–**30**), useful starting materials for the preparation of inositols, pseudosugars [93], and selectively protected derivatives of the phytoalexin resveratrol (**31**) [94]. The latter compound can be isolated from *Vitis* species and other plants, where it is elicited in response to fungal infections. Because of its important biological properties (for example, it has been reported to be highly antioxidant, a coronary vasodilator, an inhibitor of platelet aggregation, and one of the cardioprotective phenolic derivatives from red wine), it is the object of extensive investigation. As shown in Figure 6.9, regioselective derivatization of **31** at positions 3, 5, and 4′ has been achieved by a chemo-enzymatic procedure based on standard chemical acylation and/or alcoholysis/esterification catalyzed by lipase PS and CalB [95]. Some of these compounds as well as other ester derivatives have been tested as cancer cell growth inhibitors.

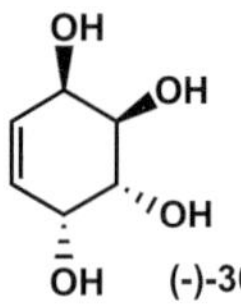

Finally, as an example of regioselective aminolysis, a recent report by Conde and Lopez-Serrano deserves to be cited. They used CalB for the regioselective monoamidation of *N*-protected aspartic acid diethyl esters with butylamine [96].

6.3.5
Rationales

In contrast to enzyme enantioselectivity, the rationalization of the regioselective outcome of reactions catalyzed by hydrolases is still a difficult task. Thus, whereas

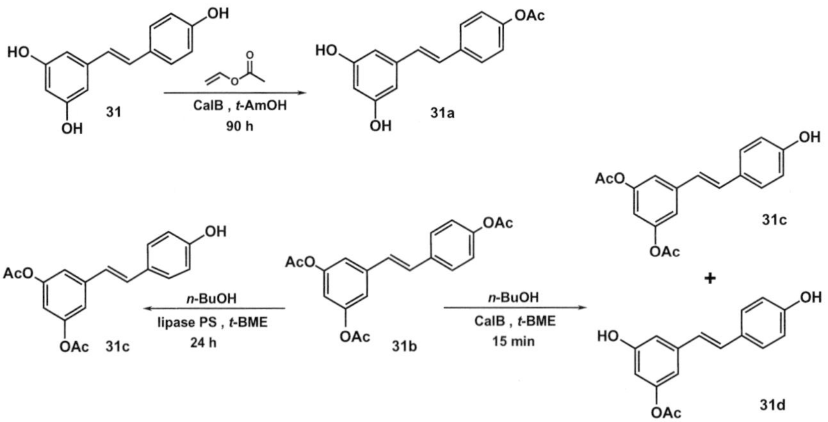

Figure 6.9 Regioselective derivatization of the phytoalexin resveratrol.

enantioselectivity can be usually explained in terms of bulkiness of the substituents at the chiral center and their interaction with the amino acids of the active site (interaction which allows the preferential accommodation of one of the two enantiomers in a productive way), it is likely that regioselectivity is mainly determined by electrostatic interactions of the different hydroxyl (or amino) groups of the nucleophile (which is "bulky" in any case) with the amino acids of the enzyme. Therefore, a correct analysis of these interactions (particularly by molecular modeling) requires a detailed knowledge of the enzyme structure and, more specifically, of its active site (better if solvated with organic solvent molecules).

In the absence of this information, early attempts at rationalization of the experimental results were based on a detailed investigation of enzymatic substrate specificity. For instance, acylation of enantiomeric methyl glycopyranosides by different lipases was focused on the characterization of the reaction outcomes (percentage of the formed regioisomers after the complete disappearance of starting materials or after a fixed reaction time), and the results obtained were interpreted on the basis of the relative orientation of hydroxyls at C-2, C-3, and C-4 [97].

In another paper, the subtilisin-catalyzed esterification of several enantiomeric benzyl and naphthyl glycopyranosides was described. The D-sugar derivatives were all good substrates, and subtilisin regioselectivity was similar for all the compounds tested. On the other hand, most of the L-glycopyranosides were transformed during longer reaction times with a lower regioselectivity. The kinetic constants of the reactions were determined by measuring the initial rates with different concentrations of sugars and ester, and a possible rationale based on the different sugar structures was suggested [98].

Similarly, a possible rationale was suggested to explain the remote control of lipase PS site- and regioselectivity observed in the acetylation of stevioside (**19**) and steviolbioside (**20**) [73]. The calculations, based on the conformational behavior of the substrates in different simulated solvents, correlated well with the experimental results and suggested that the substrate conformation might have been the main factor influencing and determining the recognition by this lipase: the less sterically hindered hydroxyl group entered the active site of the catalyst more favorably, thus directing the acetylation alternatively to the 6'-OH in **20** and the 6"-OH in **19**.

More recently, a similar approach was proposed to explain the regioselectivity displayed by *Candida antartcica* lipase A in the esterification of quinic and shikimic acid derivatives [99].

Other authors have attempted to rationalize the enzyme selectivity toward sugar derivatives by simulating the interaction of the substrates with the enzyme by molecular modeling [100], although this is not an easy task [101]. The first published example related to the regioselective acylation of sucrose [102]. However, the two observed acylation sites of sucrose were not on the same monosaccharide unit, and therefore a possible explanation of subtilisin selectivity could reside in the different steric hindrances.

The results obtained in the esterification of sucrose with two different lipases were analyzed by Ballesteros and coworkers [103]. Initially they examined computationally the conformational space available to the reacting species in the active site and evaluated the importance of the enthalpic and entropic effects [103a]. A more recent study, based on rigorous conformational sampling simulating three different dielectric constants of the "virtual" organic solvent, allowed the elucidation of the complexity of the interactions that govern the binding "landscape" and the subsequent catalytic steps [103b].

In two elegant papers [104], Kaslauskas, Gotor, and coworkers have shed light on the unusual selectivity observed in the acylation of 2-deoxynucleosides [77–78]. Computer modeling of phosphonate transition state analogs suggested that CalB favored acylation of the 5'-OH because this orientation allowed the base ring to bind in a hydrophobic pocket and the sugar residue to form a stronger key hydrogen bond [104a]. A similar approach suggested a rationale for the preferential acylation of 3'-OH by the action of lipase PS [104b].

A different methodological approach was suggested by Colombo and coworkers [105], specifically to find an explanation for the previously described selectivity of subtilisin toward a series of enantiomeric sugars [98]. Using molecular dynamics and QM/MM calculations, the molecular recognition of the substrates by subtilisin active site was investigated, and the differences in energy of the tetrahedral intermediates (TI) mimicking the acylation transition states were evaluated as well. QM/MM analysis of the energy-structure correlations of the TIs showed a clustering toward lower energy of the conformations corresponding to the major experimental products for each enantiomer.

6.4
Closing remarks

This chapter describes, to the best of author's knowledge, the "state of the art" in the enzymatic chemo- and regioselective acylaton of polyfunctionalized compounds at the end of 2006. Twenty years after the first reports by Klibanov's group, these peculiar biocatalyzed transformations have shown an unexpected efficiency and versatility.

It is possible to foresee interesting developments in the area of new materials (glycoside esters with gelating properties, new polymers, and new derivatives of natural polysaccharides), in new synthetic applications (particularly for the preparation of new derivatives of bioactive compounds), and in the development of more suitable rationales by molecular modeling.

6.5 References

1 a) P. Muller, *Pure Appl. Chem.* 1994, **66**, 1077–1184; b) IUPAC Compendium of Chemical Terminology 2nd Edition, 1997.

2 S. Riva, J. Chopineau, A. P. G. Kieboom, A. M. Klibanov, *J. Am. Chem. Soc.* 1988, **110**, 584–589.

3 P. Cesti, A. Zaks, A. M. Klibanov, *Appl. Biochem. Biotechnol.* 1985, **11**, 401–407.
4 M. Therisod, A. M. Klibanov, *J. Am. Chem. Soc.* 1986, **108**, 5638–5640.
5 M. Therisod, A. M. Klibanov, *J. Am. Chem. Soc.* 1986, **109**, 3977–3981.
6 S. Riva, A. M. Klibanov, *J. Am. Chem. Soc.* 1988, **110**, 3291–3295.
7 See, for instance, a) H. Waldmann, D. Sebastian, *Chem. Rev.* 1994, **94**, 911–937. b) M. Ferrero, V. Gotor, *Chem. Rev.* 2000, **100**, 4319–4348.
8 V. Gotor, *J. Mol. Cat. B–Enzymatic* 2002, **19–20**, 21–30.
9 a) A. L. Margolin, D. L. Delinck, M. R. Whalon, *J. Am. Chem. Soc.* 1990, **112**, 2849–2854. b) D. L. Delink, A. L. Margolin, *Tetrahedron* 1990, **31**, 3093–3096.
10 N. Chinsky, A. L. Margolin, A. M. Klibanov, *J. Am. Chem. Soc.* 1989, **111**, 386–388.
11 L. Gardossi, D. Bianchi, A. M. Klibanov, *J. Am. Chem. Soc.* 1991, **113**, 6328–6329.
12 X. C. Liu, D. S. Clark, J. S. Dordick, *Biotechnol. Bioeng.* 2000, **69**, 457–460.
13 a) S. Tawaki, A. M. Klibanov, *Biocatalysis* 1993, **8**, 3–19. b) C. Ebert, L. Gardossi, P. Linda, R. Vesnaver, M. Bosco, *Tetrahedron*, 1996, **52**, 4867–4876.
14 R. Kumar, A. Azim, V. Kumar, S. K. Sharma, A. K. Prasad, O. W. Howardt, C. E. Olsen, S. C. Jain, V. S. Parmar, *Bioorg. Med. Chem.* 2001, **9**, 2643–2652.
15 A. X. Yan, R. Y. Chan, W. S. Lau, K. S. Lee, M. S. Wong, G. W. Xing, G. L. Tian, Y. H. Ye, *Tetrahedron*, 2005, **61**, 5933–5941.
16 O. Torre, V. Gotor-Fernandez, I. Alfonso, L. F. Garcia-Alles, V. Gotor, *Adv. Synth. Catal.* 2005, **347**, 1007–1014.
17 a) A. Baldessari, L. E. Iglesias, E. G. Gros, *J. Chem. Research (S)* 1992, 204–205. b) L. E. Iglesias, A. Baldessari, E. G. Gros, *Biotechnol. Lett.* 1998, **20**, 275–277.
18 C. Hedfords, E. Ostmark, E. Malmstrom, K. Hult, M. Martinelle, *Macromolecules* 2005, **38**, 647–649.

19 a) D. H. Altreuter, J. S. Dordick, D. S. Clark, *J. Am. Chem. Soc.* 2002, **124**, 1871–1876. b) D. H. Altreuter, J. S. Dordick, D. S. Clark, *Enz. Microb. Technol.* 2002, **31**, 10–19.
20 I. A. Cotterill, J. O. Rich, *Org. Process. Res. Dev.* 2005, **9**, 818–821.
21 See, for instance, a) L. Chebil, C. Humeau, A. Falcimaigne, J.-M. Engasser, M. Ghoul, *Process Biochem.* 2006, **41**, 2237–2251. b) D. Lambusta, G. Nicolosi, A. Patti, C. Sanfilippo, *J. Mol. Catal. B–Enzymatic* 2003, **22**, 271–277. c) F. J. Plou, M. A. Cruces, M. Ferrer, G. Fuentes, E. Pastor, M. Bernabè, M. Christensen, F. Comelles, J. L. Parra, A. Ballesteros, *J. Biotechnol.* 2002, **96**, 55–66. d) S. Riva, *J. Mol. Catal. B–Enzymatic* 2002, **19–20**, 43–54. e) B. La Ferla, *Monatsh. Chem.* 2002, **133**, 351–368. f) G. Carrea, S. Riva, *Angew. Chem. Int. Ed.* 2000, **39**, 2226–2254. g) M. Ferrero, V. Gotor, *Chem. Rev.* 2000, **100**, 4319–4348. h) S. Riva in *Enzymatic reactions in organic media* (Eds. A. M. P. Koskinen, A. M. Klibanov), Blackie Academic & Professional, Glasgow, 1996, pp. 140–169.
22 For a review see. F. Theil, *Catal. Today* 1994, **22**, 517–536.
23 P. Ciuffreda, S. Casati, E, Santaniello, *Tetrahedron Lett.* 2003, **44**, 3663–3665
24 C. Pichon, M. E. Martin-Gourdel, D. Chauvat, C. Alexandre, F. Huet, *J. Mol. Catal. B–Enzymatic* 2004, **27**, 65–68.
25 A. Baldessari, C. M. Mangone, E. G. Gros, *Helv. Chim. Acta* 1998, **81**, 2407–2413.
26 a) B. Danieli, P. De Bellis, L. Barzaghi, G. Carrea, G. Ottolina, S. Riva, *Helv. Chim. Acta* 1992, **75**, 1297–1304. b) N. Armesto, M. Ferrero, S. Fernandez, V. Gotor, *J. Org. Chem.* 2002, **67**, 4978–4981. c) N. Armesto, M. Ferrero, S. Fernandez, V. Gotor, *J. Org. Chem.* 2003, **68**, 5784–5787.
27 A. Intra, A. Bava, G. Nasini, S. Riva, *J. Mol. Catal. B–Enzymatic* 2004, **29**, 95–98.
28 J. Gu, M. E. Ruppen, P. Cai, *Org. Lett.* 2005, **7**, 3945–3948.
29 S. Riva, in *Biocatalysis in the pharmaceutical and biotechnology industries* (Ed. R. N. Patel), CRC Press, Boca Raton, FL (USA), 2007, pp. 591–604.

30 S. Riva, R. Bovara, G. Ottolina, G. Carrea, *J. Org. Chem.* 1989, **54**, 3161–3164.

31 a) A. Bertinotti, G. Carrea, G. Ottolina, S. Riva, *Tetrahedron* 1994, **50**, 13165–13172. b) B. Danieli, G. Lesma, M. Luisetti, S. Riva, *Tetrahedron* 1997, **53**, 5855–5862.

32 M. M. Cruz Silva, S. Riva, M. L. Sa e Melo, *Tetrahedron* 2005, **61**, 3065–3073.

33 P. C. Michels, Y. L. Khmelnitsky, J. S. Dordick, D. S. Clark, *Trends Biotechnol.* 1998, **16**, 210–215.

34 V. M. Mozhaev, C. L. Budde, J. O. Rich, A. Y. Usyatinsky, P. C. Michels, Y. L. Khmelnitsky, J. S. Dordick, D. S. Clark, *Tetrahedron* 1998, **54**, 3971–3982.

35 Y. L. Khmelnitsky, C. L. Budde, J. M. Arnold, A. Y. Usyatinsky, J. S. Dordick, D. S. Clark, *J. Am. Chem. Soc.* 1997, **119**, 11554–11555.

36 a) C. Gao, O. Mayon, D. A. MacManus, E. N. Vulfson, *Biotechnol. Bioeng.*, 2001, **71**, 235–243. b) S. Riva, B. Danieli, M. Luisetti, *J. Nat. Prod.* 1996, **59**, 618–621. c) B. Danieli, M. Luisetti, S. Riva, A. Bertinotti, E. Ragg, L. Scaglioni, E. Bombardelli, *J. Org. Chem.* 1995, **60**, 3637–3642.

37 a) T. Maruyama, S.-I. Nagasawa, M. Goto, *J. Biosci. Bioeng.* 2002, **94**, 357–361. b) O.-J. Park, G.-J. Long, J. W. Yang, *Enzyme Microb. Technol.* 1999, **25**, 455–462.

38 a) J. Garcia, S. Fernandez, M. Ferrero, Y. S. Sanghvi, V. Gotor, *Tetrahedron-Asymmetry* 2003, **14**, 3533–3540. b) L. Lay, L. Panza, S. Riva, M. Khitri, S. Tirendi, *Carbohydr. Res.* 1996, **291**, 197–204.

39 Q. Wu, M. Wang, Z. C. Chen, D. S. Lu, X. F. Lin, *Enzyme Microb. Technol.* 2006, **39**, 1258–1263.

40 J. F. Kennedy, H. Kumar, P. S. Panesar, S. S. Marwaha, R. Goyal, A. Parmar, S. Kaur, *J. Chem. Technol. Biotechnol.* 2006, **81**, 866–876. For additional reviews see also [21c, 21e, 21h].

41 S. Cai, S. Hakomori, T. Toyokuni, *J. Org. Chem.* 1992, **57**, 3431–3437.

42 a) I. Matsuo, M. Isomura, R. Walton, K. Ajisaka, *Tetrahedron Lett.* 1996, **37**, 8795–8798. b) L. Panza, M. Luisetti, E. Crociati, S. Riva, *J. Carbohydr. Chem.* 1993, **12**, 125–130. c) L. Panza, S. Brasca, S. Riva, G. Russo, *Tetrahedron-Asymmetry* 1993, **4**, 931–932. d) M. J. Chin, G. Iacazio, D. G. Spackman, N. J. Turner, S. M. Roberts, *J. Chem. Soc., Perkin Trans. I* 1992, 661–662.

43 a) A. Rencurosi, L. Poletti, G. Russo, L. Lay, *Eur. J. Org. Chem.* 2003, 1672–1680. b) A. Rencurosi, L. Poletti, M. Guerrini, G. Russo, L. Lay, *Carbohydr. Res.* 2002, **337**, 473–483. c) B. La Ferla, L. Lay, G. Russo, L. Panza, *Tetrahedron-Asymmetry* 2000, **11**, 3647–3651. d) B. La Ferla, L. Lay, L. Poletti, G. Russo, L. Panza, *J. Carbohydr. Chem.* 2000, **19**, 331–343. e) L. Lay, L. Panza, S. Riva, M. Khitri, S. Tirendi, *Carbohydr. Res.* 1996, **291**, 197–204.

44 a) P. Simerska, A. Pisvejcova, M. Kuzma, P. Sedmera, V. Kren, S. Nicotra, S. Riva, *J. Mol. Catal. B–Enzymatic* 2004, **29**, 219–225. b) P. Simerska, M. Kuzma, A. Pisvejcova, L. Weignerova, M. Mackova, S. Riva, V. Kren, *Folia Microbiol*, 2003, **48**, 329–337. c) L. Husakova, S. Riva, M. Casali, S. Nicotra, M. Kuzma, Z. Hunkova, V. Kren, *Carbohydr. Res.* 2001, **331**, 143–148. d) L. Weignerova, P. Sedmera, Z. Hunkova, P. Halada. V. Kren, M. Casali, S. Riva, *Tetrahedron Lett.* 1999, **52**, 9297–9299.

45 a) K. Adelhorst, F. Bjorkling, S. E. Godtfredsen, O. Kirk, *Synthesis* 1990, 112–115. b) I. Perez-Victoria, J. C. Morales, *Tetrahedron* 2006, **62**, 878–886.

46 G. John, G. Zhu, J. Li, J. D. Dordick, *Angew. Chem. Int. Ed.* 2006, **45**, 4772–4775.

47 a) X. Chen, A. Johnson, J. S. Dordick, D. G. Rethwisch, *Macromol. Chem. Phys.* 1994, **195**, 3567–3578. b) Q. Wu, N. Wang, Y.-M. Xiao, D.-S. Lu, X.-F. Lin, *Carbohydr. Res.* 2004, **339**, 2059–2067.

48 a) A. Cordova, T. Iversen, K. Hult, *Macromolecules* 1998, **31**, 1040–1045. b) K. S. Bisht, F. Deng, R. A. Gross, D. L. Kaplan, G. Swift, *J. Am. Chem. Soc.* 1998, **120**, 1363–1367.

49 a) B. J. Kline, E. J. Beckman, A. J. Russell, *J. Am. Chem. Soc.* 1998, **120**, 9475–9480. b) H. Uyama, K. Inada, S. Kobayashi, *Macromol. Rapid. Commun.*

1999, **20**, 171–174. c) O.-J. Park, D.-Y. Kim, J. S. Dordick, *Biotechnol. Bioeng.* 2000, **70**, 208–216. d) D.-Y. Kim, J. S. Dordick, *Biotechnol. Bioeng.* 2001, **76**, 200–206.

50 J. Hu, W. Gao, A. Kulshrestha, R. A. Gross, *Macromolecules* 2006, **39**, 6789–6792.

51 F. F. Bruno, J. A. Akkara, M. Ayyagari, D. L. Kaplan, R. Gross, G. Swift, J. S. Dordick, *Macromolecules* 1995, **28**, 8881–8883.

52 J. Li, W. Xie, H. N. Cheng, R. G. Nickol, P. G. Wang, *Macromolecules* 1999, **32**, 2789–2792.

53 a) V. Sereti, H. Stamatis, E. Koukios, F. N. Kolisis, *J. Biotechnol.* 1998, **66**, 219–223. b) V. Sereti, H. Stamatis, C. Pappas, M. Polissiou, F. N. Kolisis, *Biotechnol. Bioeng.* 2001, **72**, 495–500.

54 L. Ferreira, R. Carvalho, M. H. Gil, J. S. Dordick, *Biomacromolecules* 2002, **3**, 333–341.

55 A. Zaks, A. M. Klibanov, *J. Biol. Chem.* 1988, **263**, 3194–3201.

56 a) M. Jackson, H. H. Mantsch, *Biochim. Biophys. Acta* 1991, **1078**, 231–235. b) J. T. Chin, S. L. Wheeler, A. M. Klibanov, *Biotechnol. Bioeng.* 1994, **44**, 140–145.

57 N. Rangel Pedersen, P. J. Halling, L. H. Pedersen, R. Wimmer, R. Matthiesen, O. R. Veltman, *FEBS Lett.* 2002, **519**, 181–184.

58 I. Perez-Victoria, J. C. Morales, *Tetrahedron* 2006, **62**, 2361–2369.

59 L. Ferreira, M. H. Gil, J. S. Dordick, *Biomaterials* 2002, **23**, 3957–3967.

60 a) M. Ferrer, M. A. Cruces, M. Bernabè, A. Ballesteros, F. J. Plou, *Biotechnol. Bioeng.* 1999, **65**, 10–16. b) M. Ferrer, M. A. Cruces, F. J. Plou, M. Bernabè, A. Ballesteros, *Tetrahedron* 2000, **56**, 4053–4061.

61 M. T. Gustavsson, P. V. Persson, T. Iversen, M. Martinelle, K. Hult, T. T. Teeri, H. Brumer III, *Biomacromolecules* 2005, **6**, 196–203.

62 a) K. Hostettmann, A. Marston, Eds., *Chemistry and pharmacology of natural products: saponins*, Cambridge University Press, Cambridge (UK), 1995. b) R. Ikan, Ed., *Naturally occurring glycosides*, John Wiley & Sons Ltd., Chichester (UK), 1999.

63 J. Koester, R. Bussmann, W. Barz, *Arch. Biochem. Biophys.* 1984, **234**, 513–521.

64 S. Gebhardt, S. Bihler, M. Schubert-Zsilavecz, S. Riva, D. Monti, L. Falcone, *Helv. Chim. Acta.* 2002, **85**, 1943–1959.

65 a) J. A. Perez, C. Boluda, H. Lopez, J. M. Trujillo, J. M. Hernadez, *Chem. Pharm. Bull.* 2004, **52**, 1123–1124. b) D. Colombo, F. Compostella, F. Ronchetti, A. Scala, L. Toma, H. Tokuda, H. Nishino, *Eur. J. Med. Chem.* 2001, **36**, 691–695, and references therein. c) A. Evidente, T. Fujii, N. S. Iacobellis, S. Riva, A. Sisto, G. Surrico, *Phytochemistry* 1991, **30**, 3505–3510.

66 D. Monti, A. Candido, M. M. Cruz Silva, V. Kren, S. Riva, B. Danieli, *Adv. Synth. Catal.* 2005, **347**, 1168–1174.

67 B. Danieli, M. Luisetti, S. Riva, A. Bertinotti, E. Ragg, L. Scaglioni, E. Bombardelli, *J. Org. Chem.* 1995, **60**, 3637–3642.

68 S. Gebhardt, S. Bihler, M. Schubert-Zsilavecz, S. Riva, D. Monti, L. Falcone, *Helv. Chim. Acta.* 2002, **85**, 1943–1959.

69 a) R. Teng, C. Aug, D. McManus, D. Amstrong, S. Mau, A. Bacic, *Helv. Chim. Acta.* 2004, **87**, 1860–1872.

70 B. Yu, G. Xing, Y. Hui, X. Han, *Tetrahedron Lett.* 2001, **42**, 5513–5516.

71 B. Danieli, M. Luisetti, S. Steuer, A. Michelitsch, W. Likussar, S. Riva, J. Reiner, M. Schubert-Zsilavecz, *J. Nat. Prod.* 1999, **62**, 670-673.

72 S. Riva, D. Monti, M. Luisetti, B. Danieli, *Proc. N. Y. Acad. Sci.* 1998, **864**, 70–80.

73 G. Colombo, S. Riva, B. Danieli, *Tetrahedron* 2004, **60**, 741–746.

74 a) B. Danieli, P. De Bellis, G. Carrea, S. Riva, *Heterocycles* 1989, **29**, 2061–2064. b) B. Danieli, P. De Bellis, G. Carrea, S. Riva, *Helv. Chim. Acta* 1990, **73**, 1837–1844. c) B. Danieli, A. Bertario, G. Carrea, B. Redigolo, F. Secundo, S. Riva, *Helv. Chim. Acta* 1993, **76**, 2981–2991. d) B. Danieli, M. Luisetti, G. Sampognaro, G. Carrea, S. Riva, *J. Mol. Cat. B-Enzymatic* 1997, **3**, 193–201. e) S. Riva, B. Danieli, M. Luisetti, *J. Nat. Prod.* 1996, **59**, 618–621.

75 a) N. Nakajama, K. Ishihara, T. Itoh, T. Furuya, H. Hamada, *J. Biosci. Bioeng.*, 1999, **87**, 105–107. b) C. Gao, P. Mayon, D. A. MacManus, E. N. Vulfson, *Biotechnol. Bioeng.*, 2001, **71**, 235–243. c) A. Kontogianni, V. Shouridou, V. Sereti, H. Stamatis, F. N. Kolisis, *J. Lip. Sci. Technol.*, 2001, **103**, 655–660. d) E. Enaud, C. Humeau, B. Piffaut, M. Girardin, *J. Mol. Catal. B–Enzymatic* 2004, **27**, 1–6.

76 See, for instance. a) Y. Duan, Z. Du, Y. Yao, R. Li, D. Wu, *J. Agric. Food Chem.* 2006, **54**, 6219–6225. b) M. Ardhaoui, A. Falcimaigne, J.-M. Engasser, P. Moussou, G. Pauly, M. Ghoul, *J. Mol. Catal. B–Enzymatic* 2004, **29**, 63–67. c) A. Kontogianni, V. Skouridou, V. Sereti, H. Stamatis, F. N. Kolisis, *J. Mol. Catal. B–Enzymatic* 2003, **21**, 59–62.

77 a) K. Nozaki, A. Uemura, J.-I. Yamashita, M. Yasumoto, *Tetrahedron Lett.* 1990, **31**, 7327–7328. b) L. F. Garcia-Alles, V. Gotor, *Tetrahedron* 1995, **51**, 307–316. c) J. Garcia, S. Fernandez, M. Ferrero, Y. S. Sanghvi, V. Gotor, *J. Org. Chem.* 2002, **67**, 4513–4519. d) X.-F. Sun, N. Wang, Q. Wu, X. F. Lin, *Biotechnol.Lett.* 2004, **26**, 1019–1022. e) J. Garcia, S. Fernandez, M. Ferrero, Y. S. Sanghvi, V. Gotor, *Tetrahedron Lett.* 2004, **45**, 1709–1712.

78 a) I. Lavandera, S. Fernandez, M. Ferrero, V. Gotor, *J. Org. Chem.* 2001, 4079–4082. b) I. Lavandera, S. Fernandez, M. Ferrero, V. Gotor, *J. Org. Chem.* 2004, **69**, 1748–1751.

79 A. Diaz-Rodriguez, S. Fernandez, Y. S. Saghvi, M. Ferrero, V. Gotor, *Org. Process Res. Dev.* 2006, **10**, 581–587.

80 K. Rege, S. Hu, J. A. Moore, J. S. Dordick, S. M. Cramer, *J. Am. Chem. Soc.* 2004, **126**, 12306–12315.

81 J. Garcia, S. Fernandez, M. Ferrero, Y. S. Sanghvi, V. Gotor, *Org. Lett.* 2004, **6**, 3759–3762.

82 J. Garcia, A. Diaz-Rodriguez, S. Fernandez, Y. S. Saghvi, M. Ferrero, V. Gotor, *J. Org. Chem.* 2006, **71**, 9765–9771.

83 a) B. K. Liu, N. Wang, Q. Wu, C. Y. Xie, X. F. Lin, *Biotechnol.Lett.* 2005, **27**, 717–720. b) M. Tamarez, B. Morgan, G. S. K. Wong, W. Tong, F. Bennett, R. Lovey, J. L. McCormick, A. Zaks, *Org. Process Res. Dev.* 2003, **7**, 951–953.

84 B. Danieli, P. De Bellis, G. Carrea, S. Riva, *Gazz. Chim. Ital.* 1991, **121**, 123–125.

85 S. K. Singh, A. P. Felse, A. Nunez, T. A. Foglia, R. A. Gross, *J. Org. Chem.* 2003, **68**, 5466–5473.

86 P. K. Vermula, J. Li, G. John, *J. Am. Chem. Soc.* 2006, **128**, 8932–8938.

87 D. C. Palmer, F. Terradas, *Tetrahedron Lett.* 1994, **35**, 1673–1676.

88 V. C. O. Njar, E. Caspi, *Tetrahedron Lett.* 1987, **28**, 6549–6552.

89 a) V. S. Parmar, A. K. Prasad, N. K. Sharma, *Pure Appl. Chem.* 1992, **64**, 1135–1139. b) V. S. Parmar, A. K. Prasad, N. K. Sharma, *Chem. Commun.* 1993, 27–29.

90 D. Lambusta, G. Nicolosi, A. Patti, C. Sanfilippo, *J. Mol. Catal. B–Enzymatic* 2003, **22**, 271–277.

91 M. Natoli, G. Nicolosi, M. Piattelli, *J. Org. Chem.* 1992, **57**, 5776–5778.

92 D. Lambusta, G. Nicolosi, A. Patti, M. Piattelli, *Synthesis* 1993, **11**, 1155–1158.

93 A. Patti, G. Sanfilippo, M. Piattelli, G. Nicolosi, *Tetrahedron-Asymmetry* 1996, **7**, 2665–2670.

94 G. Nicolosi, C. Spatafora, C. Tringali, *J. Mol. Catal. B–Enzymatic* 2002, **16**, 223–229.

95 V. Cardile, L. Lombardo, C. Spatafora, C. Tringali, *Bioorg. Chem.* 2005, **33**, 22–33.

96 S. Conde, P. Lopez-Serrano, *Eur. J. Org. Chem.* 2002, 922–929.

97 a) D. Colombo, F. Ronchetti, A. Scala, L. Toma, *J. Carbohydr. Chem.* 1992, **11**, 89–94. b) D. Colombo, F. Ronchetti, L. Toma, *Tetrahedron* 1991, **47**, 103–110. c) P. Ciuffreda, D. Colombo, F. Ronchetti, L. Toma, *J. Org. Chem.* 1990, **55**, 4187–4190.

98 a) B. Danieli, F. Peri, G. Roda, G. Carrea, S. Riva, *Tetrahedron* 1999, **55**, 2045–2060. b) B. Danieli, F. Peri, G. Carrea, D. Monti, S. Riva, *Carbohydr. Lett.* 1995, **1**, 363–368.

99 N. Armesto, S. Fernandez, M. Ferrero, V. Gotor, *Tetrahedron* 2006, **62**, 5401–5410.

100 R. J. Kazlauskas, *Science* 2001, **293**, 2277–2279.

101 N. Boissiere-Junot, C. Tellier, C. Rabiller, *J. Carbohydr. Chem.* 1998, **17**, 99–115.

102 a) J. O. Rich, B. A. Bedell, J. S. Dordick, *Biotechnol.Bioeng.* 1995, **45**, 426–434. b) J. O. Rich, J. S. Dordick, *J. Am. Chem. Soc.* 1997, **119**, 3245–3252.

103 a) G. Fuentes, M. A. Cruces, F. J. Plou, A. Ballesteros, C. S. Vendra, *ChemBioChem* 2002, **3**, 907–910. b) G. Fuentes, A. Ballesteros, C. S. Vendra, *Protein Sci.* 2007, **13**, 3092–3103.

104 a) I. Lavandera, S. Fernández, J. Magdalena, M. Ferrero, R. J. Kazlauskas, V. Gotor, *ChemBioChem* 2005, **6**, 1381–1390. b) I. Lavandera, S. Fernandez, J. Magdalena, M. Ferrero, H. Grewal, C. K. Saville, R. J. Kazlauskas, V. Gotor, *ChemBioChem* 2006, **7**, 693–698.

105 S. Pieraccini, M. Sironi, G. Colombo, *Chem. Phys. Lett.* 2006, **418**, 373–376.

7
Industrial-Scale Applications of Enzymes in Non-Aqueous Solvents

David Pollard and Birgit Kosjek

7.1
Introduction

Within the last decade, biocatalysis has become fully integrated into the synthetic toolbox of fine chemical and pharmaceutical companies producing efficient and competitive processes. Well over 100 different biocatalytic processes have been implemented on an industrial scale [1, 2]. The most notable of the large-scale processes include the synthesis of acrylamide (30 000 t/y, nitrile hydratase, [3, 4]), nicotinamide (>3500 t/y, nitrile hydratase [5]), and lactam antibiotics (>10 000 t/y, penicillin amidase, [6]). Many enzymatic processes are carried out on a scale of between 10 kg/y and 100 t/y, and in most cases the unique properties of enzymes for outstanding chemo and regio selectivity have been utilized for the manufacture of drugs and agrochemicals. This is vitally important to the pharmaceutical industry, where 80% of the molecules are chiral [7] and strict regulatory guidelines dictate the need for highly enantiopure compounds (>99.5% *ee*) [8]. In addition, the demand for specifically functionalized chiral intermediates continues to grow as drug molecules become more complex with multiple chiral centers [8, 9]. This demand has led to the exploitation of hydrolases with technology platforms for the resolution of alcohols and amines in organic solvents. Kinetic resolution provides a fast method of synthesizing chiral molecules in order to meet aggressive development timelines of industry. This is further enhanced by the development of dynamic kinetic resolution (DKR) and deracemization methods that have led to a number of efficient industrial scale processes with effective yields. Similarly, the enzymatic synthesis of cyanohydrins from aldehydes and ketones in biphasic reactions has made a significant industrial impact on the supply of chiral cyanohydrin synthons. This chapter outlines the successful industrial processes which have combined enzymes and organic solvents for the proven advantages of high substrate concentration, effective space time yields, thermodynamic equilibrium control in favor of synthesis, and ease of catalyst reuse using immobilized enzymes.

Organic Synthesis with Enzymes in Non-Aqueous Media. Edited by Giacomo Carrea and Sergio Riva
Copyright © 2008 WILEY-VCH Verlag GmbH & Co. KGaA, Weinheim
ISBN: 978-3-527-31846-9

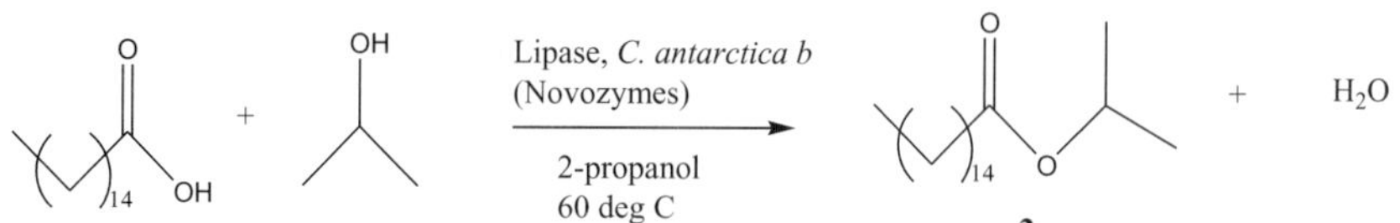

Scheme 7.1 Synthesis of isopropyl palmitate from palmitic acid and isopropanol.

7.2
Ester Synthesis by Esterification

The industrial use of hydrolases was originally dominated by proteases and amylases for the hydrolytic reactions of proteins and carbohydrates for dairy and detergent applications. The ability of hydrolyases to catalyze transesterification, the condensation of an acid with an alcohol or amine to esters and amides, has led to simple processes for triglyceride modification and ester synthesis for the industries of food, fragrance, and personal care products. Unilever and Fuji Oil manufacture several 100 t/y of cocoa butter substitutes obtained by acidolysis of palm oil fractions using the lipase from *Mucor miehei* [10, 11]. The lipase immobilized on polyacrylate resin catalyzes the 1,3 specific transesterification of cheap mid-fractions of palm oil containing a majority triglyceride 1,3-dipalmitoyl-2-monolein with stearic acid in heptane. The simple esters of isopropyl palmitate (2) and isopropyl myristate are synthesized for personal care products including cosmetics, soaps, lubricants, and greases by Uniquema (Scheme 7.1) [12, 13]. The immobilized lipase from *C. antarctica b* catalyzes the transesterification of palmitic acid (1) [3.1 M (800 g/L)] with isopropanol at 60 °C with 99% yield. The generated water is removed by azeotropic distillation, and isopropanol is continuously fed to the reaction to replace the distilled material.

Similarly, the esterification of geraniol and citronellol is carried out in hexane and catalyzed by the immobilized lipase *Mucor miehei*. The geraniol and citronellol esters are extensively used for the fragrance industry, where the use of lipases benefits mild conditions and low impurity side products [14, 15].

7.3
Resolution of Racemic Alcohols

The application of enzymatic acylation for the resolution of racemic alcohols in organic solvent has shown to be an effective method to rapidly synthesize chiral alcohols. The racemic alcohols are treated with the lipase and acylating agent; one enantiomer remains unconverted whereas the second enantiomer is esterified and easily separated by distillation (Scheme 7.2). Vinyl acetate or isopropenyl acetate are typical acylating agents, as the generated vinyl alcohol tautomerizes rapidly

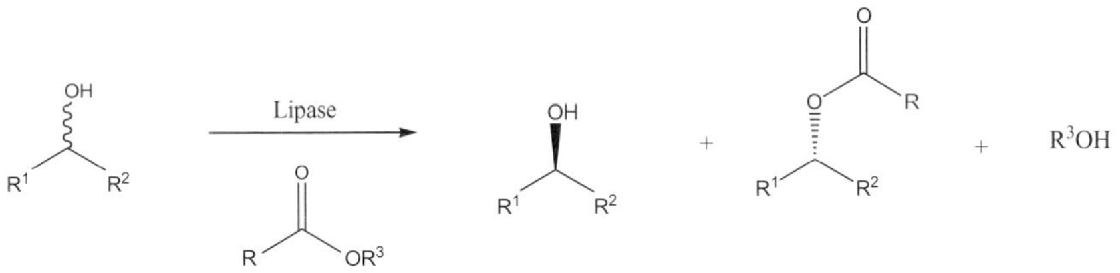

Scheme 7.2 Resolution of racemic alcohols through enzymatic acylation.

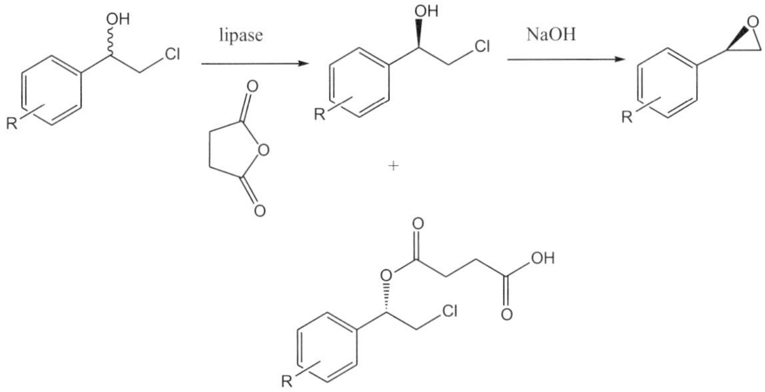

Scheme 7.3 The synthesis of styrene oxides through the enzymatic acylation of chloroalcohols.

and irreversibly to acetaldehyde, allowing the equilibrium to shift in favor of the product. BASF have shown this technology to be effective on an industrial scale for the synthesis of styrene oxides (Scheme 7.3) [7].

7.3.1
Synthesis of a Chiral R-(+) Hydromethyl Glutaryl Coenzyme A Reductase Inhibitor (Anticholesterol Drug)

The anticholesterol drug candidate, hydroxymethyl glutaryl coenzyme A (HMG CoA) reductase inhibitor (3) was resolved by diastereoselective acylation (Scheme 7.4) [16]. The drug is a potentially new anticholesterol drug that acts by inhibition of HMG Co A reductase. The lipase from *Pseudomonas cepacia* (Amano), immobilized on an accurel polypropylene support, efficiently acylated the undesired enantiomer to yield the S-(-) acetylated product (5) and unreacted desired chiral R-(+) alcohol (4) The process was demonstrated on a multikilogram scale using isopropenyl acetate in toluene, achieving a 48% yield and an *ee* of 98.5%. The immobilized catalyst was easily separated by filtration and then washed with toluene. The remaining isolation included methanol extraction followed by distillation and crystallization.

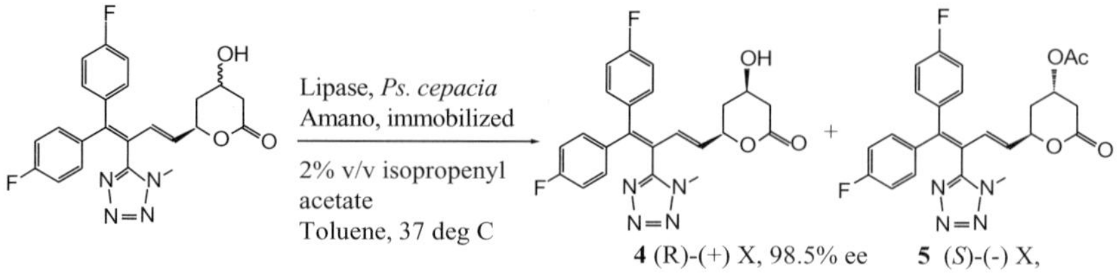

Scheme 7.4 Resolution of racemic alcohol enzymatic acylation.

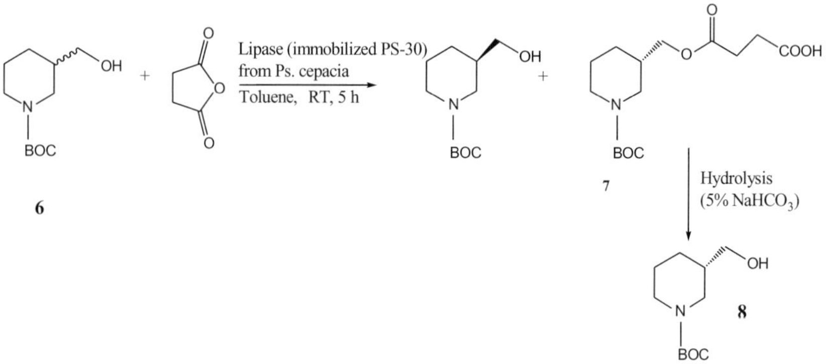

Scheme 7.5 The synthesis of (S)-N-Boc-hydroxymethylpiperidine by enzymatic resolution.

7.3.2
The Synthesis of S-N(*tert*-Butoxycarbonyl)-3-Hydroxymethylpiperidine (Tryptase Inhibitor)

(S)-N-(*tert*-butoxycarbonyl)-hydroxymethylpiperidine (**8**) is a key intermediate in the synthesis of a potent tryptase inhibitor (Scheme 7.5). It was synthesized from (R,S)-3-hydroxymethylpiperidine via fractional crystallization of the corresponding L(-)dibenzoyl tartrate salt followed by hydrolysis and acylation [17]. The lipase from *Pseudomonas cepacia* (PS-30) immobilized on polypropylene accurel PP catalyzed the esterification of racemic **6** with succinic anhydride and toluene, giving the (S)-hemisuccinate ester (**7**). This was easily separated and hydrolyzed by base to the (S)-Boc-protected 3-hydroxymethylpiperidine (**8**). Using this repeated esterification procedure gave a 32% yield (maximum theoretical yield = 50%) and 98.9% *ee*.

7.3.3
The Synthesis of (S)-[1-(Acetoxyl)-4-(3-Phenyl)Butyl]Phosphonic Acid Diethyl Ester (Anticholesterol Drug)

(S)-[1-(acetoxyl)-4-(3-phenyl)butyl]phosphonic acid diethyl ester (**10**) is a key intermediate required for the synthesis of (**11**) (Scheme 7.6), which is being

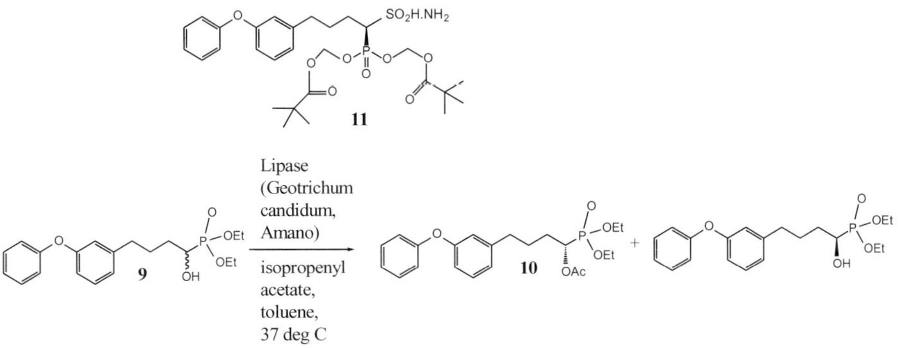

Scheme 7.6 Stereoselective acetylation of the hydroxyl diethyl ester (**9**) to the (S)-acetate as an intermediate for the synthesis of a squalene synthase inhibitor (**11**) of cholesterol synthesis.

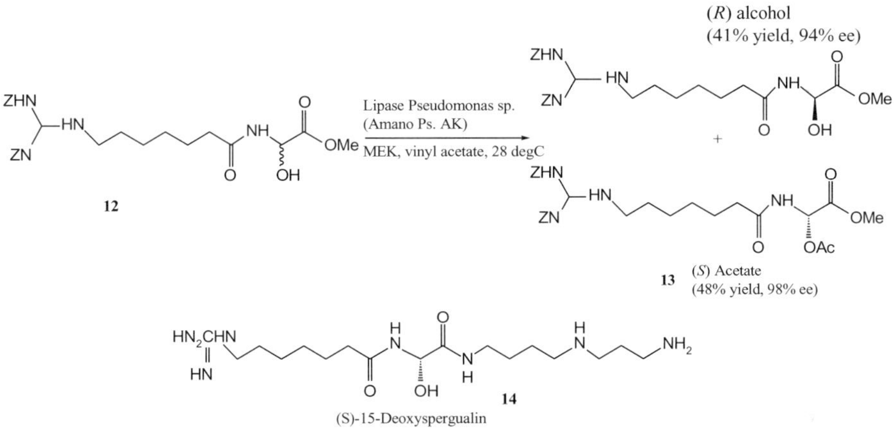

Scheme 7.7 The acylation of racemic alcohol to yield (S)-acetate, an intermediate for the synthesis of (S)-15-deoxyspergualin.

investigated as a potential squalene synthase inhibitor for anticholesterol [18]. The acetylation of racemic substrate was demonstrated using a G. candidum lipase with toluene and isopropenyl acetate as the acyl donor. A reaction yield of 38% (theoretical maximum 50%) and an ee of 95% was obtained for the desired enantiomer (**10**).

7.3.4
Synthesis of Desoxyspergualin (Immunosuppressant and Antitumor Drug)

The resolution of the hydroxyglycine derivative **12** was required as the intermediate in the synthesis of a clinical trial candidate for an immunosuppressive agent and antitumor compound (−)-15-deoxyspergualin (**14**) (Scheme 7.7). The lipase from

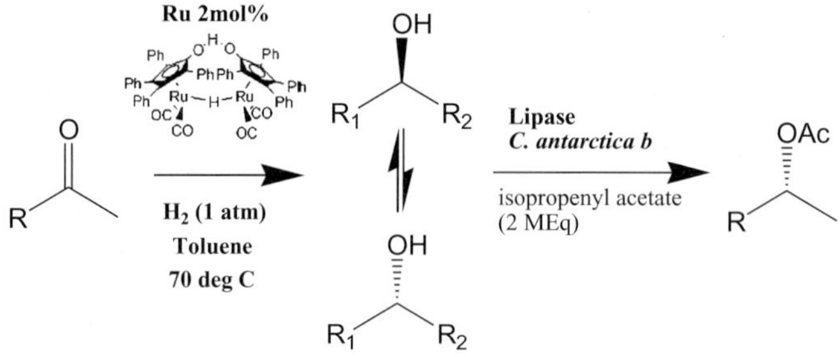

Scheme 7.8 Dynamic kinetic resolution of racemic alcohols by the combination of transition metal catalysis with enzymatic acylation.

Pseudomonas species (Amano lipase AK) catalyzed the reaction with vinyl acetate and methyl ethyl ketone. The desired (*S*)-acetate (**13**) was obtained in 48% yield (50% of theoretical maximum) and 98% *ee*. [19].

7.3.5
Dynamic Kinetic Resolution of Racemic Alcohols

The dynamic kinetic resolution of optically active alcohols by combining enzymatic resolution with transition metal redox racemization has been demonstrated on an industrial scale by DSM using methods adapted from the work of Bäckvall [20, 21] and Kim [22]. The nonacylated enantiomer is racemized *in situ* by a transition metal (Ru)-catalyzed Meerwein-Ponndorf-Verley reduction or Oppenauer oxidation and continuous removal of one of the enantiomers using stereoselective enzymatic acylation (Scheme 7.8). The methodology has been demonstrated for a range of alcohols including aromatic, heterocyclic, or alicyclic ring alcohols with effective yields and high selectivity (>99%) with simple isolation by crystallization or distillation [23–25]. It has been demonstrated on a 500 kg to 1 ton scale for the synthesis of (*S*)-3,5-trifluorofluoromethylphenyl alcohol, which is an important intermediate for the synthesis of the drug molecule EMEND, an orally active NK1 receptor antagonist for the treatment of chemotherapy-induced emesis [26].

7.4
Kinetic Resolution of Racemic Amines

Chiral amines are valuable synthons that dominate agrochemicals and pharmaceutical drug pipelines. Current methods for the preparation of amines are largely based upon resolution, and this is an excellent example of an industrial success. BASF makes a range of chiral amines by acylating racemic amines with proprietary esters, whereby one enantiomer is acylated to the amide, which can be easily separated from the unreacted amine [7]. For example, the resolution of racemic **15** in

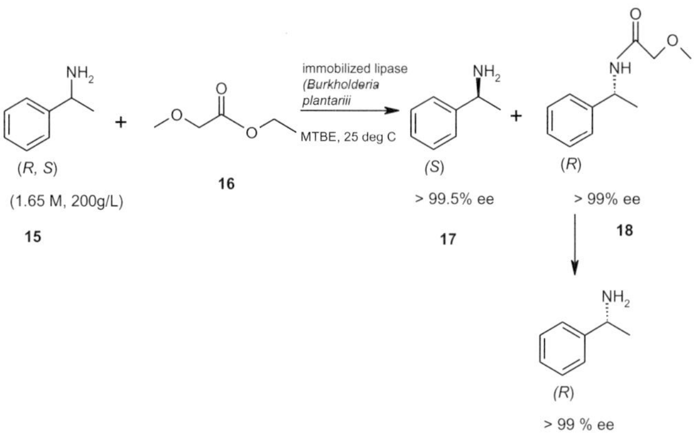

Scheme 7.9 Resolution of racemic amines by lipase-catalyzed enantioselective amide formation (BASF).

high concentration (1.65 M) is catalyzed by the lipase from *Burkholderia plantarii* using ethyl methoxyacetate (**16**) as the acyl donor. This gave 41% (R) amide (**18**) and 45% (S)-amine (**17**) both with ee >99% and E values of >500 (Scheme 7.9) [27].

The disadvantage of using alkyl ester type acyl donors is the typical slow rate of reaction. BASF made an enormous improvement by identifying an alternative acyl donor: ethyl methoxyacetate [28, 29]. This gave a 100-fold faster reaction than that with butyl acetate. This is probably because of an enhanced carbonyl activity induced by the electronegative α-substituent which accounts for the activating effect of the methoxy group. The lipase is immobilized on polyacrylate and removed by filtration. The (S)-amine is separated by distillation or extraction, and the (R) amine is released through basic hydrolysis of the (R) amide. The unconverted (S) enantiomer can be racemized using a palladium catalyst [27].

A wide variety of amines, including aryl alkyl amines, alkyl amines, and amino alcohols have been resolved on a multi-ton scale (Scheme 7.10) [7]. This method has been used with methyl ethers. For example the racemic (1-methoxy)-2-propylamine (R,S **19**) can be separated to give the (S)-**20** which is required for the synthesis of a corn herbicide (**21**) and produced at 2500 t/y [7, 30] (Scheme 7.11).

7.4.1
Resolution of Rac-2-Butylamine

More recent was the demonstration of the enzymatic resolution of *sec*-butylamine (**22**), which is an important intermediate for a number of drug compounds for Bristol Myers Squibb [31]. (S)-*sec*-butylamine (**23**) of high selectivity (99.5% ee) was obtained by *C. antarctica* lipase-catalyzed acylation with ethyl decanoate in methyl *tert*-butyl ether (Scheme 7.12). Initial studies using ethyl or vinyl butyrate gave poor selectivity. This was due to the low stereodifferentiation between the two

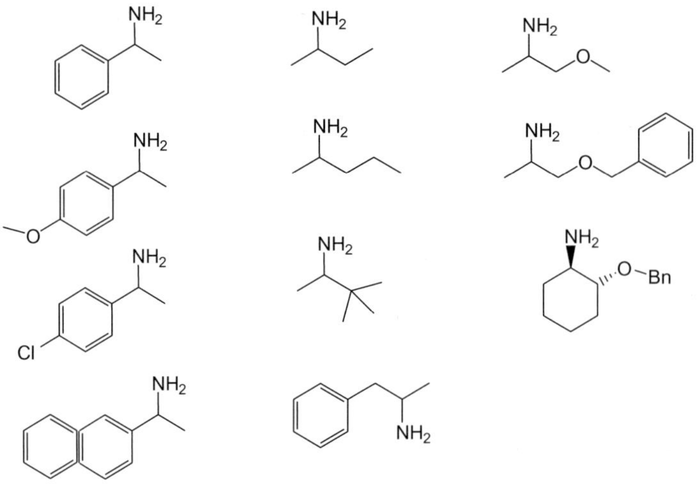

Scheme 7.10 The resolution of chiral amines: examples of the substrate range used by the BASF process demonstrated on a multi-ton scale.

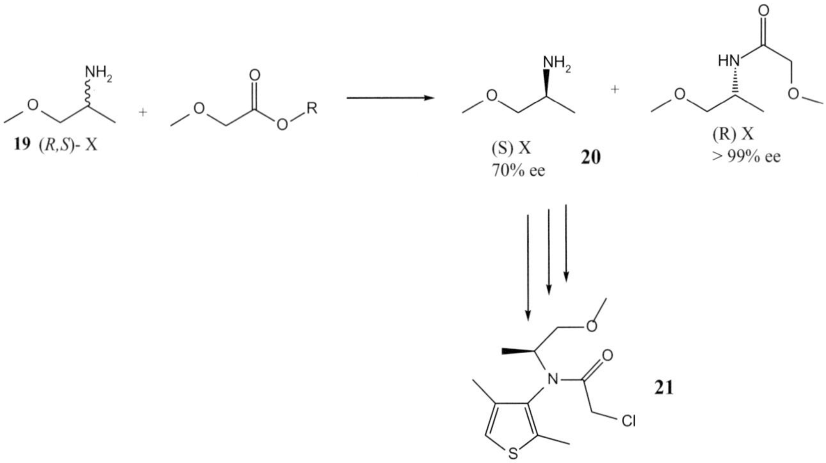

Scheme 7.11 Resolution of racemic (1-methoxy)-2-propylamine to give the (S)-amine, an intermediate in the synthesis of a chiral corn herbicide.

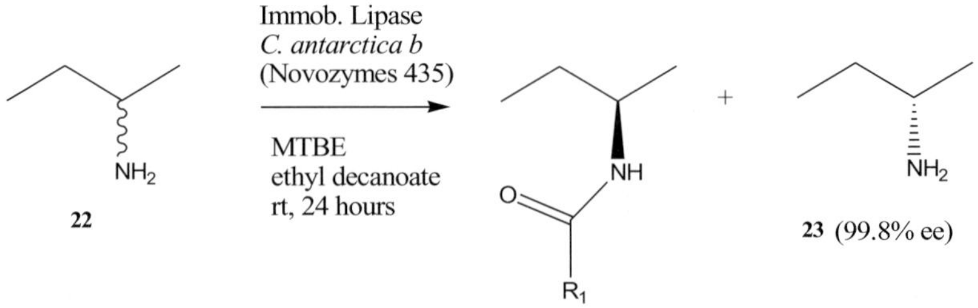

Scheme 7.12 The kinetic resolution of racemic sec-butylamine.

similar substituents, methyl and ethyl, attached to the carbon atom bearing the amino group.

Switching the acyl donor to ethyl esters of fatty acids showed an improvement in selectivity with increasing carbon chain length from C2 to C10 (ethyl decanoate). The (S)-amine **23** was synthesized after 1 day and the enzyme was recovered from the reaction by filtration, then washed and reused. The residual amine was extracted from the MTBE solution with dilute phosphoric acid (pH 2.9–4.4). The aqueous extract was concentrated to give a phosphate salt of the (S)-sec-butylamine with ee 99.7%. This was followed by a crystallization procedure in ethanol to give 99.8% ee and 50% yield.

7.4.2
Synthesis of a Selective, Non-Peptide, Non-Sulfhydryl Farnesyl Protein Transfer Inhibitor (Antitumor Agent)

Schering Plough demonstrated the kinetic resolution of a secondary amine (**24**) via enzyme-catalyzed acylation of a pendant piperidine (Scheme 7.13) [32]. The compound **27** is a selective, non-peptide, non-sulfhydryl farnesyl protein transfer inhibitor undergoing clinical trials as a antitumor agent for the treatment of solid tumors. The racemic substrate (**24**) does not contain a chiral center but exists as a pair of enantiomers due to atropisomerism about the exocylic double bond. The lipase Toyobo LIP-300 (lipoprotein lipase from *Ps. aeruginosa*) catalyzed the isobutylation of the (+) enantiomer (**26**), with MTBE as solvent and 2,2,2-trifluoroethyl isobutyrate as acyl donor [32]. The acylation of racemic **24** yielded (+) **26** at 97% ee and (−) **25** at 96.3% ee after 24 h with an E >200. The undesired enantiomer (**25**)

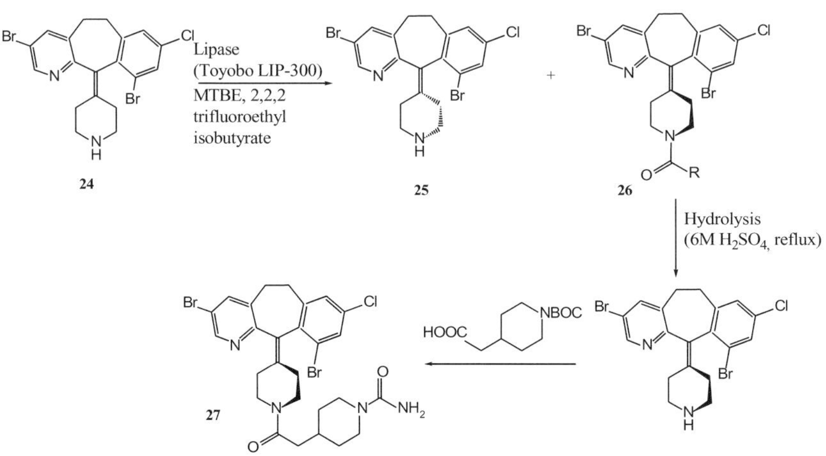

Scheme 7.13 Enzymatic resolution for the synthesis of **27** for the sythesis of selective, non-peptide, non-sulfhydryl farnesyl protein transfer inhibitor currently undergoing phase II clinical trials for the treatment of solid tumors.

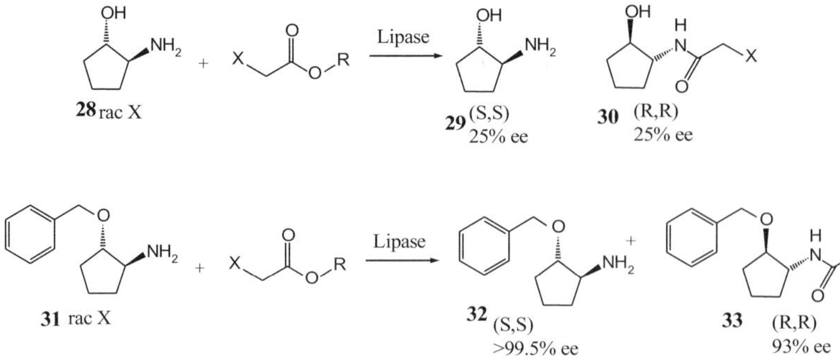

Scheme 7.14 Enzyme-catalyzed kinetic resolution of racemic trans-2-aminocyclopentanol and its benzyl ether.

can be racemized by refluxing in di(ethyleneglycol) dibutyl ether. A 65% overall yield was obtained for (+) **26** with ee 98% after three rounds of enzymatic resolution, which was completed at a 50 kg scale.

7.5
Resolution of Amino Alcohols and Methyl Ethers

The enzymatically catalyzed kinetic resolution of amino alcohols has been established on the multi-ton scale by BASF [7] (Scheme 7.14). Initial studies gave poor selectivity for the unprotected alcohols, as the resolution of *trans*-2-aminocyclopentanol (racemic **28**) gave the amine (S,S) **29** and the amide (R,R) **30** in 25% ee. When the hydroxy functionality was protected as an ether, then resolution of racemic benzyl ether **31** proceeds with high *ee* to the give the amine (S,S) **32** and the RR amide **33** with >99.5 and 93 % *ee* respectively [33, 34].

7.6
Resolution of an Ester

7.6.1
Ester Resolution for the Synthesis of Emtricitabine (Antiviral Drug)

The resolution of racemic FTC butyrate (**34**) was required for the synthesis of the antiviral drug emtricitabine (Emtriva) (Scheme 7.15): a nucleoside reverse transcriptase inhibitor targeted for treatment of human immunodeficiency virus (HIV) and hepatitis B infections [35]. The racemic FTC butyrate ester (**34**) was treated with immobilized cholesterol esterase, which cleaved the required isomer to the corresponding alcohol (−) **35** with 91% *ee* and 52% conversion [36]. The product was isolated as the hydrochloride salt to give 31% yield (98% *ee*) from the 8 kg demonstration. The esterase was immobilized by precipitation onto an accurel polypropylene support using acetone followed by cross linking with glutaralde-

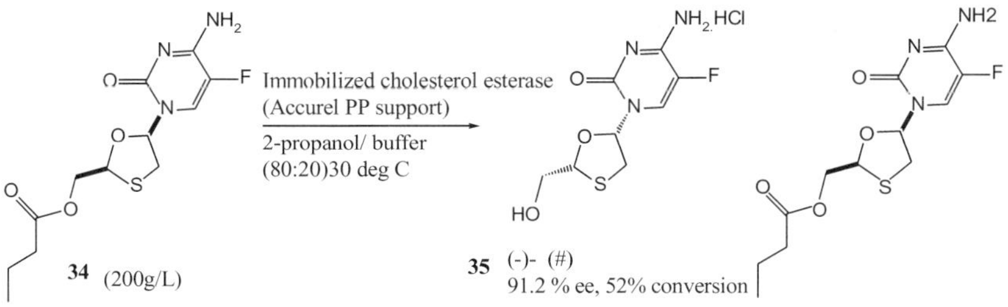

Scheme 7.15 The resolution of an FTC butyrate.

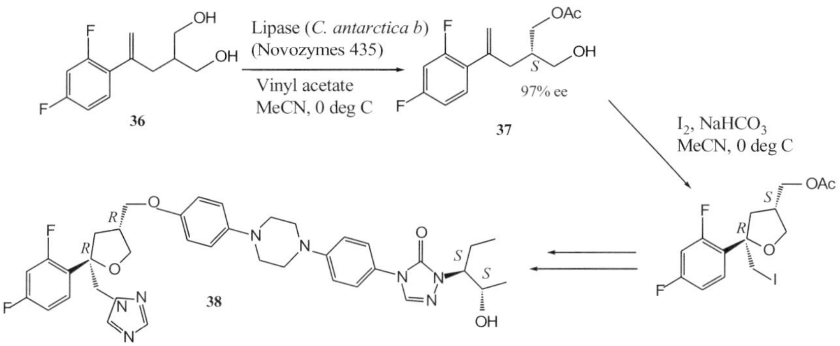

Scheme 7.16 Desymmetrization of 2-substituted 1,3-propanediol to (S)-monoacetate by enzymatic hydrolysis in 80% organic solvent.

hyde. This allowed the enzyme to tolerate an 80% v/v isopropanol solution, enabling a high substrate concentration (200 g L^{-1}) to be achieved. The enzyme recycling was demonstrated up to fourteen times.

7.7
Desymmetrization by Transesterification

7.7.1
Synthesis of an Antifungal Azole Derivative

The desymmetrization of prochiral esters, monoacetates, or diols by hydrolysis or acylation has been an industrially useful asymmetric tool for generating chiral intermediates. The desymmetrization of 2-substituted 1,3-propanediol (36) to chiral monoacetate (37) by asymmetric transesterification was demonstrated by Schering Plough (Scheme 7.16). The 2R4S phenylsulfonate (36) is a key intermediate for the synthesis route to an antifungal azole derivative (38) for the treatment of systemic *Candida* and *Aspergillus* infections [37]. The highly selective synthesis of monoacetate (37) (97% ee) was catalyzed by lipase cal b with vinyl acetate in acetonitrile at high substrate loading (200 g L^{-1}). The reaction was carried out at

0 °C in order to improve selectivity and minimize the undesired over-hydrolysis to diacetate which occurred to an extent of up to 17% [38]. The subsequent iodocyclization step was carried out in the same solvent at 0 °C after catalyst removal by filtration. A batch of 30 kg was delivered and catalyst reusability demonstrated for 6 cycles without major loss of activity.

7.8
Regioselective Acylation

7.8.1
Regioselective Acylation of Drug Intermediate for an Antileukaemic Agent

A number of industrial examples have used lipases to catalyze the regioselective acylation of important drug intermediates. The regioselective acylation at the 5′ position of a powerful anti-leukaemic agent (**39**) [39] from Glaxo Wellcome was required to improve the compound's water solubility and bioavailability [40] (Scheme 7.17). All of the chemical synthetic approaches used, including selective acylation or deacylation of the corresponding triacetate, showed poor selectivity, requiring chromatography to remove the undesired acetate. However the lipase cal b was found to catalyze the acylation with vinyl acetate and 1,4-dioxane using an effective substrate concentration (100 g L^{-1}). Optimization of the solvent type and temperature led to minimizing the undesired impurities of 3′-mono and diacetates to <0.5%. The isolation of product (**40**) was a simple filtration step to remove the enzyme followed by catalyst washing with methanol and distillation. A scaleable process was demonstrated, giving an 85% yield [40].

7.8.2
Regioselective Acylation of Ribavirin: Antiviral Agent

Another example of regioselective acylation was required for a ribavirin antiviral agent (**44**) used in combination with α-2-β-interferon to treat hepatitis C (Scheme

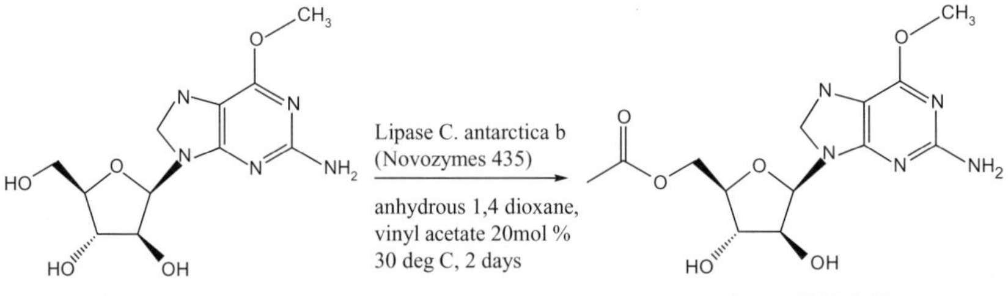

Scheme 7.17 Enzymatic acylation of 2-amino-9-β-D-arabinofuranosyl-6-methoxy-9H-purine (Glaxo Wellcome).

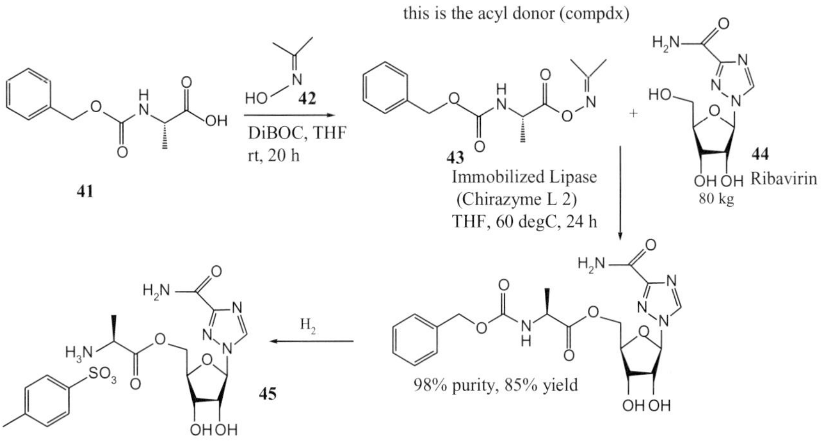

Scheme 7.18 Regioselective acylation to synthesize the alanine ester of ribavirin, an antiviral agent.

7.18). The alanine ester of ribavirin (**45**) was required in order to improve the bioavailability [41]. Chemical acylation attempts of the unprotected ribavirin (**44**) proved unsuccessful as it resulted in a complex mixture of mono-, di- and triacylated products. Pursuing an enzymatic route seemed feasible, as lipase-catalyzed acylation of 5′-hydroxyl has been shown for several nucleosides with amino acid derivates. The Schering Plough team found the amino acid α-alanine was accepted by the lipase chiralzyme L2 (**41**). Therefore a suitable donor for the enzymatic acylation was synthesized by coupling L-Cbz-Ala (**41**) with acetone oxime (**42**) in the presence of di-*tert*-butyl dicarbonate in THF to give a 96% yield. The reaction mixture was then diluted with THF, ribavirin was added, and the acylation reaction was initiated by addition of the chiralzyme lipase. After 24 h at 60 °C, the reaction was complete without impurities and the enzyme was removed by filtration followed by distillation and precipitation with MTBE [41]. This one-pot synthesis route was used to synthesize 80 kg of the intermediate.

7.9
Asymmetric Ring Opening of Racemic Azlactone

7.9.1
Synthesis of (S)-Benzyl-L-*tert*-Leucine Butyl Ester

The amino acid (S)-*tert*-leucine (**48**) has been shown to be an important intermediate for a wide variety of pharmaceutically active molecules in the areas of antitumor, anti-inflammatory, and antiviral activity [42]. A number of routes to (S)-*tert*-leucine (**48**) have been developed, in particular the aqueous-based process from Degussa using asymmetric reductive amination of the prochiral keto acid with dehydrogenase and cofactor recycling [43, 44]. An industrial alternative was

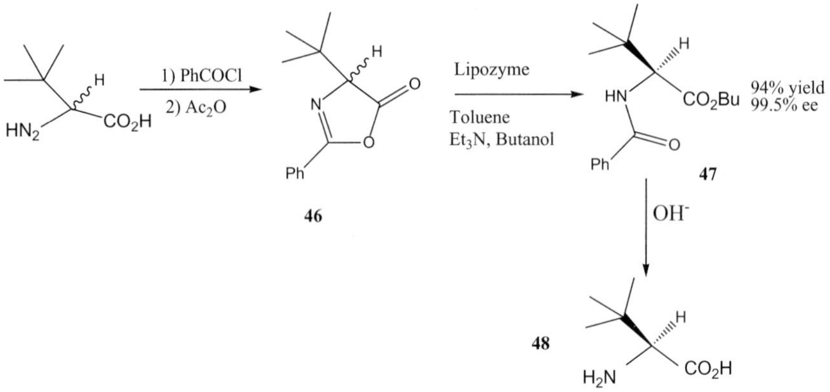

Scheme 7.19 The synthesis of homochiral L-(S)-tert-leucine via a lipase-catalyzed dynamic resolution process.

demonstrated by Chiroscience which used the lipase from *Mucor miehei* to catalyze the asymmetric ring opening of the racemic azlactone 2-phenyl-4-*tert*-butyloxazolin-5(4H)-one (**46**) in anhydrous toluene with 2 equivalents of triethylamine [45, 46] (Scheme 7.19). The N-benzoyl-L-*tert*-leucine butyl ester (**47**) was synthesized in 94% yield, 99.5% ee, and standard conditions for the hydrolysis of the butylester gave the L-(S)-*tert*-leucine (**48**) in 80% yield and 99.5% ee. The azlactone (**46**) was easily synthesized via cyclodehydration of the corresponding N-benzoyl amine acid. The ring opening allowed a dynamic resolution process because of the low pKa (8.9) of the C-4 proton, which enabled the unreactive enantiomer of the racemic azlactone to undergoe facile epimerization.

7.9.2
Synthesis of (S)-γ-Fluoroleucine Ethyl Ester

Fluorinated amino acids and their derived peptides are important pharmaceutical agents because of their broad biological properties as enzyme inhibitors, receptor antagonists, and lipophilicity enhancing agents [47]. However, the asymmetric synthesis of γ-fluoro-α-amino acids remains a significant challenge. The stereoselective incorporation of the γ-fluoro-containing side chains have mostly been obtained by either a chiral auxiliary-directed diastereoselective alkylation or a chiral phase transfer-catalyzed alkylation of the N-protected precursors, where only modest stereoselectivities <40 *de* or *ee* are obtained [48, 49]. Recently, Merck & Co. Inc. developed an elegant route to (S)-γ-fluoroleucine ethyl ester (**50**) via dynamic kinetic resolution of an azlactone (**49**) catalyzed by immobilized lipase cal B (Novozymes 435) (Scheme 7.20) [49]. The azlactone substrate (**49**) was easily synthesized with high yield from the cyclodehydration of the amide acid using EDCI and CH_2Cl_2. The lipase rapidly catalyzed the ring opening of 2-(3-butenyl)azlactone (**49**) in methyl *tert*-butyl ether (MTBE) with ethanol and triethylamine, which established the stereochemistry of the (S)-amide ester (**50**) at 84% *ee* and 80% yield [50]. The system operates by the dynamic kinetic mechanism, with spontaneous

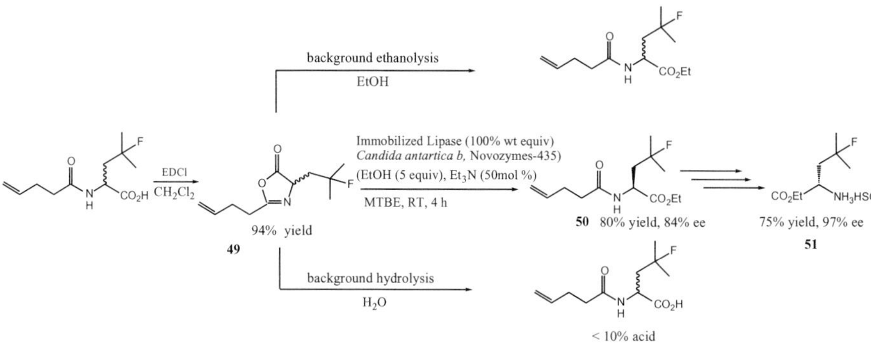

Scheme 7.20 Ring opening of azlactone with immobilized lipase Cal b to the (S)-γ-fluoroleucine ethyl ester.

racemization of the azlactone via enol tautomerization, allowing the yield of (S)-γ-fluoroleucine ethyl ester (**50**) to be greater than 50%. The competing reactions of selective nucleophilic alcohol addition and the non-selective background hydrolysis can erode product yield and ee. This was minimized by process optimization of solvent type and reaction conditions. The readily available immobilized catalyst allowed ease of operation, and isolation was by simple filtration and solvent washes. The free amine (**51**) was liberated by oxidative removal of the pentenoyl group and isolated as its hydrogen sulfate salt in 97% ee and 75–90% yield. This provided a simple and efficient route for multi-kilogram deliveries.

7.10
Cyanohydrin Formation

The formation of enantiomerically enriched cyanohydrins from aldehydes and ketones is an increasingly important industrial tool. Cyanohydrin synthons provide significant utility, as both the nitrile and alcohol functional groups can be effectively derivatized without a resultant loss in optical purity. Such transformations have led to a wide variety of products, including α-hydroxyketones, β-hydroxyamines, α-aminonitriles, and α-hydroxyesters [51, 52]. DSM has pioneered the large-scale production of both (R) and (S)-cyanohydrins using the hydroxynitrile lyase (HNL) biocatalysts with a wide range of aliphatic, aromatic, and oxygen/sulphur-containing heteroaromatic aldehydes and ketones [51, 53] using biphasic systems of >50–60 vol% organic phase.

7.10.1
Synthesis of (R)-2-Chloromandelic Acid: Intermediate for Clopidogrel

(R)-2-chloromandelic acid (**53**) is a key intermediate for an antidepressant and for the production of the platelet aggregation inhibitor clopidogrel (**55**) (Plavix) (Scheme 7.21) [54]. The hydroxynitrile lyase (HNL) from almond (*Prunus amygdalus*)

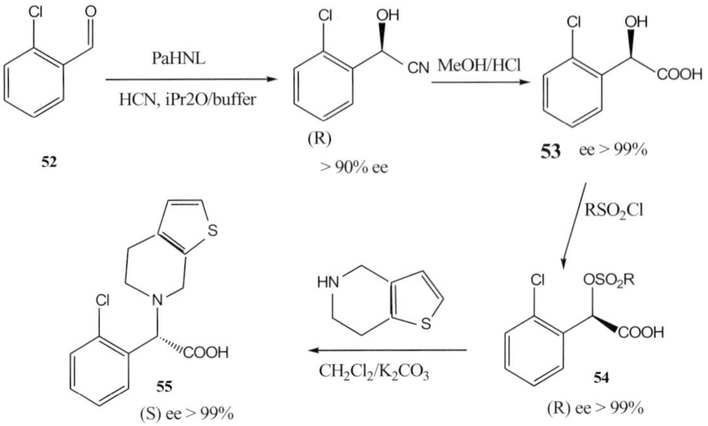

Scheme 7.21 Stereoselective addition of hydrogen cyanide to chlorobenzaldehyde catalyzed by hydroxynitrile lyase.

catalyzes the stereoselective addition of hydrogen cyanide to the chlorobenzaldehyde (**52**) in a biphasic system [53]. Originally the process used almond flour extract or immobilized enzyme on avicel microcrystalline cellulose. The enzyme can now be readily over-expressed in the yeast *Pichia* host, which provides a readily available enzyme supply [55, 56]. The cell lysate is used in the reaction with 60% MTBE and citric acid buffer at pH 5.8 and 15 °C. The high solvent concentration enables a 200 g L^{-1} substrate concentration to be achieved, giving a reaction duration of 2 h with a space time yield (STY) of 2.1 mol L^{-1} h^{-1} [53, 57]. The combination of low pH, biphasic solvent, and low temperature is used to suppress the spontaneous non-enzymatic addition of hydrogen cyanide leading to racemic cyanohydrins. The HCN is a weak acid that is essentially undissociated at neutral to acidic pH values and distributes well between the organic and the aqueous phase, with a slight preference for the organic phase. The almond source of the HNL was found to act as a surfactant, so intense mixing is important to maximize contact of the two-phase system to form an emulsion. After the reaction, the emulsion is broken and the phases separated. The aqueous phase containing the enzyme is recycled and the organic MTBE phase is stabilized with acid and processed through a wiped film evaporator to remove the MTBE, HCN, and residual water [53]. The cyanohydrin is then transformed into the corresponding α-hydroxy-carboxylic ester (**54**), which can be reacted with the tetrahydrothienylpyridine after activation with phenyl sulfonyl chloride to give clopidogrel (**55**).

7.10.2
Synthesis of Chiral Phenoxybenzaldehyde Cyanohydrin: Intermediate for the Synthesis of Pyrethroid Insecticides

The (S)-cyanohydrins and (S)-hydroxycarboxylic acids syntheses are catalyzed by HNLs from *Hevea brasilisensis* and *Manihot esculenta* in the form of economically

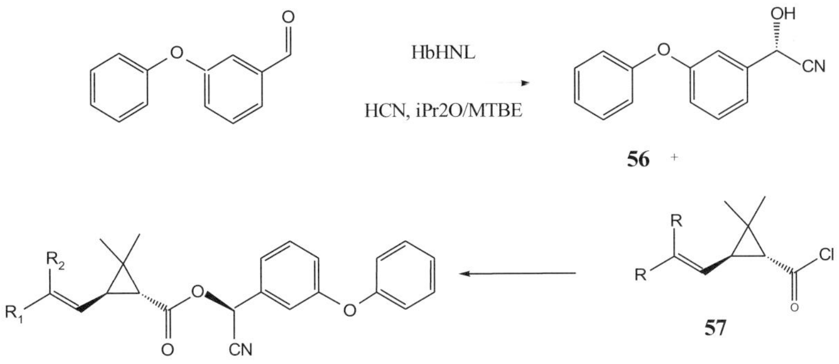

Scheme 7.22 Synthesis of pyrethroid insecticides via (S)-phenoxybenzaldehyde cyanohydrin.

R_1	R_2	
Br	Br	Deltamethrin
Cl	Cl	Cypermethrin
Br$_3$C	Br	Tralomethrin

attractive recombinant enzyme preparations expressed in E. coli [58]. The largest commercial production of chiral cyanohydrin is that of (S)-3-phenoxybenzaldehyde cyanohydrin (**56**) (Scheme 7.22), which is used for the production of synthetic pyrethroids [53, 59, 60]. The chiral pyrethrum acid chloride derivative (**57**) is coupled with the (S)-cyanohydrin (**56**) to yield a range of neurotoxin insecticides for household and crop protection applications. The cyanohydrin formation is carried out in a biphasic 60% MTBE system similar to that described for the chloromandelic acid reaction. The cyanohydrin (**56**) is 98% ee with 98% yield and STY 2.1 mol L^{-1} h^{-1}, and the enzyme can be recycled five times.

7.10.3
Hydrocyanation of 3-Pyridinecarboxyaldehyde

The formation of enantiomerically enriched cyanohydrins has been successful for a wide range of aliphatic, aromatic, and oxygen/sulfur containing heteroaromatic aldehydes and ketones. However nitrogen-containing heteroaryl carboxaldehydes such as 3-pyridinecarboxyaldehyde (**58**) have shown to be a significant challenge. For example, low selectivity has been shown for the hydrocyanation of 3-pyridinecarboxyaldehyde via monometallic bifunctional BINOLAM catalysts to produce (S)-cyanohydrin [61] or by HNL-catalyzed reactions [62, 63].

The 3-pyridinecarboxyaldehyde **58** is highly water soluble, and so the spontaneous cyanide addition to give racemic cyanohydrin cannot be suppressed unless the aqueous pH is lowered below 3.5, which is not tolerated by the enzymes. The only available option is to operate in a 100% organic solvent system. This was recently made possible by the availability of the cross linked enzyme aggregate particles (CLEAs), which can tolerate organic solvents [64]. The individual precipitated protein molecules are chemically bonded to one another through the formation

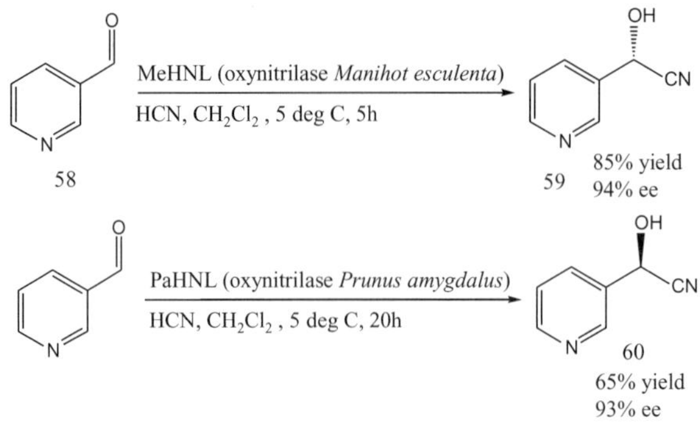

Scheme 7.23 Synthesis of chiral cyanohydrins from 3-pyridinecarboxyaldehyde in organic solvent using commercially available CLEA particles of hydroxynitrile lyases.

of imines via the amine functionality of the protein and a polyaldehyde cross linker. A number of enzyme CLEA systems are readily available for industrial use [65]. Recently the Merck research laboratories demonstrated the synthesis of both (*R*) and (*S*)-cyanohydrins (**59, 60**) with >93% *ee* through hydrocyanation of 3-pyridinecarboxyaldehyde (**58**) in >65% yield using dichloromethane (Scheme 7.23) [66]. The commercially available CLEA particles of hydroxynitrile lyases (HNLs) from cassava and almond were used to produce both enantiomers. The use of dichloromethane, free hydrogen cyanide, and low temperatures was important for improving the cyanohydrin stereoselectivity and suppressing the background reaction. The aggregate particles were easily separated from the reaction by filtration followed by washing, and up to ten times reuse was demonstrated.

7.11
Outlook

This chapter demonstrates that using enzymes in organic solvents has made a considerable impact on a range of industries employing commodity chemicals, agrochemicals, and pharmaceutical compounds with outstanding chemo and regioselectivity. Many of these processes are highly efficient, with effective substrate concentrations (1–2 M) and reaction times (STY >100 g L^{-1} h^{-1}). Classical racemate resolution will continue to have its place in industry for the rapid synthesis of enantiomerically pure building blocks. This will be further enhanced by the increased use of dynamic kinetic resolution technology. The integration of enzyme catalysts into chemical synthesis will continue to grow owing to the increased use of isolated enzymes for established chemistries such as ketone reductions [67–69]. This will expand to emerging catalyst libraries such as transamination, hydroxylation, and enoate reductions [8]. A number of industrial-scale ketone reductions are currently completed with ketoreductases in the presence of 5–20% v/v organic solvent and high substrate concentrations up to 2 M [70–72]. The application of catalyst improvement via directed evolution will drive the ability of these types of reactions to be carried out at higher organic solvent concentrations [73–76]. A recent industrial example showed the reduction of ethyl 4-chloroacetoacetate by an evolved ketoreductase using 45% v/v butyl acetate [76]. The expansion of available catalysts and novel catalyst forms, such as CLEAs, in combination with the ability to design the catalyst to fit the process, will promote the continued impact of enzyme catalysis upon industrial chemical synthesis in non-aqueous systems.

7.12 References

1 A. Liese, K. Seelback, C. Wandrey, *Industrial Biotransformations*, Wiley-VCH, Weinheim, 2006.
2 A. Bommarius, B. R. Riebel, *Biocatalysis*, Wiley-VCH, Weinheim, 2004.
3 H. Yamada, M. Kobayashi, *Biosci. Biotechnol. Biochem.* 1996, **60**, 1391.
4 J. Han, Y. Zhang, J. Xue, J. Huang, X. Chen, *Shanghai Huagong*, 1995, **20**, 7.
5 C. Chassin, *Specialty Chem.* 1996, **16**, 3, 102–105.
6 P. Cheetham, *Applied Biocatalysts*, Harwood Acad. Press, Amsterdam, 2000.
7 B. Hauer et al., *Angew. Chem. Int. Ed.* 2004, **43**, 788–824.
8 D. J. Pollard, J. Woodley, TIBTECH 2007, **25**, 2, 66–73.
9 D. Yazbeck, R. Daniel, C. A. Martinez, J. Tao, *Tetrahedron Asymm.* 2004, **15**(18), 2757–2763.
10 M. H. Coleman, A. R. MacRae, Unilever NV, DE-B 2705608, 1977.
11 T. Matsuo, N. Sawamura, Y. Hashimoto, W. Hashida, Fuji Oil Co., EP 0035883A2, 1977.
12 G. A. Hills, A. R. Macrae, Unichema Chemie BV, EP 0383405, 1990.
13 P. A. Kemp, A. R. Macrae, Unichema Chemie BV, EP 0506159, 1992.
14 E. Anderson, K. Larsson, O. Kirk, *Biocatal. Biotrans.* 1998, **16**, 181–204.
15 F. M. Fonteyn et al., *Biotechnol. Lett.* 1994, **16**, 693–696.
16 R. N. Patel, *Appl. Microbiol. Biotechnol.* 1992, **38**, 56–60.
17 A. Goswami et al., *Org. Process Res. Dev.* 2001, **5**, 415–420.
18 R. N. Patel, A. Banerjee, L. J. Szarka, *Tetrahedron Asymm.* 1997, 8, 7, 1055–1059.
19 R. N. Patel, A. Banerjee, L. J. Szarka, *Tetrahedron. Asymm.* 1997, **8** (*11*), 1767–1771.
20 J. E. Bäckvall, *Chem. Int. Ed.* 2004, **43**, 6535.
21 O. Pamies, J-E. Bäckvall, *Curr. Opin. Biotechnol.* 2003, **14**, 407–413.
22 P. Kim, P. Park, *Chem. Eur. J.* 2006, **12**, 225.
23 T. Dax, M. Stanek, P. Pochlauer, WO 095628, 2005.
24 G. Verzijl, J. Vries, Q. Broxterman, DSM patent, WO 01/90396, 2001.
25 G. Verzijl, J. Vries, Q. Broxterman, *Tetrahedron Asymm.* 2005, **16**, 1603–1610.
26 K. M. Brands et al., *J. Am. Chem. Soc.* 2003, **125**, 2129–2135.
27 F. Balkenhohl et al., *J. Prakt. Chem.* 1997, **339**, 381–384.
28 H. Riechers, J. Simon, A. Hohn, A. Kramer, F. Funke, W. Siegel, C. Nubling, WO 2000047546, 1999.
29 M. Breuer et al., *Angew. Chem. Int. Ed.* 2004, **43**, 788–824.
30 E. Ladner, *Chim. Oggi* 1999, **17**, (*7*), 51–55.
31 A. Goswami, Z. Guo, W. Parker, R. Patel, *Tetrahedron Asymm.* 2005, **16**, 1715–1719.
32 B. Morgan, A. Zaks, D. R. Dodds et al., *J. Org. Chem.* 2000, **65**, 18, 5451–5459.
33 F. Balkenhohl, K. Ditrich, C. Nubling, BASF AG WO 9623894, 1995.
34 K. Nubling, C. Ditrich, C. Dully, DE 19837745, 1999.
35 D. Liotta, W. Choi, PCT Int Appl., WO 91252418, 1991.
36 A. Osborne, D. Brick, G. Ruecroft, I. Taylor, *Org. Proc. Res. Dev.* 2006, **10**, 670–672.
37 K. Saksena, *Tetrahedron Lett.* 1996, **37**, 5657–5660.

38 B. Morgan, D. R. Dodds, A. Zaks, D. Andrews, R. Klesse, *J. Org. Chem.* 1997, **62**, 7736–7743.
39 C. U. Lambe, et al., *Cancer Res.* 1997, **55**, 3352–3356.
40 M. Mahmoudian, J. Eaddy, M. Dawson, *Biotechnol. Appl. Biochem.* 1999, **99**, 229–233.
41 M. Tamarez, B. Morgan, G. Wong, W. Tong, F. Bennett, R. Lovey, *Org. Proc. Res. Dev.* 2003, **7**, 951–953.
42 A. S. Bommarius, K. Drauz, W. Hummel, M. R. Kula, C. Wandrey, *Biocatalysis* 1994,**10**(1–4),37–47.
43 A. S. Bommarius, M. Schwarm, K. Drauz, *J. Mol. Cat. B. Enzym.* 1998, **5**, 1–11.
44 A. Menzel, H. W. Werner, J. Altenbuchner, H. Groger, *Eng. Life Sci.* 2004, **4**, 6, 573–576.
45 N. J. Turner, J. R. Winterman, *Tetrahedron Lett.* 1995, **36**, 7, 1113–1116.
46 R. McCague, S. J. C. Taylor, in *Chirality In Industry II* (A. N. Collins, G. N. Sheldrake, J. Cosby, Eds.) John Wiley & Sons, NY, 1997, pp. 201–203.
47 Y. Filler, Y. Kobayashi, L. M. Yagupolskii, *Organofluorine compounds in medicinal chemistry and biomedical applications*, Elsevier Biomedical Press, Amsterdam, 1993.
48 G. Haufe, K. W. Laue, M. U. Triller, *Tetrahedron*, 1998, **54**, 5929–5938.
49 J. Limanto, et al., *J. Org. Chem.* 2005, **70**, 2372–2375.
50 P. Devine, J. Limanoto, A. Shafiee, V. Upadhyay, US patent 0234128 A1, 2005.
51 M. Sharma, N. N. Sharma, T. C. Bhalla, *Enzym. Microbiol. Technol.* 2005, **37**, 279–294.
52 J. M. Brunel, I. P. Holmes, *Angew. Chem. Int. Ed.* 2004, **43**, 2752.
53 P. Poechlauer, W. Skranc, M. Wubbolts, *Asymmetric Catalysis on Industrial Scale*, Wiley VCH, Weinheim, Germany, 2004.
54 A. Bousquet, A. Musolino, Sanofi-Synthelabo, EP 1021449 1998.
55 A. Glieder et al., *Angew. Chem. Int. Ed.* 2003, **42**, 4815–4818.
56 R. Weiss et al., *J. Mol. Cat. B. Enzym.* 2004, **29**, 211–218.
57 P. Pochlauer, R. Neuhofer, H. Griengl, R. Reintjens, H. J. Wories, DSM, EP 0927766, 1999.
58 F. Effenberger et al., DSM Fine Chemicals, Austria, EP 969095, 2000.
59 H. Griengl et al., *Tetrahedron* 1998, **54**, 14477.
60 H. Semba, A. Maschio, Nippon Shokubai Co., EP 1016712 B1, 1999.
61 A. Baeza, J. Casa, C. Najera, J. M. Sansano, J. M. Saa, *Eur. J. Org. Chem.* 2006, **8**, 1949.
62 M. Schmidt, S. Herve, N. Klemper, H. Griengl, *Tetrahedron* 1996, **52**, 7833.
63 S. Nanda, Y. Kato, Y. Asano, *Tetrahedron* 2005, **61**, 10908.
64 L. Cao, F. Van Rantwijk, P. A. Sheldon, *Org. Lett.* 2000, **2**, 361.
65 W. R. K. Schoevart et al., Int. Appl. WO 046965, 2006.
66 C. Roberge, F. Fleitz, D. Pollard, P. Devine, *Tetrahedron Lett.* 2006, accepted.
67 D. Pollard, M. Truppo, J. Pollard, C. Chen, J. Moore, *Tetrahedron Asymm.* 2006, **17**(4), 554–559.
68 M. Truppo et al., *J. Mol. Cat. B. Enzym.* 2006, **38**(3–6), 158–162.
69 B. Kosjek, D. M. Tellers, M. Biba, R. Farr, J. C. Moore, *Tetrahedron Asymm.* 2006, **17**, 2798–2803.
70 A. Gupta, M. Bobkova, A. Zimmer, A. Schwarz, WO patent 045598 A1, 2006.
71 H. Groger et al., *Angew. Chem. Int. Ed.* 2006, **45**, 5677–5681.
72 M. Truppo, D. Pollard, P. Devine, *Org. Lett.* 2006, accepted.
73 U. T. Bornscheuer, C. Bessler, R. Srinivas, S. H. Krishna, *TIBTECH*. 2002, **20**, 10, 433–437.
74 J. Sylvester, H. Chautard, F. Cedrone, M. Delcourt, *Org. Proc. Res. Dev.* 2006, **10**, 562–571.
75 G. W. Huisman, D. Gray, *Curr. Opin. Biotechnol.* 2002, **13**, 352–358.
76 C. Davis, J. Grate, D. Gray, *Codexis*. WO 04015132, 2004.

Part Three Biocatalysis in Biphasic and New Reaction Media

8
Biocatalysis in Biphasic Systems: General
Pedro Fernandes and Joaquim M. S. Cabral

8.1
Introduction

The concept of biphasic systems in biocatalysis typically encompasses the use of two immiscible liquid phases, where one of the phases (aqueous based) provides a protective environment for the biocatalyst, whereas the second phase is a substrate and/or product pool. The validity of this approach for the production of relevant compounds using enzymes and whole cells was clearly highlighted in the late 1970s and early 1980s, and since then the array of applications has consistently increased and diversified. In the original (and still more common) design of a biphasic system, the second phase is based on a water-immiscible biocompatible organic solvent with high solubilization capability for substrate and/or product. The biphasic system may be alternatively formed by two aqueous-phase systems, the organic solvent being replaced by a second aqueous phase, polymer based. In a third option, which has been gaining relevance in recent years, the organic solvent is replaced by an ionic liquid. The present chapter aims to provide an overview of the concept and applications of organic-aqueous two-liquid-phase systems in enzyme biocatalysis. To limit the scope, processes involving the use of living cells will not be considered.

8.2
Organic-Aqueous Biphasic Systems: General Considerations

The introduction of an organic solvent into an otherwise aqueous-phase system is the most immediate approach to overcoming the poor water solubility of many of the chemicals that can be used for as substrates for the production of key molecules, either as building blocks or end products, in a wide array of relevant fields, including the pharmaceutical, food and feed, agricultural, and soaps and detergent sectors [1]. In most of these systems biocatalysis takes place in the bulk aqueous phase, although interfacial biocatalysis may occur, where the enzyme acts while positioned in the organic-aqueous interface. Schemes of biocatalytic organic-aqueous biphasic systems are given in Figure 8.1.

Organic Synthesis with Enzymes in Non-Aqueous Media. Edited by Giacomo Carrea and Sergio Riva
Copyright © 2008 WILEY-VCH Verlag GmbH & Co. KGaA, Weinheim
ISBN: 978-3-527-31846-9

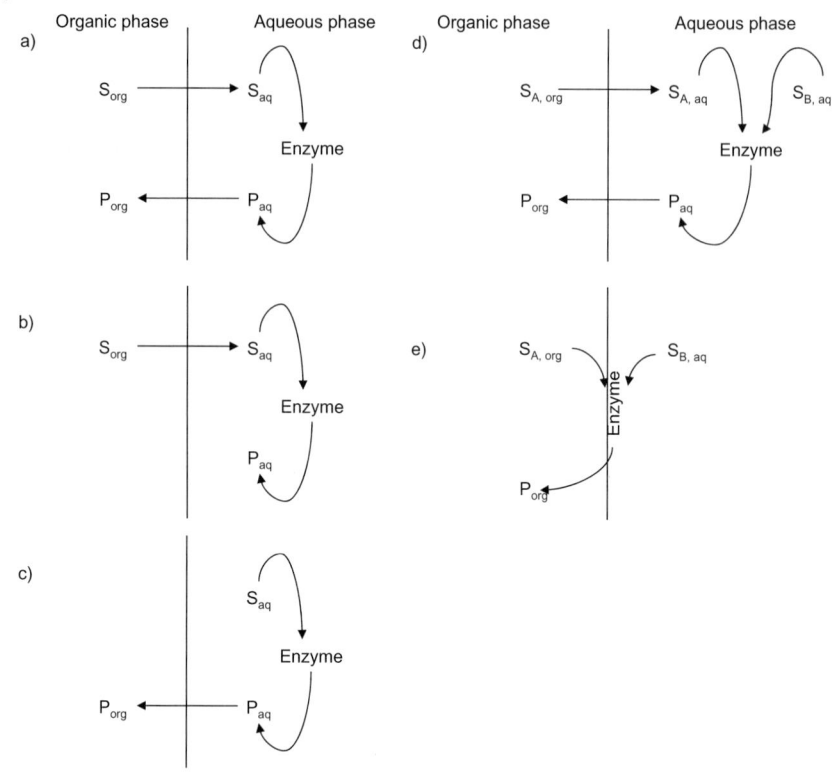

Figure 8.1 Schematic representation of some biocatalytic systems commonly found in organic-aqueous biphasic systems: (a) hydrophobic substrate and product, (b) hydrophobic substrate and hydrophilic product, (c) hydrophilic substrate and hydrophobic product, (d) and (e) hydrophobic substrate A, hydrophilic substrate B, hydrophobic product. In systems (a) to (d) catalysis occurs in the bulk aqueous phase; in system (e) catalysis occurs in the interface.

Given the overall high substrate concentrations that can be achieved in this way, high volumetric productivities can be envisaged in comparison with conventional aqueous-based systems. This approach also makes feasible reactions that are thermodynamically unfavored in aqueous media. The products formed, also often of a hydrophobic nature, partition back to the organic phase, a feature that allows *in situ* process recovery. Isolation of a given compound from an organic solvent is a relatively simple task. Furthermore, the potential inhibitory role of substrate/product toward the biocatalyst is minimized, since they are only slightly soluble in the aqueous phase where the enzyme tends to remain, and their partition to the aqueous phase is unfavored. Nevertheless, the amount of water in biphasic systems is above the minimum required for enzyme molecules to retain catalytic activity [2]. Biphasic systems are not without some drawbacks. The introduction of the organic phase increases the complexity of the reaction system and causes disturbance of downstream processing should emulsification take place. Vigorous agitation is required to increase mass transfer, bringing deleterious shear effects on biocatalysts. Enzyme inactivation at the aqueous-organic interface, in particular, is a commonly reported disadvantage of these systems [3]. On the other hand, however, some enzymes are less sensitive to interfacial inactivation, namely hydroxynitrile lyases and epoxide hydrolases [4], whereas a few enzymes, i.e. lipases and phospholipases, may actually require an aqueous-organic interface to display the desired activity (interfacial activation) as a result of conformational changes upon adsorption to the interface [5–7]. Along with those previously mentioned, chymotrypsin, dehydrogenases, laccases, penicillin acylases, proteases, subtilisin, and trypsin are other enzymes commonly used in organic-aqueous biphasic systems.

8.3 Classification of the Systems: Macro- and Microheterogeneous Systems

Macroheterogeneous systems are composed of distinguishable organic and aqueous phases. These systems are widely disseminated and are particularly useful when hydrophilic and hydrophobic substrates are simultaneously used; substrate and product are insoluble in the same media, and/or co-factors and the remaining components are insoluble in the same media. Biphasic systems are particularly useful when biocatalytic systems involving at least one hydrophobic species and requiring co-factor regeneration is used, since the latter is performed in aqueous media [8] (Figure 8.2). An extreme case of two liquid phase systems occurs when the aqueous phase is decreased in such manner that it is no longer distinguishable from the continuous organic phase. The resulting microheterogeneous system thus consists of water-in-oil dispersions stabilized by surfactants. These amphiphilic molecules lower the interfacial tension, typically to 10^{-5} dyne cm^{-1}[5], therefore reducing the positive free energy change of dispersion related to surface formation. The microemulsions thus formed are isotropic, thermodynamically

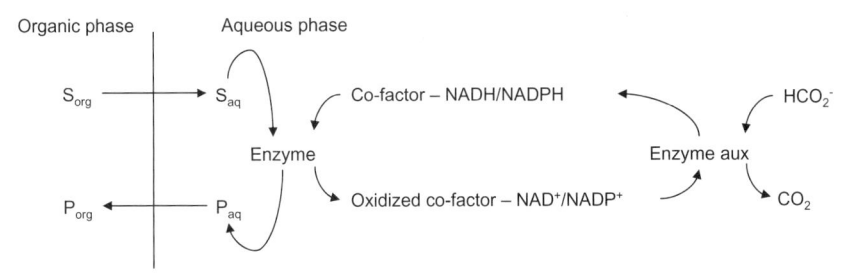

Figure 8.2 Schematics of a biocatalytic system in an organic-aqueous biphasic system involving co-factor (e.g., NADH, NADPH) regeneration, which occurs in the bulk aqueous phase. Substrate migrates from the bulk organic phase into the bulk aqueous phase, and the resulting product diffuses back into the bulk organic phase.

stable solutions, which can even occur spontaneously if the right mixture of surfactant, oil, and water are allowed to homogenize [9]. A particular structure of microemulsion yields reverse micelles, quite widespread in enzymatic biocatalysis.

Reversed micelles consist of water droplets, where enzyme and water-soluble molecules are encapsulated, surrounded by the amphiphilic molecules of a surfactant, which stabilizes the interface area, to yield a thermodynamically stable, isotropic, colloidal dispersion of water in oil (where oil is a general term for an apolar organic solvent). The polar heads of the surfactant are directed toward the aqueous core, whereas the hydrophobic tails are directed to the bulk non-polar organic phase. Reversed micelles can thus be seen as microreactors, with high interfacial areas (10 to $100 \, m^2 cm^{-3}$). The use of reverse micelles is particularly attractive when low water contents are needed, a typical requirement of synthetic processes, since these are favored by the shift of the thermodynamic equilibrium, and unwanted side reactions can be controlled. Reverse micelles have a rigid interior core that results from immobilization of the water present through the hydration of the hydrophilic moiety of the surfactants; in systems where the amount of water is enough to allow for mobile or free water in the core after the hydration requirements of the hydrophilic head groups of the surfactant, the result is microheterogeneous systems, generally termed microemulsions [9]. The use of microemulsions for laccase applications in biocatalysis provides a recent illustrative example [10]. These microemulsion systems present some drawbacks, mainly due to the presence of surfactants in relatively high concentrations, since this feature makes the separation of products difficult, and surfactants have denaturating effects on enzymes [11].

In order to ease product recovery and simultaneously allow enzyme reuse, gelled microemulsions have been effectively tested. The microemulsion is mixed with a hydrogelation mixture based on gelatin, gelatin-silica, k-carrageenan, or agar, which, upon gelation, yields a rigid, stable, and catalytically active gel in the presence of non-polar solvents [12]. Micro-emulsions can alternatively be formed by ultrasonication, thus avoiding the use of surfactants, but ultrasonic intensity has to be limited within a given range so as to promote emulsification while avoiding enzyme denaturation. An ultrasonic intensity of 106 W was proved the best compromise when the hydrolytic activity of lipase in a water/isooctane system was assessed. Still, the addition of oleic acid was required to improve the stability of the enzyme, when compared with a conventional stirred system, although activity was improved with stirred and reversed-micelle systems [13]. Extensive details on microemulsion systems can be found elsewhere [3, 11, 14].

8.4
Mechanisms of Enzyme Inactivation

Organic solvents can have a deleterious effect on enzymes, and work has been performed to gain a deeper insight into their role in enzyme inactivation, which ultimately can potentially be ascribed to three mechanisms [15–17]. Solvent molecules dissolved in the aqueous phase containing the biocatalyst may interact with non-polar groups of the enzyme and eventually disrupt hydrophobic interactions. Solvent molecules may compete with the protein molecule for the essential water required by the latter for proper polypeptide conformation, and ultimately strip the essential water layer, a phenomenon that can be mostly observed when water-miscible solvents are used, since the non-polarity of hydrophobic solvents prevents the disruption of the interaction of the enzyme molecule with the essential water; although solvent molecules of polar water miscible solvents may replace water molecules in the essential layer without structural effects, the same is not true when the replacement is made with molecules of low-polarity water-miscible organic solvents, because interaction with the protein molecule is quite different from that with water. Adsorption of enzyme molecules onto the aqueous–organic interface is possibly the more relevant mechanism for enzyme inactivation in biphasic systems, and its toxic effect is apparently cumulative, with molecular toxicity effects due to the organic solvent molecules dissolved in the aqueous phase [18]. The interface disturbs electrostatic, hydrogen-bond, and hydrophobic interactions of the protein, particularly those established between the hydrophobic core of the enzyme and the hydrophobic interface, resulting in irreversible denaturation, since the protein is unable to refold correctly after desorption from the interface [17, 19, 20, 21]. This inactivation mechanism presents a particular problem in biphasic systems, since increasing the interfacial area increases enzyme inactivation, although, on the other hand, it favors mass transfer [15, 16, 22]. Enzymes are not equally susceptible to interfacial inactivation; those with a more stable internal structure are likely to be less affected then those with a more flexible internal structure [17]. However, their behavior is difficult to predict, since several factors are to be considered, either related to the properties of the protein, such as adiabatic compressibility, secondary structure, surface hydrophobicity, and thermostability, or to the solvent polarity, pH, and ionic strength of the aqueous phase and interfacial tension and operational parameters such as stirring rate and organic-to-aqueous phase volume ratio [17, 22, 23].

8.5
Approaches to Protection of the Enzyme

Immobilization onto a solid support, either by surface attachment or lattice entrapment, is the more widely used approach to overcome enzyme inactivation, particularly interfacial inactivation. The support provides a protective microenvironment which often increases biocatalyst stability, although a decrease in biocatalytic activity may occur, particularly when immobilization is by covalent bonding. Nevertheless, this approach presents drawbacks, since the complexity (and cost) of the system is increased, and mass transfer resistances and partition effects are enhanced [24]. For those applications where enzyme immobilization is not an option, wrapping up the enzyme with a protective cover has proved promising [21].

This approach is based in the development of an enzyme-polymer conjugate. The polymer reacts with groups located in the enzyme surface, in such a manner that does not interfere with biocatalyst activity, to provide a shell that does not disturb the interaction of the modified enzyme with small molecules (substrate), but prevents its interaction with hydrophobic interfaces. Using as a model system three different enzymes, glucose oxidase, D-amino acid oxidase, and trypsin, each conjugated with dextran-aldehyde, conjugates were yielded that showed considerable stability in the presence of both organic-aqueous and air-liquid interfaces, as compared to unmodified enzymes. The incorporation of a sacrificial protein in the bioconversion that would be preferentially adsorbed at the interface and thus minimize the interfacial adsorption of the biocatalytic enzyme has also been suggested [23].

8.6
Solvent Selection

In the process of choosing an adequate water-immiscible solvent for the implementation of biphasic systems, several issues have to be addressed: the solvent has to present high affinity to substrates and/or products, be cheap, non-biodegradable, and non-toxic to humans, and must have a relatively high boiling point (though preferably lower then water) and relatively low volatility. Along with these features, it should be pointed out that the organic solvent may influence the enantioselectivity, regioselectivity, and chemoselectivity of the enzymatic reaction. However, by far the most important feature is the biocompatibility toward the biocatalyst. Extensive efforts have been made to correlate enzyme activity and stability in non-aqueous media, and particularly in organic-aqueous two-phase systems. These are discussed in this chapter, with given solvent parameters, and predictive equations are developed. Among such physical parameters, some of the more extensively used are the dielectric constant, the Hildebrand solubility, the three-dimensional solubility parameter, the Dimroth-Reichardt E_T parameter, and the Hansch parameter log P_{oct} (logarithm of the partition coefficient in a standard octanol–water two-phase system). None of the above has proved totally satisfactory, but the latter parameter is the most commonly used tool to provide guidance in solvent selection for biphasic biocatalysis, particularly since log P_{oct} values can be easily obtained or estimated (see http://www.syrres.com/eSc/est_kowdemo.htm, assessed 19th January 2007) [25]. This model suggests that solvents with a log P_{oct} value below 2 are toxic, whereas those with a log P_{oct} value above 4 are biocompatible, leaving a transitional region within. Its predictive application should be carefully assessed, as since it tends to prove misleading when matching solvents with different functionalities [26–28]. It has been suggested that the mechanisms involving enzyme/solvent interaction in biphasic systems may be too complex for the solvent biocompatibility to be predicted by a single physical parameter [27, 29]. The chemical functionality of the solvent has been efficiently used to correlate alcohol dehydrogenase activity in the presence of several water-immiscible organic solvents, a model system where the log P_{oct}-based predictive approach was not effective [27]. The chemical functionality of the solvent may include a set of molecular properties including the presence or absence of given residues or heteroatoms, the nature of the molecular orbitals, and concomitant effects on relevant macroscopic properties of the solvent, such as solvation energy, solubility, and surface tension [27]. The use of a relatively complex tool, multiple linear and non-linear regression analysis, may thus be required to quantitatively characterize and predict solvent biocompatibility [29, 30]. This approach requires the evaluation of the role played by a likely wide array of potentially significant physical parameters, as demonstrated in the work by Barberis and co-workers [29] and by Nurok and co-workers [30]. The former group assessed solvent-accessible non-polar saturated area (NPSA), solvent-accessible non-polar unsaturated area (NPUA), solvent-accessible polar surface area (PSA), polarizability, dipole moment, log P_{oct}, density, molecular volume, dipolarity-polarizability, hydrogen-bond donor ability, and hydrogen-bond acceptor ability, when studying the effect of organic solvents on the stabilitity of subtilisin and papain in organic solvents. Nurok and co-workers [30] were able to show that log P_{oct}, either individually or combined with a representative physical parameter, in their case NPUA, yielded regression models to predict initial rate/specificity, which nevertheless failed to cover in both cases the whole range of solvents tested. Barberis and co-workers [29], assessed solvent dipolarity/polarizability, solvent hydrogen-bond acid or acidity, solvent hydrogen-bond base or basicity, molar volume, Hildebrand cohesive energy density, dielectric constant, refraction index, dipole moment, Dimroth-Reichardt polarity indicator, log P_{oct}, Drago solvent polarity scale, and dielectric parameter. Although these authors were successful in correlating the stability of each of three phytoproteases to key physicochemical parameters, they highlighted that extrapolation of data to other biocatalysts is not feasible since each biocatalyst requires individual evaluation. A similar suggestion on the difficulties in developing a widely applicable predictive equation correlating the properties of organic solvents with enzyme activity/stability is also hinted at in the work of Nurok and co-workers [30]. Nevertheless there is a far better understanding of the mechanisms underlying biocatalyst/organic solvent interaction.

Efforts have also been made to correlate solvent properties, namely hydrophobicity, dipole moment, and dielectric constant with enantioselectivity, but such correlations have only been demonstrated in some systems [31–33].

8.7 Operational Parameters

Overall biotransformation rate in biphasic systems depends on both mass transfer and biochemical reaction. In this section, we focus on the parameters affecting mass transfer, which is influenced by interfacial area, mass transfer coefficients, and concentration driving force [34]. Each of these factors, on the other hand, depends on particular parameters. Thus, interfacial area depends on phase ratio and phase composition, which affect phase densities and interfacial tension, and intensity and nature of mixing, which influence the degree of dispersion of the two phases. Mass transfer coefficients are also influenced by phase composition, which controls diffusivity and brings interfacial turbulence, intensity and nature of mixing, which affect film thickness and interfacial turbulence, and physical properties of the systems. The concentration driving force depends on the solute concentration in the bulk of the two phases and on the partition coefficient, which governs the distribution of the substrate between the two phases. Some more detail on this is given in the following text. The particular field of microemulsions will not be addressed here since it has been extensively reviewed recently [14].

8.7.1
Interfacial Area and Volume Phase Ratio

The most common approach in organic-aqueous biphasic systems is to operate with a continuous aqueous phase, which prevents the occurrence of extreme pH spots, with a volume phase ratio up to 0.4 [35]. An adequate selection of the organic-aqueous volume phase ratio allows for high interfacial areas. The influence of the volume phase ratio in the interfacial area (a) can be predicted according to Equation 1.

$$a = \frac{6\phi}{d_{32}} \quad (1)$$

where ϕ is volume fraction of the dispersed phase and d_{32} is the Sauter mean droplet diameter, which can be estimated by

$$d_{32} = z_1 f(\phi) We^{z_0} d_i \quad (2)$$

where d_i is the impeller diameter, z_0 is a constant, usually assumed as -0.6, and z_1 is a constant with values that range between 0.04 and 0.4, and We is the Weber number, $We = \rho_c N^2 d_i^3 \sigma^{-1}$, where ρ_c is the density of the continuous phase, N is the impeller speed, and σ is the interfacial tension of the dispersed phase [36–38]. The function $f(\phi) = 1 + z_2 \phi$, if $\phi < 0.2$, and $2 < z_2 < 9$, whereas for $\phi > 0.3$, $f(\phi) = [\ln(z_3 + z_4 \phi)(\ln z_3)^{-1}]^{-3/5}$, where $z_3 \approx 0.011$ and z_4, which depends on the ratio between coalescence and dispersion coefficients, is typically close to 1.0. However, z_4, as well as other constant parameters, is characteristic of a given system and must be determined empirically if high accuracy is required, since they depend on the nature of each of the two liquid phases, diameter and nature of the impeller [37, 38]. Seader and Henley [36] provide a more general approach for the determination of z_1 and $f(\phi)$ in Equation 2. Thus, for We < 10000, $z_1 = 0.052$ and $f(\phi) = \exp(4\phi)$, whereas for We above 10000, $z_1 = 0.39$ and $f(\phi) = 1$, since d_{32} is almost independent of hold-up up to $\phi = 0.5$ [36]. In order to extend the use of d_{32} predictive correlation over a wider range of viscosities, a correction factor for viscosity can be added to Equation 2, where z_5 is a constant, depending on the system [39], leading to Equation 3.

$$d_{32} = z_1 f(\phi) We^{z_0} d_i \left(\frac{\mu_d}{\mu_c}\right)^{z_5} \tag{3}$$

More detailed information on the values used for parameters termed z_i can be found elsewhere [40].

In systems used in biphasic biocatalysis, maximum interfacial area is observed for a volume fraction of the organic phase close to that observed for phase inversion, which occurs within a 0.4 to 0.7 range. Extreme values of the volume fraction of the organic phase, under 0.1 or over 0.9 should also be avoided, since phase separation is poor and makes product recovery difficult.

It should be pointed out that the increase in the volume fraction of the dispersed phase may not always lead to a larger interfacial area, mainly in a high range of Weber numbers, roughly above 6000 [37]. In order to minimize the risk of uncontrolled phase inversion while still allowing for high interfacial area, several predictive models have been developed and tested for modeling phase inversion in biphasic systems [40–42]. One of these is given in Equation 4

$$\frac{\phi_{cr}}{1-\phi_{cr}} = \frac{d_{32}^{o/w}}{d_{32}^{w/o}} \tag{4}$$

where ϕ_{cr} represents the volume fraction of the organic phase at inversion, $d_{32}^{o/w}$ is the Sauter mean droplet diameter in the oil-water dispersion, and $d_{32}^{w/o}$ is the Sauter mean droplet diameter in the aqueous-organic dispersion [39].

Biphasic liquid systems may exist in two different states, phase separation state and emulsion state. The transition between these occurs at given hydrodynamic conditions, namely a given stirring speed in a stirred vessel. Irrespective of the state, increasing the stirring speed leads to enhanced mass transfer rates, since convection is favored. In the phase separation state, only convective mechanisms influence mass transfer since the interfacial area is constant, whereas in the emulsion state, further turbulence, along with the increased interfacial area (also positively influenced by increased stirring speed), enhances convective mixing. Since increased stirring enhances shear stress, the possibility of this effect leading to the inactivation of enzyme molecules was investigated. Early papers focusing on protein denaturation due to shear were based on data gathered through the use of the Couette viscometer [43, 44]. Such experimental work ruled out shear forces as having a relevant (if any) action on enzyme molecules.

Later work, carried out using mechanically stirred vessels, corroborated earlier conclusions [23, 45, 46]. Mechanical stirring was actually shown to enhance irreversible lysozyme inactivation (and aggregation) but does not directly bring it about, since it does not induce cleavage of enzyme molecules. It was evidenced that interfaces, either physical (air-liquid, glass-liquid, liquid-liquid) or molecular (as an outcome of the exposed hydrophobic surface of denaturated enzyme molecules that avoid aqueous media and form aggregates) induce and increase enzyme inactivation in stirred vessels. An increase in stirring rate enhances the rate of surface renewal, and concomitantly interfacial deleterious interaction is favored [23]. The pH value of the aqueous phase was also shown to affect interfacial inactivation, since it influences the adsorption of the enzyme molecule onto the interface [17].

8.7.2
Mass Transfer Coefficients and Process Modeling

A commonly used mass transfer reaction model is presented in Figure 8.1a, where the reaction occurs in the bulk aqueous phase [35, 47]. It is assumed that the substrate dissolved in the organic phase diffuses into the aqueous phase, reaching equilibrium. In the absence of reaction, once equilibrium is achieved, apparent mass transfer ceases. Given the presence of active enzyme, depletion of substrate in the aqueous phase occurs, and the system moves into a new equilibrium. Thus, the overall reaction rate depends both on reaction and mass transfer.

$$\frac{dS_{org}}{dt} = -J_s \frac{A}{V_{org}} \tag{5}$$

where S_{org} is the substrate concentration in the organic phase, t is the time, A is the area, V_{org} is the volume of organic phase and J_s is the flux of substrate, defined as

$$J_s = k_s (S_{org} - S_{org,eq}) \tag{6}$$

where k_s is the substrate mass transfer coefficient and $S_{org,eq}$ is the substrate concentration in the organic phase at equilibrium.

A similar equation can be written for the aqueous phase:

$$\frac{dS_{aq}}{dt} = J_s \frac{A}{V_{aq}} - r(S) \tag{7}$$

where S_{aq} is the substrate concentration in the aqueous phase, V_{aq} is the volume of organic phase, and r(s) is the reaction rate.

As for the product,

$$\frac{dP_{aq}}{dt} = J_P \frac{A}{V_{aq}} + \left(\frac{M_P}{M_S}\right) r(S) \tag{8}$$

where P_{aq} is the product concentration in the aqueous phase, M_P and M_S are the molecular weights of product and substrate, respectively, and J_P is the flux of product, defined as

$$J_P = k_P (P_{org} - P_{org,eq}) \tag{9}$$

where k_p is the product mass transfer coefficient and P_{org} and $P_{org,eq}$ are the product concentrations in the organic phase and in the organic phase at equilibrium, respectively.

And finally,

$$\frac{dP_{org}}{dt} = J_P \frac{A}{V_{org}} \tag{10}$$

Partition coefficients relate the concentrations in each of the two phases at equilibrium:

$$m_s = \frac{S_{o,eq}}{S_{aq,eq}} \tag{11}$$

$$m_p = \frac{P_{o,eq}}{P_{aq,eq}} \tag{12}$$

The model provides a good approach for the biotransformation system and highlights the main parameters involved. However, prediction of mass transfer effects on the outcome of the process, through evaluation of changes in the mass transfer coefficients, is rather difficult. A similar mass transfer reaction model, but based on the two-film model for mass transfer for a transformation occurring in the bulk aqueous phase as shown in Figure 8.3, could prove quite useful. Each of the films presents a resistance to mass transfer, but concentrations in the two fluids are in equilibrium at the interface, an assumption that holds provided surfactants do not accumulate at the interface and mass transfer rates are extremely high [36].

The overall process is described by the following equations:

$$\frac{dS_{org}}{dt} = -\frac{A}{V_{org}} J_{b,org} \tag{13}$$

where $J_{b,org}$ is the flux of substrate in the bulk organic phase.

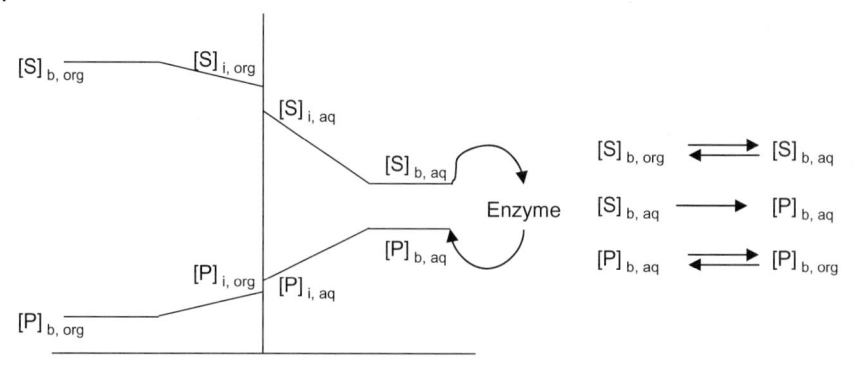

Figure 8.3 Two-liquid film model adapted to a biotransformation system where substrate and product are hydrophobic and reaction occurs in the bulk aqueous phase.

Substrate migrates from the bulk organic phase into the interface:

$$J_{s,org} = k_d(S_{b,org} - S_{i,org}) \quad (14)$$

where k_d is the mass transfer coefficient in the dispersed phase (here assumed as the organic phase), $S_{b,org}$ and $S_{i,org}$ are the substrate concentrations in the bulk organic phase and in the interface.

A mass balance to the aqueous phase yields

$$\frac{dS_{aq}}{dt} = \frac{A}{V_{aq}} J_{s,aq} - r(S) \quad (15)$$

where $J_{s,aq}$ in the flux of substrate in the aqueous phase, defined as

$$J_{s,aq} = k_c([S]_{i,aq} - [S]_{b,aq}) \quad (16)$$

where k_c is the mass transfer coefficient in the continuous phase (here assumed as the aqueous phase), and $S_{b,aq}$ and $S_{i,aq}$ are the substrate concentrations in the bulk aqueous phase and in the interface.

The partition coefficient which relates equilibrium concentrations in the interface is given by

$$m_s = \frac{[S]_{i,aq}}{[S]_{i,org}} \quad (17)$$

The product formed partitions back to the organic phase, according to

$$\frac{dP_{aq}}{dt} = \frac{A}{V_{aq}} J_{p,aq} + r(S)\left(\frac{M_p}{M_s}\right) \quad (18)$$

where $J_{p,aq}$ is the flux of product in the aqueous phase, defined as

$$J_{p,aq} = k_c([P]_{b,aq} - [P]_{i,aq}) \tag{19}$$

where $P_{b,aq}$ and $P_{i,aq}$ are the product concentrations in the bulk aqueous phase and in the interface.

And finally

$$J_{p,org} = k_d([P]_{i,org} - [P]_{b,org}) \tag{20}$$

where $J_{p,org}$ is the flux of product in the aqueous phase and $P_{b,org}$ and $P_{i,org}$ are the product concentrations in the bulk organic phase and in the interface.

A partition coefficient which relates concentrations for the product at the interface can also be defined by

$$m_p = \frac{[P]_{i,aq}}{[P]_{i,org}} \tag{21}$$

Correlations for the mass transfer coefficients in the continuous (k_c) and disperse phase (k_d) can be found in the literature. An expression commonly used was developed by Calderbank and Moo-Young [38]:

$$k_c = 0.13(P\mu_c V_c^{-1} \rho_c^{-2})^{0.25} (\mu_c \rho_c^{-1} D^{-1})^{-0.67} \tag{22}$$

where D is the diffusivity of the compound in the continuous phase, V_c is the volume of the continuous phase, μ_c is the viscosity of the continuous phase, ρ_c is the density of the continuous phase, and P is the power dissipated by the agitator, given by

$$P = Po\rho_c N^3 d_i^5 \tag{23}$$

where Po is the power number, d_i is the impeller diameter and N is stirring rate.

On the other hand, k_d can be estimated through [36]

$$k_d = 2\pi^2 d_i (3d_{32})^{-1} \tag{24}$$

provided the droplets can be considered rigid spheres.

A more elaborate relation for determination of k_d can be found elsewhere [48].

The overall mass transfer coefficients based on either the continuous (K_c) or dispersed (K_d) phases can be written as

$$(K_c)^{-1} = \frac{m}{k_d} + (k_c)^{-1} \tag{25}$$

and

$$(K_d)^{-1} = \frac{m}{k_d} + (mk_c)^{-1} \tag{26}$$

Irrespective of the model used, numerical integration is required. Stirring has a direct effect on the intrinsic mass transfer coefficients, and thus adequate mixing is required to ensure intimate contact between the two phases in order to effectively transfer substrate(s) and/or products from one phase to the other. Mass transfer limitations may also occur in isolated enzyme-catalyzed reactions if the solvent is highly viscous or if the biocatalyst has a high specific activity [35]. Again, biocatalyst immobilization often brings about internal mass transfer resistances, but these are strongly related to the nature of the immobilization matrix [49].

Although many biochemical reactions take place in the bulk aqueous phase, there are several, catalyzed by hydroxynitrile lyases, where only the enzyme molecules close to the interface are involved in the reaction, unlike those enzyme molecules that remain idly suspended in the bulk aqueous phase [6, 50, 51]. This mechanism has no relation to the interfacial activation mechanism typical of lipases and phospholipases. Promoting biocatalysis in the interface may prove fruitful, particularly if substrates are dissolved in both aqueous phases, provided that interfacial stress is minimized. This approach was put into practice recently for the enzymatic epoxidation of styrene [52]. By binding the enzyme to the interface through conjugation of chloroperoxidase with polystyrene, a platform that protected the enzyme from interfacial stress and minimized product hydrolysis was obtained. It also allowed a significant increase in productivity, as compared to the use of free enzyme, and simultaneously allowed continuous feeding, which further enhanced productivity.

A particular model, the adsorbed enzyme model, was developed, as it often provides a more realistic approach to such a system. However, both models (mass transfer and adsorbed enzyme) may coexist simultaneously, a matching of the quantitative mathematical descriptions being required to rule out one of the models [6].

The use of the Hatta number has been suggested to assess whether the reaction occurs in the bulk phase or in the interface. This parameter relates the transfer rate due to catalysis to the transfer rate due to diffusion. For a given substrate, the Hatta number can be given by

$$Ha = \frac{1}{k_{aq}} \left(\frac{2k_c DS_{aq,eq}^{n-1}}{n+1} \right)^{0.5} \tag{27}$$

where D is the diffusivity of the substrate in the aqueous phase, n is the order of the reaction, and k_c is the reaction rate constant. Further details regarding this matter can be found elsewhere [53].

On the basis of the Hatta number, the transformations carried out in biphasic systems can be described as slow (Ha < 0.3), intermediate (with a kinetic–diffusion regime 0.3 < Ha < 3.0), and fast (Ha > 3). These are diffusion limited and take place near the interface (within the diffusion layer). Slow transformations are under kinetic control and occur mostly in a bulk phase, so that the amount of substrate transformed in the boundary layer in negligible. When diffusion and reaction rate are of similar magnitude, the reaction takes place mostly in the diffusion layer, although extracted substrate is also present in the continuous phase, where it is transformed at a rate depending on its concentration [38, 50, 54].

8.8
Reactor Types

Stirred tank reactors are widely used in organic-aqueous biocatalysis. Membrane reactors are also used, the membrane creating a physical boundary that contains the enzyme in a given phase, while simultaneously allowing for phase separation, a major asset for the simplification of downstream processing, and allowing high interfacial areas, from 500 to 5000 $m^2 m^{-3}$ [54]. Typical reactor configurations are given in Figure 8.4. Within the concept of stirred reactors, emulsion reactors and Lewis cell reactors can be used. In emulsion reactors, stirring leads to the dispersion of one phase in the bulk of the second phase, in the form of tiny droplets. Intimate contact between the two phases is thus achieved. Interfacial area varies according to stirring intensity. Lewis cells, on the other hand, provide a constant interfacial area of contact between the two phases. These reactors are cylindrical vessels divided into two independently stirred compartments in which each of the two phases is contained. An interfacial plate, containing a central hole that allows contact between the two phases, divides the two compartments. The diameter of this hole defines the interfacial area. Stirring intensity requires careful monitoring to avoid excessive rippling, which disturbs the interface. These reactors are therefore mostly used for exact characterization of biphasic systems by fixing a given interfacial area. In emulsion reactors, droplet lifetime and size distribution, and hence interfacial area, are difficult to establish, which introduces further complexity in the system.

Membrane reactors can be considered passive or active according to whether the membrane plays the role of a simple physical barrier that retains the free enzyme molecules solubilized in the aqueous phase, or it acts as an immobilization matrix binding physically or chemically the enzyme molecules. Polymer- and ceramic-based micro- and ultrafiltration membranes are used, and particular attention has to be paid to the chemical compatibility between the solvent and the polymeric membranes. Careful, fine control of the transmembrane pressure during operation is also required in order to avoid phase breakthrough, a task that may sometimes prove difficult to perform, particularly when surface active materials are present or formed during biotransformation. Silicone-based dense-phase membranes have also been evaluated in whole-cell processes [55, 56], but

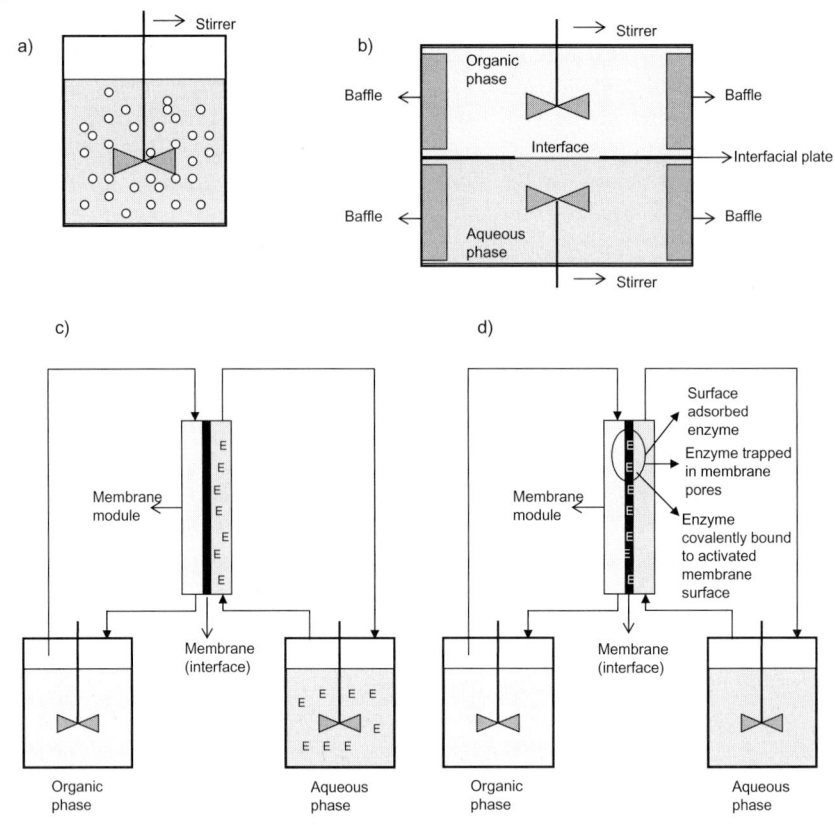

Figure 8.4 Reactor types used in organic-aqueous biphasic systems: (a) Emulsion reactor, (b) Lewis cell, (c) passive membrane reactor, (d) active membrane reactor. "E" represents enzyme molecules.

mass transfer rates through dense membranes are rather low compared with microporous membranes [57], which considerably limits the feasibility of such an approach.

8.9
Downstream Processing

The first step in downstream processing is the separation of the product-rich phase from the second phase and the biocatalyst. This may be simplified if the enzyme is immobilized or if a membrane module is included in the experimental set-up. In the case of emulsion reactors, centrifugation for liquid phase separation is a likely separation process [58], although the small size of droplets, the possibility of stable emulsion formation during the reaction, particularly if surface-active compounds are involved, and a low phase density difference (less then 3% difference between the two phases) [59] may render this approach ineffective [35]. Thermal processing, either by heating or cooling, may help to break the emulsion and allow for a clear organic phase [58]. An alternative to phase separation is the use of a microporous membrane downstream of the reactor.

8.10
Recent Applications

Biphasic systems have been effectively used in several enzyme-catalyzed reactions, including peptide and alkyl glycosides synthesis, esterification and transesterification, alcoholysis, hydrolysis, and enantiomeric resolution [2, 24, 60]. Although application of this particular bioconversion system has been used for final products, it is mostly used in the production of intermediate compounds, particularly optically active ones, that can be used as building blocks in the pharmaceutical and food sectors [61–64]. Updated reviews have addressed this matter [2, 4, 24, 60–63], and examples of some representative recent applications of this methodology are given in Table 8.1).

8.11
Conclusions

Biocatalysis in organic-aqueous systems provides a feasible approach for the development and implementation of biotransformation processes with high volumetric productivities, involving the use of hydrophobic substrates and/or products, or combinations of hydrophobic and hydrophilic compounds. This methodology has been extensively and successfully used in the production of optically active compounds, where it clearly supasses its chemical counterpart and has been shown to allow for the expression of biocatalytic activity not apparent in aqueous media. Efforts to further understand, predict, and hence model the interaction between organic solvents and enzyme are continuously being carried out, contributing to a faster, cheaper, and more effective development of biphasic systems. It is also noted that there is a trend toward further development of the characterization of engineering aspects in biphasic biocatalysis, which certainly contributes to more efficient and reproducible systems, enables effective scale-up, and leads to the implementation of integrated systems, with considerable cost advantages.

Acknowledgments

P. Fernandes acknowledges grants SFRH/BPD/20416/2004 from Fundação para a Ciência e a Tecnologia, Portugal.

Table 8.1 Some recent examples of biphasic systems in biocatalysis.

Biocatalyst	System	Reference
Alcohol dehydrogenase (Regeneration of co-factor is performed by the same enzyme, using 2-propanol as co-factor)	Reduction of acetophenone to (R)-phenylethanol in phosphate buffer/ methyl tert-butyl ether	[27]
Alcohol dehydrogenase (coupled to formate dehydrogenase for co-factor regeneration)	Reduction of ketones in phosphate buffer/n-heptane	[65]
Chloroperoxidase	Oxidation of primary alcohols to aldehydes in acetate or citrate buffer/hexane or ethyl acetate	[66]
Epoxide hydrolase	Kinetic resolution of epoxides in Tris buffer/octane	[18]
β-Glycosidase and β-galactosidase	Synthesis of alkyl-b-glucosides in acetate buffer/1-hexanol or 1-octanol	[67]
Hydrolase	Resolution of (R,S)-ethyl mandelate and (R,S)-ethyl 2-chloromandelate in phosphate buffer/iso-octane	[68, 69]
Hydroxynitrile lyase	Synthesis of (R)-mandelonitrile in sodium phosphate buffer/methyl tert-butyl ether	[70]
Laccase	Oxidation of ferulic acid in sodium phosphate buffer/ethyl acetate to yield colorants	[71]
Lipase	Resolution of (R)-isomer and (S)-isomer of 2-hydroxy octanoic acid in phosphate buffer/iso-octane	[72]
Penicillin acylase (immobilized)	Penicillin G hydrolysis in countercurrent system, in water (low pH, 4–5.6)/butyl acetate	[73]
Pyruvate decarboxylase	Production of (R)-phenylacetylcarbinol, a precursor of ephedrine in MES-KOH buffer/1-pentanol	[28]
Thermolysine	Syntheses of N-(benzyloxycarbonyl)-L-alanyl-L-phenylalanine methyl ester and N-(benzyloxycarbonyl)-L-aspartyl-phenylalanine methyl ester in HEPES or MES-NaOH buffers/ethyl acetate	[74]
Thermolysine	Synthesis of ZAlaPheOMe in a membrane contactor using Tris-HCl/ethyl acetate	[54]

References

1 A. J. J. Straathof, S. Panke, A. Schmid, Curr. Opin. Biotechnol. 2002, 13, 548–556.
2 S. H. Krishna, Biotechnol. Adv. 2002, 20, 239–266.
3 N. L. Klyachko, A. V. Levashov, (Bioorganic synthesis in reverse micelles and related systems) Curr. Opin. Colloid Interface Sci. 2003, 8, 179–186.

4 S. Panke, M. Wubbolts, (Enzyme technology and bioprocess engineering) *Curr. Opin. Biotechnol.* 2002, **13**, 111–116.
5 H. Stamatis, A. Xenakis, F. N. Kolisis, *Biotechnol. Adv.* 1999, **17**, 293–318.
6 A. J. J. Straathof, *Biotechnol. Bioeng.* 2003, **83**, 371–375.
7 A. Aloulou, J. A. Rodriguez, S. Fernandez, D. van Oosterhout, D. Puccinelli, F. Carrière, *Biochim. Biophys. Acta* 2006, **1761**, 995–1013.
8 M. Eckstein, T. Daussmann, U. Kragle, *Biocatal. Biotrans.* 2004, **22**, 89–96.
9 S. P. Moulik, B. K. Paul, *Adv. Colloid Interface Sci.* 1998, **78**, 99–195.
10 J. Rodakiewicz-Nowak, N. N. Pozdnyakova, O. V. Turkovskaya, *Biocatal. Biotrans.* 2005, **23**, 271–279.
11 C. M. L. Carvalho, J. M. S. Cabral, *Biochemie* 2000, **82**, 1063–1085.
12 H. Stamatis, A. Xenakis, *J. Mol. Catal. B: Enzymatic* 1999, **6**, 399–406.
13 M. M. R. Talukder, M. M. Zaman, Y. Hayashi, J. C. Wu, T. Kawanishi, *Biocatal. Biotrans.* 2006, **24**, 189–194.
14 C. M. L. Carvalho, J. M. S. Cabral (Reversed micellar bioreaction systems), in *Multiphase Bioreactor Design*, J. M. S. Cabral, M. Mota, J. Tramper (eds.), Taylor & Francis, London, pp. 181–223, 2001.
15 A. S. Ghatorae, G. Bell, P. J. Halling, *Biotechnol. Bioeng.* 1994a, **43**, 331–336.
16 A. S. Ghatorae, M. J. Guerra, G. Bell, P. J. Halling, *Biotechnol. Bioeng.* 1994b, **44**, 1355–1361.
17 A. C. Ross, G. Bell, P. J. Halling, *Biotechnol. Bioeng.* 2000, **67**, 498–503.
18 H. Baldascini, K. J. Ganzeveld, D. B. Janssen, A. A. C. M. Beenackers, *Biotechnol. Bioeng.* 2001, **73**, 44–54.
19 A. Sadana, *Bioseparation* 1993, **3**, 297–320.
20 H. Wu, Y. Fan, J. Sheng, S. F. Sui, *Eur. Biophys. J.* 1993, **22**, 201–205.
21 L. Betancor, F. López-Gallego, A. Hidalgo, N. Alonso-Morales, M. Fuentes, R. Fernández-Lafuente, J. M. Guisán, *J. Biotechnol.* 2004, **110**, 201–207.
22 K.-K. Fan, P. Ouyang, X. Wu, Z. Lu, *Enzyme Microb. Technol.* 2001, **28**, 3–7.
23 H. Baldascini, D. B. Janssen, *Enzyme Microb. Technol.* 2005, **36**, 285–293
24 P. Fernandes, J. M. S. Cabral (Biotransformations) in *Basic Biotechnology* 3rd edn, Cambridge University Press, Cambridge, 579–626, 2006.
25 R. F. Rekker, R. Mannhold, *Calculation of drug lipophilicity: the hydrophobic fragmentat constant approach*, VCH, Weinheim, Germany, 1992.
26 M. Vermuë, J. Tramper, *Pure Appl. Chem.* 1995, **67**, 345–373.
27 M. V. Filho, T. Stillger, M. Müller, A. Liese, C. Wandrey, *Chem. Int. Ed.* 2003, **42**, 2993–2996.
28 B. Rosche, M. Breuer, B. Hauer, P. L. Rogers, *Biotechnol. Bioeng.* 2004, **86**, 788–794.
29 S. Barberis, E. Quiroga, S. Morcelle, N. Priolo, J. M. Luco, *J. Mol Catal. B: Enzymatic* 2006, **38**, 95–103.
30 D. Nurok, R. M. Kleyle, B. B. Muhoberac, M. C. Frost, P. Hajdu, D. H. Robertson, S. V. Kamat, A. J. Russell, *J. Mol. Catal. B: Enzymatic*, 1999, **7**, 273–282.
31 G. Carrea, G. Ottolina, S. Riva, *Trends Biotechnol.* 1995, **13**, 63–70.
32 M. Persson, D. Costes, E. Wehtje, P. Adlercreutz, *Enzyme Microb. Technol.* 2002, **30**, 916–923.
33 G. de Gonzalo, G. Ottolina, F. Zambianchi, M. W. Fraaije, G. Carrea, *J. Mol. Catal. B: Enzymatic* 2006, **39**, 91–97.
34 T. C. Lo (Commercial liquid-liquid extraction equipment) in *Handbook of Separation Techniques for Chemical Engineers*, Philip A. Schweitzer, eds., McGraw-Hill Professional, New York, pp. 1-449–1-518.
35 G. J. Lye, J. M. Woodley, (Advances in the selection and design of two liquid phase biocatalytic reactors), in *Multiphase Bioreactor Design*, J. M. S. Cabral, M. Mota, J. Tramper (eds.), Taylor & Francis, London, pp. 115–134, 2001.
36 J. D. Seader, E. J. Henley, *Separation Process Principles* 1998, John Wiley & Sons, New York.
37 P. A. Quadros, C. M. S. G. Baptista, *Chem. Eng. Sci.* 2003, **58**, 3935 – 3945.
38 A. Campanella, M. A. Baltanás, *Chem. Eng. J.* 2006, **118**, 141–152.

39 L. Y. Yeo, O. K. Matar, E. S. P. de Ortiz, G. F. Hewitt, *Chem. Eng. Sci.* 2002, **57**, 1069–1072.
40 B. Hu, P. Angelia, O. K. Matar, G. F. Hewitt, *Chem. Eng. Sci.* 2005, **60**, 3487–3495.
41 L. Y. Yeo, O. K. Matar, E. S. P. de Ortiz, G. F. Hewitt, *J. Colloid Interface Sci.* 2002, **248**, 443–454.
42 B. Hu, L. Liu, O. K. Matar, P. Angeli, G. F. Hewitt, E. S. P. de Ortiz, *Tsinghua Sci. Technol.* 2006, **11**, 202–206.
43 C. R. Thomas, P. Dunnill, *Biotechnol. Bioeng.* 1979, **21**, 2279–2302.
44 C. R. Thomas, A. W. Nienow, P. Dunnill, *Biotechnol. Bioeng.* 1979, **21**, 2263–2278.
45 S. Colombié, A. Gaunand, B. Lindet, *Enzyme Microb. Technol.* 2001, **28**, 820–826.
46 S. Colombié, A. Gaunand, B. Lindet, *J. Mol. Catal. B: Enzymatic* 2001, **11**, 559–565.
47 H. J. Chae, Y. J. Yoo, *J. Chem. Technol. Biotechnol.* 1997, **70**, 163–170.
48 R. E. Treybal, *Mass-transfer operations*, 3rd edn., McGraw-Hill International Student Edition, Singapore, 1981.
49 B. S. Ferreira, P. Fernandes, J. M. S. Cabral, (Design and modelling of immobilized biocatalytic reactors), in *Multiphase Bioreactor Design*, J.M.S. Cabral, M. Mota, J. Tramper (eds.), Taylor & Francis, London, pp. 85–114, 2001.
50 M. Bauer, R. Geyer, H. Griengl, W. Steiner, *Food Technol. Biotechnol.* 2002, **40**, 9–19.
51 L. G. Cascão-Pereira, A. Hickel, C. J. Radke, H. W. Blanch, *Biotechnol. Bioeng.* 2003, **83**, 498–501.
52 G. Zhu, P. Wang, *J. Biotechnol.* 2005, **117**, 195–202.
53 K. R. Westerterp, W. P. M. van Swaij, A. A. C. M. Beenackers, *Chemical Reactor Design and Operation*, John Wiley and Sons, Chichester, 1984.
54 A. Trusek-Holownia, *Biochem. Eng. J.* 2003, **16**, 69–77.
55 S. D. Doig, A. T. Boam, D. I. Leak, A. G. Livingston, D. C. Stuckey, *Biotechnol. Bioeng.* 1998, **58**, 587–594.
56 S. D. Doig, A. T. Boam, A.G. Livingston, D. C. Stuckey, *Biotechnol. Bioeng.* 1999, **63**, 601–611.
57 R.-S. Juang, S.-Y. Tsai, *J. Membrane Sci.* 2006, **282**, 484–492.
58 A. Schmid, A. Kollmer, R. G. Mathys, B. Witholt, *Extremophiles* 1998, **2**, 249–256.
59 W. L. McCabe, J. C. Smith, P. Harriott, *Unit Operations in Chemical Engineering*, 5th edn., McGraw-Hill Higher Education, New York, 1993.
60 G. Carrea, S. Riva, *Chem. Int. Ed.* 2000, **39**, 2226–2254.
61 W. A. Loughlin, *Bioresource Technol.* 2000, **74**, 49–62.
62 J. Schrader, M. M. W. Etschmann, D. Sell, J.-M. Hilmer, J. Rabenhorst, *Biotechnol. Lett.* 2004, **26**, 463–472.
63 S. Panke, M. Wubbolts, *Curr. Opin. Chem. Biol.* 2005, **9**, 188–194.
64 S. Riva, *Trends Biotechnol.* 2006, **24**, 219–226.
65 H. Gröger, W. Hummel, C. Rollmann, F. Chamouleau, H. Hüsken, H. Werner, C. Wunderlich, K. Abokitse, K. Drauzd, S. Buchholz, *Tetrahedron*, 2004, **60**, 633–640.
66 E. Kiljunen, L. T. Kanerva, *J. Mol. Catal. B: Enzymatic* 2000, **9**, 163–172.
67 S. Papanikolaou, *Bioresource Technol.* 2001, **77**, 157–161.
68 P.-Y. Wang, S.-W. Tsai, *Enzyme Microb. Technol.* 2005, **37**, 266–271.
69 P.-Y. Wang, S.-W. Tsai, *Enzyme Microb. Technol.* 2006, **39**, 930–935.
70 W. F. Willeman, P. J. Gerrits, U. Hanefeld, J. Brussee, A. J. J Straathof, A. van der Gen, J. J. Heijnen, *Biotechnol. Bioeng.* 2002, **77**, 239–247.
71 R. Mustafa, L. Muniglia, B. Rovel, M. Girardin, *Food Res. Int.* 2005, **38**, 995–1000.
72 K. Sabaki, S. Hara, N. Itoh, *Desalination*, 2002, **149**, 247–252.
73 J. L. den Hollander, M. Zomerdijk, A. J. J. Straathof, L. A. M. van Der Wielen, *Chem. Eng. Sci.* 2002, **57**, 1591–1598.
74 M. Miyanaga, M. Ohmori, K. Imamura, T. Sakiyama, K. Nakanishi, *J. Biosci. Bioeng.* 2000, **90**, 43–51.

9
Biocatalysis in Biphasic Systems: Oxynitrilases
Manuela Avi and Herfried Griengl

9.1
Introduction

In the last decades, cyanohydrins have become versatile chiral building blocks, not only for laboratory synthesis, but also for a range of pharmaceuticals and agrochemicals. Several methods for the enantioselective preparation of these compounds have been published [1, 2]. The most important synthetic approaches are catalysis by oxynitrilases, also termed hydroxynitrile lyases (HNLs), wording used in this chapter, [3] and by transition metal complexes [4], whereas the relevance of cyclic dipeptides as catalysts is decreasing [2].

HNLs are found in over 3000 higher plant species, like *Rosaceae, Euphorbiaceae, Linaceae* and *Filitaceae*, in several bacteria, and in some insects [5, 6]. For cyanohydrin synthesis, mainly five HNLs have been used: the (R)-selective HNLs from *Prunus amygdalus* (PaHNL) and *Linum usitatissimum* (LuHNL) and the (S)-selective ones from *Hevea brasiliensis* (HbHNL), *Manihot esculenta* (MeHNL), and *Sorghum bicolor* (SbHNL) [1].

The first synthesis of mandelonitrile using hydroxynitrile lyases was published by Rosenthaler in 1908 [7], but only fifty years later this research field regained attention for synthetic purposes by Becker et al. [8–10]. However, because of the spontaneous chemical addition of HCN in the aqueous and alcoholic medium used, only low enantiomeric excesses (*ee*) were achieved. In addition, the low solubility of many substrates in the buffer made this reaction unsuitable for laboratory scale and industrial application.

The breakthrough came with the employment of water-immiscible organic solvents, where the undesired chemical reaction is more or less suppressed [11]. Since then, several publications dealing with liquid/liquid biphasic systems or the application of an immobilized enzyme in pure organic solvent have appeared.

In this chapter, the use of HNL-catalysis in such systems to provide increased yield and enantiomeric excess is discussed.

Organic Synthesis with Enzymes in Non-Aqueous Media. Edited by Giacomo Carrea and Sergio Riva
Copyright © 2008 WILEY-VCH Verlag GmbH & Co. KGaA, Weinheim
ISBN: 978-3-527-31846-9

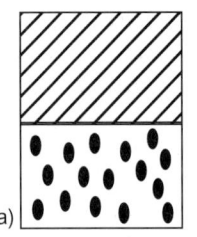

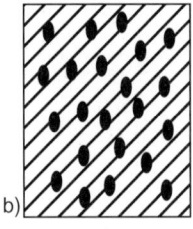

Figure 9.1 Representation of some biphasic systems: (a) liquid/liquid systems, and (b) liquid/solid system. Dashed area: organic phase; white area: water phase; black dots: biocatalyst [12].

9.2
Biphasic Systems

Over the last twenty years, biocatalysis in organic solvents has emerged as an important area of research and has led to the widespread industrial use of enzymes. In an aqueous medium, with which enzymes are naturally associated, most substrates are poorly soluble, and this leads to low reaction rates and high reaction volumes. In addition, in water, undesired side reactions and degradation of organic compounds often occur. Also, the thermodynamic equilibrium is unfavorable and product recovery can be difficult. Working in organic solvent avoids these problems, and enzymes often show beneficial behavior in such systems [12–14].

Organic solvent systems include mixtures of water and water-miscible solvents, biphasic systems of water and water-immiscible solvents, pure organic solvents, and reverse micellar systems.

In a liquid/liquid biphasic system (Figure 9.1a), the enzyme is in the aqueous phase, whereas the hydrophobic compounds are in the organic phase. In pure organic solvent (Figure 9.1b) a solid enzyme preparation is suspended in the solvent, making it a liquid/solid biphasic system. In a micellar system, the enzyme is entrapped in a hydrated reverse micelle within a homogeneous organic solvent [12]. Reverse micellar systems have not found application in the HNL-catalyzed synthesis of cyanohydrins and are therefore beyond the scope of this review.

9.2.1
Liquid/Liquid Biphasic Systems

9.2.1.1 Buffer/Organic Solvent
In an aqueous system, the only possible way to suppress the unwanted chemical HCN addition to carbonyl compounds is to work at low pH to reduce the concentration of the cyanide ion – the reactive species – and at low temperature to increase the selectivity. However, in an acidic medium, most HNLs lose activity very fast [15, 16], and it is therefore necessary to find reaction conditions to avoid these problems.

One method is to run the reaction in an aqueous buffer/organic solvent biphasic system. This makes it possible to work at high substrate and product concentrations and at the pH-optimum of the enzyme. In addition, in water immiscible solvents the non-enzymatic addition of HCN to the carbonyl group is non-existent or extremely slow. Possible disadvantages are enzyme deactivation at the interface and the presence of organic solvent dissolved in the aqueous phase [15, 17, 18].

The first use of such biphasic systems was reported for a transhydrocyanation process with acetone cyanohydrin as HCN donor [19]. In the years following this discovery, the addition of HCN in buffer/organic solvent was extensively investigated, optimizing the reaction conditions by empirical approaches. The group of van der Gen [20] examined the interplay of a number of reaction parameters for cyanohydrin synthesis with hydroxynitrile lyase of *P. amygdalus* (*Pa*HNL). They concluded that the nature of the organic solvent has a large effect on the reaction. *tert*-Butyl methyl ether (*t*BME) was clearly to be preferred over ethyl acetate. Varying the phase volume ratio in favor of the buffer increased the conversion but decreased the *ee*, which implies that the non-enzymatic reaction rate increases with larger buffer volumes. Through the optimization of these parameters, an industrially attractive process was developed. (*R*)-Mandelonitrile (9.5 kg) was produced in a total volume of 100 L in 20 min using 15 g of *Pa*HNL [20].

Similar results were found by Griengl and co-workers [21] for *Hb*HNL catalysis. Ethers, such as diisopropyl ether (DIPE) or *t*BME, were found to be the most suitable solvents. The transformation proceeds most efficiently at temperatures between 5 and 15 °C, and the formation of a stable emulsion seems to be of importance. A series of aldehydes were converted by this method (Figure 9.2). Compared to transformations in aqueous buffer medium [22], higher conversions and *ee* were achieved (Table 9.1) [21].

The influence of the solvent and other reaction parameters (pH, temperature) on enzyme stability were studied in more detail by different groups, who confirmed the previous results [17, 18, 23]. Hickel et al. studied the adsorption of *Pa*HNL at the interface, determining the tension between the two phases. Apolar solvents show a large increase in the interfacial pressure. The enzyme exhibits a high initial activity, but the loss of catalytic activity occurs within short time. *Pa*HNL is quickly adsorbed and unfolded by such hydrophobic solvents, losing its activity. Using relatively polar solvents such as DIPE and *t*BME, no changes in the interfacial tension are measured. The initial activity is much lower, but is constant over several hours. The adsorption of the enzyme at the interface is weaker, and no conformational changes of the active site occur [17, 23].

Bauer et al. [18] showed that, for *Hb*HNL also, ethers are the solvents of choice to maintain enzyme activity. The stability in the presence of the substrates was also investigated, indicating that the enzyme is already destabilized at low

Figure 9.2 Enzymatic cyanohydrin formation. R = aryl, alkyl, cycloalkyl, heteroaryl.

Table 9.1 Comparison of the conversion of aldehydes (RCHO) in an aqueous buffer [22] and a biphasic system [21].

R-	Conversion[a] and (ee)[b] [%] in aqueous citrate buffer at pH 4.5[c]	Conversion[a] and (ee)[b] [%] in buffer/tBME at pH 5.5
Ph	67 (99)	97 (99)
(E)-PhCH=CH	0 (0)	93 (98)[d]
3-PhOPh	9 (99)	99 (99)
PhOCH$_2$OCH$_2$	n.d. (0)	92 (12)
c-C$_6$H$_{11}$	94 (99)	95 (99)
2-furyl	55 (98)	95 (98)
3-fury	61 (99)	98 (98)
3-thienyl	67 (98)	98 (99)
CH$_3$(CH$_2$)$_4$	n.d. (84)	97 (98)
CH$_2$=CH	38 (94)	92 (98)
(E)-CH$_3$(CH$_2$)$_4$CH=CH	62 (99)	96 (99)
CH$_3$(CH$_2$)$_2$C≡C	43 (80)	62 (98)

a Determined by GC of the corresponding acetates or UV-spectroscopy.
b Determined by chiral GC of the acetates.
c Data from [22].
d Determined by chiral HPLC of the tBDMS ethers.
n.d. Not determined.

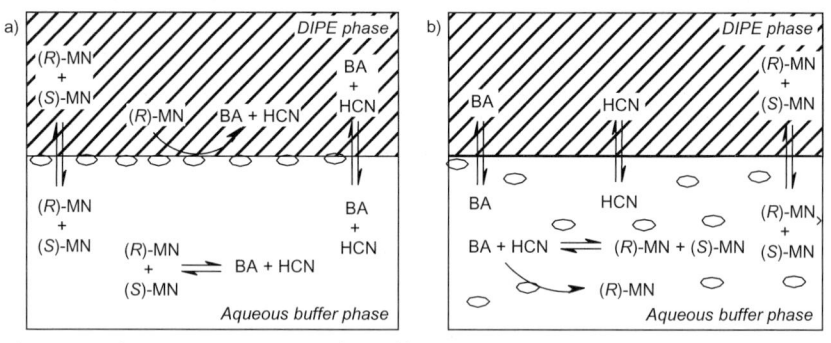

Figure 9.3 Schematic representation of possible reaction sites for PaHNL-catalyzed cleavage/synthesis of mandelonitrile (MN): (a) adsorbed enzyme model [24] and (b) mass transfer model [27]. BA: benzaldehyde; white dots: enzyme.

concentrations of benzaldehyde and HCN. Therefore high concentrations of substrates have to be avoided.

The mechanism of the reaction is still under discussion. Hickel et al. [24] proposed that mandelonitrile cleavage by PaHNL occurs at the organic solvent/water interface and not in the aqueous bulk phase (Figure 9.3a). In their recycling reactor, the buffer phase is stagnant, whereas the organic phase is mixed by recycling through an optical cell. Under these conditions no mass transfer limitation is observed [24]. Evidence for an interfacial reaction is given by the linear dependence of the reaction rate on interfacial area, no dependence on the aqueous phase volume, and the increase of product formation rate with increasing enzyme concentration until the interface is saturated by enzyme. Additionally, no reaction occurs if denatured enzyme adsorbs onto the interface prior to the reaction [24, 25].

Willeman et al. [26] modeled the enzyme-catalyzed cyanohydrin synthesis in a stirred batch tank reactor. Assumption of a mass transfer limitation (Figure 9.3b) is made, which results in a low concentration of substrate in the aqueous phase, thus suppressing the non-enzymatic reaction. In a well-stirred biphasic system the enzyme concentration was varied, keeping the phase ratio constant. A maximum rate of conversion is reached at the concentration where mass transfer of the substrate becomes limiting. Further increase of enzyme concentration does not enhance the reaction rate [27]. The different results achieved by the two groups are explained by the different process strategies. No mass transfer limitation could be detected by Hickel et al. because the stirring rate in the aqueous phase was not varied [26].

A comparison of the qualitative results of the two mathematical models shows strong similarities between the consequences of the mass transfer limitation model and the adsorbed enzyme model [28]. Experimental findings show that the reaction mode is strongly dependent on the process conditions, and the different models cannot be adopted for other conditions [29].

For preparative purposes, the use of biphasic solvent systems consisting of an aqueous phase and a water-immiscible organic phase for *Pa*HNL and *Hb*HNL catalysis has proven to have a broad applicability, also including, for example, pyrrole derivatives [30] (see Table 9.3, Section 9.2.2.3) and to be suitable for industrial scale. DSM established enzymatic hydrocyanation processes, e.g., for the production of (S)-*m*-phenoxymandelonitrile [31, 32] and large-scale production of (R)-2-(2-furyl)-2-hydroxyacetonitrile [33].

9.2.1.2 Buffer/Ionic Liquids

Less widely explored are ionic liquids (ILs) as the second phase. ILs have emerged as promising replacements for organic solvents. They are highly polar, have a negligible vapor pressure, and possess high thermal and chemical stability [34]. In recent years they have also found application in biocatalysis [35, 36]. Initial experiments by Griengl and co-workers with HNL immobilized on Celite® in neat ILs containing 1% water afforded only racemic products [37]. A biphasic system of IL/aqueous buffer 1/1 (v/v) for the synthesis of long-chain aldehydes gave higher conversions than the *t*BME/aqueous buffer system. Excellent *ee*-values were achieved with *Pa*HNL, whereas using *Hb*HNL catalysis a decrease in the *ee* was observed [37]. Zong and co-workers used a DIPE/aqueous buffer system. In *Pa*HNL-catalyzed transhydrocyanation, the addition of low amounts of IL (up to 10%) improved the *ee* and the enzyme activity. In the case of *Me*HNL catalysis, the addition of IL resulted in a decrease in both the initial rate and the *ee*, indicating that this enzyme is less tolerant of the presence of ILs [38].

9.2.2
Liquid/Solid Biphasic Systems

The occurrence of the undesired non-enzymatic cyanohydrin reaction in water/organic solvent systems has driven the research toward pure organic solvent systems. Most proteins are insoluble in organic solvents. This makes it necessary to work with crystalline, lyophilized, or immobilized enzymes in solid/liquid biphasic systems. The stability of hydroxynitrile lyases in dry organic solvents is very high, depending mainly on the substrate and the water content [39]. The necessity for small amounts of water for the catalytic activity of HNLs [40, 41] requires immobilization of the enzyme. Furthermore, besides enhancing the stability, immobilization also facilitates the separation from the reaction mixture. The most common techniques for immobilization are adsorption onto an inert carrier, covalent binding, inclusion of the protein in a polymeric gel, membrane, or capsule, and cross-linking with polyfunctional reagents [42].

9.2.2.1 Crude Enzyme Preparations
The simplest and cheapest way to produce supported *Pa*HNL is to use almond meal itself. The step of purification of the enzyme is avoided [43], but low loading of enzyme is the main disadvantage. Other HNLs have also been employed using this crude type of enzyme preparation, such as apple, apricot, cherry, plum [44, 45], peach and loquat [46] meal. However, these catalysts were applied to a minor extent compared with almond meal.

By the application of almond meal in organic solvent containing just the amount of buffer needed to maintain catalytic activity, a large number of (*R*)-cyanohydrins have been successfully synthesized, particularly from those aldehydes where the enzymatic reaction cannot compete with the non-enzymatic reaction in aqueous medium. The solvents of choice are again ethers and surprisingly ethyl acetate [47], which proved to be unsuitable for liquid/liquid biphasic systems [20]. Different groups have optimized the amount of water required for optimum reactivity, which ranges from 4% [48] to 10% [49] (v/v) of the reaction volume. This quantity of buffer is easily absorbed by the biocatalyst, which is freely dispersed in the organic solvent. A larger amount of buffer first causes coagulation of the enzyme and then leads to the formation of a second liquid phase, with a decrease in reaction rate and *ee* [48].

Lin and co-workers [40] have developed a so-called "micro-aqueous" organic reaction system. In contrast to former preparations of almond meal, the almond kernels are soaked in water prior to grinding. After the defatting step, the meal contains 8–10% water (w/w), making it unnecessary to add the amount of water needed for enzyme activity. The reactions are carried out in buffer-saturated organic solvents to avoid a possible drying effect of the solvent on the biocatalyst. Further addition of water to the reaction results in lower conversions and *ee* values,

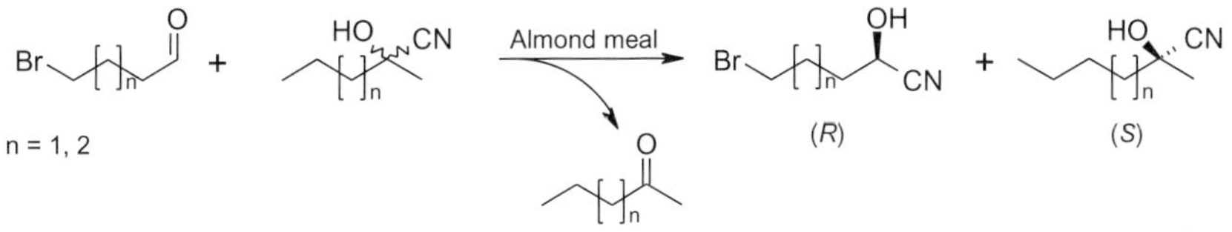

Figure 9.4 Enzymatic transcyanation of ω-bromoaldehydes and racemic ketone cyanohydrins [53].

this being caused by coagulation of the crude enzyme [40, 46]. Besides a great variety of substrates converted, such as fluorinated benzaldehydes [50] and heteroaryl carboxaldehydes [51] (see Table 9.3 for examples), a continuous process for the production of cyanohydrins has also been developed using the "micro-aqueous" organic solvent system [52].

The Gotor group succeeded in using almond meal for the enantioselective cleavage of ketone cyanohydrins, which served additionally as the cyanide source in the subsequent asymmetric addition of HCN to ω-bromoaldehydes in one pot (Figure 9.4) [53].

The presence of the aldehyde as cyanide acceptor is essential to achieve the resolution of ketone cyanohydrins with good enantioselectivities. The unreacted (S)-ketone cyanohydrin and the (R)-ω-bromoaldehyde cyanohydrin formed could be isolated in moderate to good enantiopurity (50 to 95% ee and 75 to 92% ee respectively) [53].

The conversion of ω-hydroxyalkanals to the corresponding cyanohydrins in moderate enantioselectivities could also be accomplished by transhydrocyanation with acetone cyanohydrin as the cyanide source. These substrates are considered difficult because of their high solubility in water. Through the employment of an almond meal preparation in a "micro-aqueous" organic reaction system, the ee-values could be significantly improved [54].

As well as almond meal, *Sorghum bicolor* shoots have also found application in the synthesis of aromatic cyanohydrins [55]. The enantiomeric purity obtained in transhydrocyanation experiments with acetone cyanohydrin as the cyanide source suffers from the high water content (>14% v/v) necessary for the decomposition of acetone cyanohydrin. In contrast, the application of HCN allows the use of low amounts of water (2% v/v), leading to yields and optical purities comparable with those obtained by the isolated enzymes.

The application of crude enzyme preparations also constitutes a convenient method to find new sources for hydroxynitrile lyases [56–58].

9.2.2.2 Inert Carriers

Adsorption on solid matrices represents a quite simple and inexpensive method for enzyme immobilization. Enzyme dispersion is improved, reducing the diffusion limitations and favoring the accessibility of substrate to the enzyme [12]. On the other hand, because of the weak binding, the system can suffer from catalyst leaching, and there is little stabilization of the enzyme. The most common carriers

Table 9.2 Conversion of aldehydes (RCHO) in water/EtOH and in ethylacetate/cellulose [11].

R-	Conversion (ee)[a] [%]	
	H$_2$O/EtOH	Ethyl acetate/Avicel
Ph	99 (86)	95 (99)
3-PhOPh	99 (11)	99 (98)
2-furyl	86 (69)	88 (98.5)
3-pyridyl	78 (6.7)	89 (14)
CH$_3$CH=CH	68 (76)	68 (97)
Ph(CH$_2$)$_2$	82 (27)	95 (40)
CH$_3$S(CH$_2$)$_2$	87 (60)	97 (80)
(CH$_3$)C	56 (45)	78 (73)
C$_3$H$_7$	75 (69)	75 (96)

a As (R)-(+)-MTPA-derivative.

are synthetic organic polymers (polyamides, polyesters, polysiloxanes), biopolymers (cellulose, starch), inorganic materials (sand, metal oxides, clays), proteinaceous materials (albumin, gelatine, collagen, silk), and synthetic materials (glass, silica, zeolites) [59].

First experiments were carried out by Effenberger and co-workers using *Pa*HNL immobilized on different carriers in ethyl acetate, cellulose being the most suitable support [11]. In comparison to a water/ethanol system, the enantioselectivity toward many substrates could be increased (Table 9.2). The application of DIPE as solvent improves the reaction even more [60]. Higher selectivities are achieved and the enzyme activity remains unchanged over several weeks [61]. Crystalline cellulose also proved to be a good carrier for other HNLs [62, 63], and the conversion of a range of substrates, like benzaldehyde derivatives [64], pivaldehyde and its derivatives [65], and heteroaryl carbaldehydes [62] could be accomplished. In comparison, other supports, such as ion exchange resins, are not appropriate, because conversion and *ee* decreases significantly, whereas the application of Celite® gives chemical and optical yields comparable to those using cellulose [65]. For the immobilization of *Me*HNL [66] and *Lu*HNL [67], Effenberger et al. found nitrocellulose to be more suitable.

Costes et al. investigated in detail the enzyme stability of *Hb*HNL immobilized on Celite®. In dry solvent the stability was found to be very high, whereas in buffer-saturated solvents it was severely decreased [39]. The optimum water content was found to be higher than the water solubility of the solvents, ranging between 1 and 1.5% (v/v). At lower amounts the enzyme cannot exhibit its highest activity because of insufficient hydration, and an excess of water has no positive effect [68]. Further investigations showed a strong influence of the amount of HCN on the selectivity of the enzyme. Increasing the HCN concentration above 800 mM resulted in a dramatic decrease of the *ee*. Furthermore, the higher HCN concentration in the water phase surrounding the enzyme caused a faster enzyme deactivation [68]. The same cyanohydrin transformation catalyzed by different HNL immobilized on Celite® under optimized reaction conditions gave products with quite different enantiopurity, indicating that HNLs have a limited intrinsic enantiospecificity toward certain substrates, which is caused by the imperfect fitting of the substrate in the active site [68]. The catalysis by *Pa*HNL isoenzyme#5 and by *Hb*HNL on Celite® in different solvents was investigated in detail. Both enzymes show good to excellent performance in different aromatic solvents. Furthermore, dichloromethane turned out to be a suitable solvent for *Pa*HNL on Celite®, whereas ethers gave no satisfactory results. In contrast, *Hb*HNL exhibited good activity in ethers [69].

A transhydrocyanation reaction catalyzed by *Hb*HNL on Celite® was investigated by Hanefeld and co-workers [41]. To shift the unfavorable equilibrium, the reaction was coupled with an irreversible acylation step. Problems occurred due to hydrolysis of the acyl donor by the water needed for enzyme activity and subsequent deactivation of the *Hb*HNL by the acid formed. *In situ* derivatization in the cyanohydrin synthesis catalyzed by *Pa*HNL on Celite® was investigated, and a one-pot chemoenzymatic synthesis of protected cyanohydrins was developed using ethyl cyanoformate as both HCN donor and protecting reagent [70].

9.2.2.3 Cross-Linking of Hydroxynitrile Lyases

Immobilization by chemical cross-linking affords a biocatalyst with high volumetric activity, because no inert carrier or matrix is needed. The tight binding stabilizes the enzyme and facilitates the recycling. However, direct cross-linking of the enzyme leads to low activity and poor mechanical properties. Recently cross-linked enzyme crystals (CLECs) and cross-linked enzyme aggregates (CLEAs) have been reported to exhibit the desired properties without the need of a support [59]. The most commonly used cross-linking agent is glutaraldehyde. With some enzymes the use of dextran polyaldehyde as cross-linker is advantageous, resulting in lower loss of activity [71].

The first cross-linked HNLs were reported by Costes et al. [72]. They compared *Me*HNL-CLECs with Celite®-immobilized *Me*HNL. By cross-linking, the stability of the enzyme was improved, particularly in polar organic solvents. Furthermore, the cross-linked crystals could be reused without significant loss of activity. After six consecutive batches, 70% of the initial activity was retained, whereas the immobilized enzyme showed virtually no remaining activity (<1%). Nevertheless, crystallization and cross-linking cause a considerable loss of activity compared to the immobilization on Celite® [72].

The major disadvantage of CLECs is the need to crystallize the enzyme. This can be avoided by simply precipitating the enzyme from an aqueous solution and cross-linking the resulting aggregates, leading to cross-linked enzyme aggregates (CLEA®) [73]. The Sheldon group has shown that the *Pa*HNL CLEA is a stable and recyclable biocatalyst. In a biphasic DIPE/buffer medium the CLEA was less active than the free enzyme, but only small effect was observed on the enantiomeric purity. On the other hand, using a micro-aqueous system (2% v/v buffer) the CLEA showed itself to be a superior biocatalyst, particularly for the conversion of slow reacting aldehydes (Table 9.3). No performance loss was observed on reusing the enzyme preparation 10 consecutive times [74].

The cross-linked aggregates of MeHNL also showed themselves to be highly active and robust catalysts. Optimized procedures give MeHNL CLEAs with activity recoveries up to 93% measured by a synthetic assay [75]. As observed earlier for PaHNL CLEAs [74], this result is in contrast with the photometric assay, indicating that a fast assay severely underestimates the recovery of initial activity because of rate-limiting diffusion [75].

Furthermore, unlike PaHNL CLEA, which requires 2% of water [74], the CLEAs of MeHNL maintain their activity even in completely dry solvent. The remaining limiting factors in applying such systems are the reaction equilibrium and the intrinsic enantiomeric preference of the biocatalyst. 2-Propenal is obtained under these conditions with 96% conversion, but the inherent *ee* of 57% could not be increased [75]. The major advantage of the employment of dry organic solvent is the possibility of *in situ* derivatizations of the cyanohydrins, thus having the chance of shifting the equilibrium [76].

Enantiocomplementary processes have been developed for the conversion of 3-pyridinecarboxaldehyde using commercially available HNL CLEAs. This otherwise difficult substrate was obtained with up to 85% yield and 94% *ee* after reaction optimization [77].

The major advantage offered by the use of CLEAs is the possibility of bienzymatic reactions by the application of combi-CLEAs, as shown by Mateo et al. [78]. The utilization of apparently incompatible enzymes with respect to the reaction conditions is thereby possible.

9.2.3
Other Biphasic Systems

Some other systems, which cannot be assigned unambiguously, have also been investigated for HNL catalysis. The driving force for the investigations was the attempt to improve enzyme stability and to suppress the undesired non-enzymatic side reaction.

Boy et al. [79] used lyotropic liquid crystals (LC) for the immobilization of HbHNL. The solid LC phase is not used because of the high viscosity. Therefore, the LC is used in a biphasic system consisting of the LC and an organic solvent. Such biphasic liquid crystal systems consist of organic solvent, water, and surfactant, where poorly soluble substrates and products are dissolved in the organic solvent and the liquid crystal matrix, which contains the enzyme, has a protective effect on it. By optimization and by virtue of the immobilization, it is possible to establish an extractive continuous process [79].

The immobilization of hyroxynitrile lyase from *Sorghum bicolor* on a porous membrane incorporated in a standard pump-around reactor was patented by Andruski and Goldberg [80]. A solution of aldehyde and HCN in different organic

Table 9.3 Examples of cyanohydrin formation in different biphasic systems.

Aldehyde	Time [h]	Conv. (ee) [%]	Source[a]	Method	Ref.
benzaldehyde	10	72 (92)	Pa (R)	Transcyanation, buffer/ether	19
	48	100 (>99)	Pa (R)	Microaqueous, almond meal/IPE, 12°C	40
		91 (97)	Sb (S)	Avicel-cellulose/DIPE	63
	42	97 (97)	Sb (S)	Shoots, 2% H2O v/v	55
	7	100 (98)	Me (S)	Nitrocellulose/DIPE	66
3-PhO-PhCHO		93 (96)	Sb (S)	Avicel-cellulose/DIPE	63
cinnamaldehyde	24	36.5 (51.5)	Pa (R)	Microaqueous, almond meal/IPE, 30°C	40
	4	96 (99)	Pa (R)	CLEA, 2% H2O v/v, 0°C	74
	27	20 (10)	Lu (R)	Nitrocellulose/DIPE	67
	6.2	86 (87)	Me (S)	CLEA	75
2-halobenzaldehyde X=Cl	1	96 (91)	Pa (R)	Almond meal/DIPE, 4% H2O v/v	48
	2	98 (95)	Pa (R)	CLEA, 2% H2O v/v, 0°C	74
	8.7	100 (92)	Me (S)	Nitrocellulose/DIPE	66
X=F	24	96 (84)	Pa (R)	Microaqueous, almond meal/IPE, 20°C	50
furfural (2-furaldehyde)	24	100 (99)	Pa (S)[b]	Microaqueous, almond meal/IPE, 4°C	40
	4	96 (99)	Pa (S)[b]	Avicel-cellulose/DIPE	62
		95 (99.4)	Hb (R)[b]	Buffer/tBME, large scale	33
	9	80 (80)	Sb (R)[b]	Avicel-cellulose/DIPE	62
3-furaldehyde	4	96 (99)	Pa (R)	Avicel-cellulose/DIPE	62
	33	88 (87)	Sb (S)	Avicel-cellulose/DIPE	62
	6.5	98 (92)	Me (S)	Nitrocellulose/DIPE	66
5-methylfurfural	17	70 (99)	Pa (S)[b]	Almond meal/EE, 5% H2O v/v	43
		60 (97)	Pa (S)[b]	Microaqueous, almond meal/IPE	51
2-thiophenecarboxaldehyde		70 (99)	Pa (S)[b]	Microaqueous, almond meal/IPE	51
	5	71 (99)	Pa (S)[b]	Avicel-cellulose/DIPE	62
	8	64 (91)	Sb (R)[b]	Avicel-cellulose/DIPE	62
	6	85 (96)	Me (R)[b]	Nitrocellulose/DIPE	66
3-thiophenecarboxaldehyde	6	95 (99)	Pa (R)	Avicel-cellulose/DIPE	62
	27	95 (98)	Sb (S)	Avicel-cellulose/DIPE	62
	4	98 (98)	Me (S)	Nitrocellulose/DIPE	66
pyrrole-2-carbaldehyde R = MOM		33 (81)	Pa (R)	Microaqueous, almond meal/IPE	51
R = Me	8d	11 (75)	Pa (R)	Buffer/tBME	30
R = Bn	19d	7 (51)	Hb (S)	Buffer/tBME	30
pyridine-3-carbaldehyde		99.5 (50)	Pa (R)	Microaqueous, almond meal/IPE	51
		60 (51)	Me (S)	CLEA/toluene	77
butanal	41	95 (89)	Pa (R)	Almond meal/EE	43
	19	100 (95)	Pa (R)	Transcyanation/almond meal	49
	2	91 (98)	Lu (R)	Nitrocellulose/DIPE	66
	6.5	91 (95)	Me (S)	Nitrocellulose/DIPE	67
crotonaldehyde	41	73 (99)	Pa (R)	Almond meal/EE	43
	48	99 (98)	Pa (R)	Almond meal/EE	47
	9.5	21 (99)	Lu (R)	Nitrocellulose/DIPE	67

a Source and configuration in brackets.
b Priority replacement according to CIP rules.

solvents is pumped through the membrane in a closed cycle. The membrane can be reutilized in several consecutive batches without significant loss of enantioselectivity. *m*-Phenoxybenzaldehyde was obtained in 90% conversion with an enantiomeric ratio of 93.7/6.3 (S)/(R).

9.2.3.1 Encapsulation in Sol-Gel Matrices

Enzymes can also be entrapped within the pores of a matrix or network. Small molecules can diffuse in and out of the matrix, whereas the macromolecular enzyme is maintained within the network. Precise control of the pore size is not possible, and therefore mass transfer limitations and enzyme leaching always cause problems [59].

Gröger and Vorlop [81] developed an immobilization method which is used in a buffer/organic solvent biphasic system. Here, a liquid/liquid/solid system is formed, which is not affected by enzyme leaching. The HNL from *Prunus amygdalus* is first cross-linked to increase the molecular weight and afterwards immobilized in a hydrogel matrix based on polyvinyl alcohol (PVA). Lens-shaped particles are obtained, which are macroscopic, well-defined, highly flexible, and highly active. These capsules give results in the synthesis reaction of (R)-mandelonitrile that are comparable with those using the free enzyme (93% yield, 94% *ee*). In addition, this biocatalyst shows no catalyst leaching and is recyclable without loss of activity. The high mechanical stability makes it suitable for technical reactors [81, 82].

Also, the entrapment of enzymes within hydrophobic silica sol-gel matrices, reported by Reetz [83] and others [84], results in versatile biocatalysts. The sol-gel process allows chemical inert glasses and ceramics with a desired shape to be synthesized. They are highly porous and have high mechanical and thermal resistance. If a silica matrix is generated in the presence of an aqueous enzyme solution by treating the silane precursors with a catalytic amount of sodium fluoride, the enzyme is entrapped within the gel matrix. Typically, precursors are tetraalkoxysilanes, alkyl alkoxysilanes, and mixtures of both [83]. If the resulting aquagels are not dried, the enzyme is located in the aqueous buffer inside the pores, forming a liquid/liquid/solid system. In the case of lipases, the activity in organic solvent could even be enhanced in comparison to the corresponding lipase powders under the same conditions. Optimization of the activity could be achieved by varying the ratio of the precursor mixture toward the hydrophobic silane, by increasing the alkyl chain length, and by using polymeric additives [83, 85].

Because of the structural similarity of the HNL of *Hevea brasiliensis*, the Hanefeld group assumed that the encapsulation of *Hb*HNL should proceed in a similar manner to that investigated with lipases [86]. It was found that the production procedure is crucial to obtain active sol-gels. The removal of the released methanol and submerging the formed gel in buffer are essential to obtain aquagels with at least 65% activity left. In contrast to lipases, the use of PVA as an additive and the variation of the methyltrimethoxysilane/tetramethoxysilane ratio showed no effect on the enzyme activity, but a total loss of activity was observed by drying of the aquagels to xerogels. Since a low water amount is needed for *Hb*HNL activity, the aquagels were employed directly. The buffer is encapsulated inside the pores, and no separate water phase is formed in the reaction mixture. The product *ee* for the conversion of some selected aldehydes with sol-gel HbHNL is comparable with that of other systems previously reported [86].

9.2.3.2 Comparison of CLEAs and sol-gel HNLs

The Sheldon group [87] prepared aquagels of different HNLs and compared them in the synthesis reaction of different cyanohydrins with the CLEAs and the free enzymes. The activity recovery for the aquagels and CLEAs measured by a photometric assay were quite low. Using the same loadings, the aquagels turned out to be much faster than the free enzyme. This confirms the underestimation of the recovery of activity by fast assays due to diffusion problems, as reported earlier [74, 75]. The stability and the catalytic performance of the immobilized HNLs are strongly influenced by the solvent, the immobilization method, and the enzyme source.

The conversion of benzaldehyde by the encapsulated HNLs afforded mandelonitrile in 96–98% yield and 97–99% *ee* for all three enzymes. Free and entrapped HbHNL catalyzed the conversion of hexanal with 94% *ee*, whereas *ee*-values of only 85 and 87% could be achieved with MeHNL and PaHNL preparations, respectively, limited by the intrinsic enantioselectivity of the enzymes (Table 9.4). Furthermore, the CLEAs from HbHNL and PaHNL suffered from activity loss under the reaction conditions in contrast to MeHNL CLEA, indicating that the cross-linked aggregates from MeHNL are particularly robust and necessitate only traces of water to keep the catalytic activity [87].

9.3 Conclusion

The enzymatic synthesis of chiral cyanohydrins has reached a high stage of development. The different reaction systems give the possibility to convert a great

Table 9.4 Conversion and *ee* with different enzyme praparations [87].

Aldehyde	Conversion (ee) [%]/time [h]							
	HbHNL		MeHNL			PaHNL		
	Free	Aqua gel	Free	Aqua gel	CLEA	Free	Aqua gel
BA	97(97)/4	97(99)/0.5	97(98)/4	96(99)/0.5	96(97)/2	98(97)/4	97(97)/2
Furfural	89(94)/0.5	89(94)/0.5	90(95)/0.5	95(87)/0.5	94(94)/0.5	91(96)/0.5	95(82)/0.5
Hexanal	91(94)/4	92(94)/2	93(84)/3	96(84)/3	92(81)/3	91(87)/3	88(85)/3
m-PhOBA	45(82)/72	92(98)/72	14(75)/72	98(97)/45	81(83)/72	68(99)/72	97(99)/72

variety of aldehydes and ketones in excellent yields and enantiopurities. Furthermore, the immobilization techniques make it possible to establish continuous processes, making these enzymes "workhorses" for industry.

References

1 Gregory, R. J. H. *Chem. Rev.* 1999, **99**, 3649–3682.
2 North, M. *Tetrahedron: Asymmetry* 2003, **14**, 147–176.
3 Fechter, M. H.; Griengl, H. *Food Technol. Biotechnol.* 2004, **42**, 287–294.
4 Brunel, J.-M.; Holmes, I. P. *Angew. Chem.* 2004, **116**, 2810–2837.
5 Zagrobelny, M.; Bak, S.; Rasmussen, A. V.; Jorgensen, B.; Naumann, C. M.; Moller, B. L. *Phytochemistry* 2004, **65**, 293–306.
6 Sharma, M.; Sharma, N. N.; Bhalla, T. C. *Enzyme Microb. Technol.* 2005, **37**, 279–294.
7 Rosenthaler, L. *Biochem. Z.* 1908, **14**, 238–253.
8 Becker, W.; Benthin, U.; Eschenhof, E.; Pfeil, E. *Biochem. Z.* 1963, **337**, 156–166.
9 Becker, W.; Freund, H.; Pfeil, E. *Angew. Chem., Int. Ed.* 1965, **4**, 1079.
10 Becker, W.; Pfeil, E. *Biochem. Z.* 1966, **346**, 301–321.
11 Effenberger, F.; Ziegler, T.; Foerster, S. *Angew. Chem.* 1987, **99**, 491–492.
12 Carrea, G.; Riva, S. *Angew. Chem., Int. Ed.* 2000, **39**, 2226–2254.
13 Klibanov, A. M. *Nature* 2001, **409**, 241–246.
14 Torres, S.; Castro, G. R. *Food Technol. Biotechnol.* 2004, **42**, 271–277.
15 Hickel, A.; Graupner, M.; Lehner, D.; Hermetter, A.; Glatter, O.; Griengl, H. *Enzyme Microb. Technol.* 1997, **21**, 361–366.
16 Bauer, M.; Geyer, R.; Boy, M.; Griengl, H.; Steiner, W. *J. Mol. Catal. B: Enzym.* 1998, **5**, 343–347.
17 Hickel, A.; Radke, C. J.; Blanch, H. W. *Biotechnol. Bioeng.* 2001, **74**, 18–28.
18 Bauer, M.; Griengl, H.; Steiner, W. *Enzyme Microb. Technol.* 1999, **24**, 514–522.
19 Ognyanov, V. I.; Datcheva, V. K.; Kyler, K. S. *J. Am. Chem. Soc.* 1991, **113**, 6992–6996.
20 Loos, W. T.; Geluk, H. W.; Ruijken, M. M. A.; Kruse, C. G.; Brussee, J.; van der Gen, A. *Biocatal. Biotransform.* 1995, **12**, 255–266.
21 Griengl, H.; Klempier, N.; Pöchlauer, P.; Schmidt, M.; Shi, N.; Zabelinskaja-Mackova, A. A. *Tetrahedron* 1998, **54**, 14477–14486.
22 Schmidt, M.; Hervé, S.; Klempier, N.; Griengl, H. *Tetrahedron* 1996, **52**, 7833–7840.
23 Hickel, A.; Radke, C. J.; Blanch, H. W. *J. Mol. Catal. B: Enzym.* 1998, **5**, 349–354.
24 Hickel, A.; Radke, C. J.; Blanch, H. W. *Biotechnol. Bioeng.* 1999, **65**, 425–436.
25 Pereira, L. G. C.; Hickel, A.; Radke, C. J.; Blanch, H. W. *Biotechnol. Bioeng.* 2002, **78**, 595–605.
26 Willeman, W. F.; Gerrits, P. J.; Hanefeld, U.; Brussee, J.; Straathof, A. J. J.; Van der Gen, A.; Heijnen, J. J. *Biotechnol. Bioeng.* 2002, **77**, 239–247.
27 Gerrits, P. J.; Willeman, W. F.; Straathof, A. J. J.; Heijnen, J. J.; Brussee, J.; van der Gen, A. *J. Mol. Catal. B: Enzym.* 2001, **15**, 111–121.
28 Straathof, A. J. J. *Biotechnol. Bioeng.* 2003, **83**, 371–375.
29 Cascao-Pereira, L. G.; Hickel, A.; Radke, C. J.; Blanch, H. W. *Biotechnol. Bioeng.* 2003, **83**, 498–501.
30 Purkarthofer, T.; Gruber, K.; Fechter, M. H.; Griengl, H. *Tetrahedron* 2005, **61**, 7661–7668.
31 Pöchlauer, P. *Chim. Oggi* 1998, **16**, 15–19.
32 Gröger, H. *Adv. Synth. Catal.* 2001, **343**, 547–558.
33 Purkarthofer, T.; Pabst, T.; van der Broek, C.; Griengl, H.; Maurer, O.; Skranz, W. *Org. Process Res. Dev.* 2006, **10**, 618–621.
34 Welton, T. *Chem. Rev.* 1999, **99**, 2071–2083.

35 Kragl, U.; Eckstein, M.; Kaftzik, N. *Curr. Opin. Biotechnol.* 2002, **13**, 565–571.
36 Sheldon, R. A.; Lau, R. M.; Sorgedrager, M. J.; van Rantwijk, F.; Seddon, K. R. *Green Chem.* 2002, **4**, 147–151.
37 Gaisberger, R. P.; Fechter, M. H.; Griengl, H. *Tetrahedron: Asymmetry* 2004, **15**, 2959–2963.
38 Lou, W.-Y.; Xu, R.; Zong, M.-H. *Biotechnol. Lett.* 2005, **27**, 1387–1390.
39 Costes, D.; Rotcenkovs, G.; Wehtje, E.; Adlercreutz, P. *Biocatal. Biotransform.* 2001, **19**, 119–130.
40 Han, S.; Lin, G.; Li, Z. *Tetrahedron: Asymmetry* 1998, **9**, 1835–1838.
41 Hanefeld, U.; Straathof, A. J. J.; Heijnen, J. J. *J. Mol. Catal. B: Enzym.* 2001, **11**, 213–218.
42 Bornscheuer, U. T. *Angew. Chem., Int. Ed.* 2003, **42**, 3336–3337.
43 Zandbergen, P.; van der Linden, J.; Brussee, J.; van Der Gen, A. *Synth. Commun.* 1991, **21**, 1387–1391.
44 Kiljunen, E.; Kanerva, L. T. *Tetrahedron: Asymmetry* 1997, **8**, 1225–1234.
45 Kiljunen, E.; Kanerva, L. T. *Tetrahedron: Asymmetry* 1997, **8**, 1551–1557.
46 Lin, G.; Han, S.; Li, Z. *Tetrahedron* 1999, **55**, 3531–3540.
47 Warmerdam, E. G. J. C.; van den Nieuwendijk, A. M. C. H.; Kruse, C. G.; Brussee, J.; Van der Gen, A. *Recl. Trav. Chim. Pays-Bas* 1996, **115**, 20–24.
48 van Langen, L. M.; van Rantwijk, F.; Sheldon, R. A. *Org. Process Res. Dev.* 2003, **7**, 828–831.
49 Huuhtanen, T. T.; Kanerva, L. T. *Tetrahedron: Asymmetry* 1992, **3**, 1223–1226.
50 Han, S.; Chen, P.; Lin, G.; Huang, H.; Li, Z. *Tetrahedron: Asymmetry* 2001, **12**, 843–846.
51 Chen, P.; Han, S.; Lin, G.; Huang, H.; Li, Z. *Tetrahedron: Asymmetry* 2001, **12**, 3273–3279.
52 Chen, P.; Han, S.; Lin, G.; Li, Z. *J. Org. Chem.* 2002, **67**, 8251–8253.
53 Menendez, E.; Brieva, R.; Rebolledo, F.; Gotor, V. *J. Chem. Soc., Chem. Commun.* 1995, 989–990.
54 de Gonzalo, G.; Brieva, R.; Gotor, V. *J. Mol. Catal. B: Enzym.* 2002, **19–20**, 223–230.
55 Kiljunen, E.; Kanerva, L. T. *Tetrahedron: Asymmetry* 1996, **7**, 1105–1116.
56 Solis, A.; Luna, H.; Perez, H. I.; Manjarrez, N.; Sanchez, R.; Albores-Velasco, M.; Castillo, R. *Biotechnol. Lett.* 1998, **20**, 1183–1185.
57 Solis, A.; Luna, H.; Perez, H. I.; Manjarrez, N. *Tetrahedron: Asymmetry* 2003, **14**, 2351–2353.
58 Hernandez, L.; Luna, H.; Ruiz-Teran, F.; Vazquez, A. *J. Mol. Catal. B: Enzym.* 2004, **30**, 105–108.
59 Lalonde, J.; Margolin, A. in *Enzyme Catalysis in Organic Synthesis, 2nd Ed.*, Vol. 1, Drauz, K.; Waldmann, H. (Eds.), Wiley-VCH, Weinheim, 2002, pp. 163–184.
60 Effenberger, F.; Ziegler, T.; Förster, S., Ger. Offen. DE 3701383 A1, 1988, 4 pp.
61 Effenberger, F. *Angew. Chem.* 1994, **106**, 1609–1619.
62 Effenberger, F.; Eichhorn, J. *Tetrahedron: Asymmetry* 1997, **8**, 469–476.
63 Effenberger, F.; Hoersch, B.; Foerster, S.; Ziegler, T. *Tetrahedron Lett.* 1990, **31**, 1249–1252.
64 Ziegler, T.; Hoersch, B.; Effenberger, F. *Synthesis* 1990, 575–578.
65 Effenberger, F.; Eichhorn, J.; Roos, J. *Tetrahedron: Asymmetry* 1995, **6**, 271–282.
66 Förster, S.; Roos, J.; Effenberger, F.; Wajant, H.; Sprauer, A. *Angew. Chem.* 1996, **108**, 493–494.
67 Trummler, K.; Roos, J.; Schwaneberg, U.; Effenberger, F.; Förster, S.; Pfizenmaier, K.; Wajant, H. *Plant Sci.* 1998, **139**, 19–27.
68 Costes, D.; Wehtje, E.; Adlercreutz, P. *Enzyme Microb. Technol.* 1999, **25**, 384–391.
69 Purkarthofer, T., PhD thesis, Graz University of Technology, Graz, Austria, 2004.
70 Purkarthofer, T.; Skranc, W.; Weber, H.; Griengl, H.; Wubbolts, M.; Scholz, G.; Pochlauer, P. *Tetrahedron* 2004, **60**, 735–739.
71 Mateo, C.; Palomo, J. M.; Van Langen, L. M.; Van Rantwijk, F.; Sheldon, R. A. *Biotechnol. Bioeng.* 2004, **86**, 273–276.
72 Costes, D.; Wehtje, E.; Adlercreutz, P. *J. Mol. Catal. B: Enzym.* 2001, **11**, 607–612.
73 Sheldon, R. A.; Schoevaart, R.; van Langen, L. M. *Biocatal. Biotransform.* 2005, **23**, 141–147.

74 van Langen, L. M.; Selassa, R. P.; Van Rantwijk, F.; Sheldon, R. A. *Org. Lett.* 2005, **7**, 327–329.

75 Chmura, A.; van der Kraan, G. M.; Kielar, F.; van Langen, L. M.; van Rantwijk, F.; Sheldon, R. A. *Adv. Synth. Catal.* 2006, **348**, 1655–1661.

76 Hanefeld U. personal communication, 2007.

77 Roberge, C.; Fleitz, F.; Pollard, D.; Devine, P. *Tetrahedron Lett.* 2007, **48**, 1473–1477.

78 Mateo, C.; Chmura, A.; Rustler, S.; van Rantwijk, F.; Stolz, A.; Sheldon, R. A. *Tetrahedron: Asymmetry* 2006, **17**, 320–323.

79 Boy, M.; Voss, H. *J. Mol. Catal. B: Enzym.* 1998, **5**, 355–359.

80 Andruski, S. W.; Goldberg, B., U.S. US 5177242 A, 1993, 7 pp.

81 Gröger, H.; Vorlop, K.; Capan, E., PCT Int. Appl. WO 0138554 A2, 2001, 38 pp.

82 Gröger, H.; Capan, E.; Barthuber, A.; Vorlop, K.-D. *Org. Lett.* 2001, **3**, 1969–1972.

83 Reetz, M. T.; Simpelcamp, J.; Zonta, A., Ger. Off. DE 4408152 A1, 1995, 12 pp.

84 Gill, I.; Ballesteros, A. *J. Am. Chem. Soc.* 1998, **120**, 8587–8598.

85 Reetz, M. T.; Zonta, A.; Simpelkamp, J. *Angew. Chem., Int. Ed.* 1995, **34**, 301–303.

86 Veum, L.; Hanefeld, U.; Pierre, A. *Tetrahedron* 2004, **60**, 10419–10425.

87 Cabirol, F. L.; Hanefeld U.; Sheldon, R. A. *Adv. Synth. Catal.* 2006, **348**, 1645–1654.

10
Ionic Liquids as Media for Enzymatic Transformations
Roger A. Sheldon and Fred van Rantwijk

10.1
Introduction

Ionic liquids are substances that are entirely composed of ions and are liquid at or close to room temperature. Interest in these compounds, often heralded as the green, high-tech media of the future, is increasing exponentially [1, 2]. It is based on the expectation that, because of their near-zero vapor pressure [3] and thermal stability [4] combined with their unconventional miscibility behavior, they will largely replace volatile organic solvents and revolutionize chemical process technology. Furthermore, properties such as polarity and hydrophobicity are amenable to fine tuning through an appropriate choice of cation and anion. Their synthesis, physicochemical properties, and major fields of application have been reviewed [5], and many types of ionic liquids are now commercially available in high purity [6].

Biocatalysis in ionic liquids was first reported in 2000 [7, 8, 9]. The early work involved ionic liquids composed of a 1,3-dialkylimidazolium or N-alkylpyridinium cation and a weakly-coordinating anion (Figure 10.1). More recently, attention is shifting toward new structural types. A number of reviews of this rapidly expanding subject have appeared [10, 11, 12, 13, 14].

Although enzymes function admirably in water (their natural medium), some reactions cannot easily be performed in water owing to insolubility of hydrophobic reactants or equilibrium limitations, e.g., in esterifications and amidations.

10.2
Solvent Properties of Ionic Liquids

Impurities, such as water, halides, unreacted organic salts, and organics, easily accumulate in ionic liquids [15] and may influence their solvent properties [3, 16] and/or interfere with the biocatalyst. For example, small amounts of chloride ion caused a severe deactivation of two lipases [17]. The irreproducibility of some early

Organic Synthesis with Enzymes in Non-Aqueous Media. Edited by Giacomo Carrea and Sergio Riva
Copyright © 2008 WILEY-VCH Verlag GmbH & Co. KGaA, Weinheim
ISBN: 978-3-527-31846-9

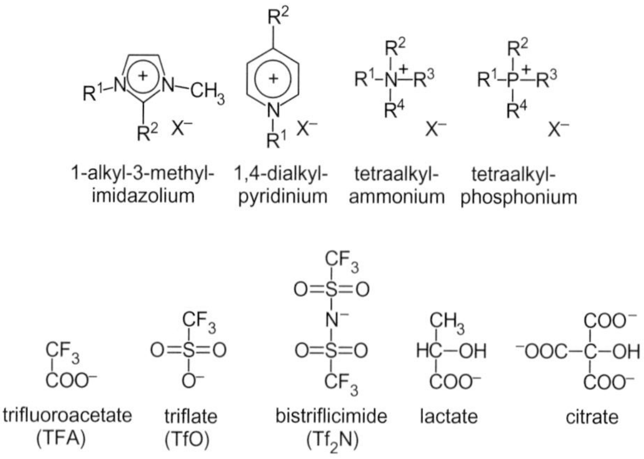

Figure 10.1 Structures of typical ionic liquids.

results with biocatalysis in ionic liquids was probably due to the presence of such impurities. Water is a common contaminant, as even water-immiscible ionic liquids are hygroscopic and readily absorb a few % of water [16]. Moreover, adventitious water in ionic liquids that contain [BF_4] or [PF_6] ions may cause partial hydrolysis of these anions with formation of HF, which inhibits many types of enzymes.

Solvent polarity, which should not be confused with hydrophilicity, is a complex concept [18], and the subject of ionic liquid polarity has been addressed by a variety of methodologies [19]. The propensity of solvents to stabilize a charge is usually determined from the absorption maximum of a solvatochromic dye [19, 20, 21] or by using a fluorescent probe [22]. By this measure, the polarity of common ionic liquid types, such as the archetypical [BMIm][BF_4] (Figure 10.1), is in the range of the lower alcohols [18, 23, 24, 25] or formamide [26]. Alternatively, solvent polarity can be gauged by measurement of keto-enol equilibria, as these are known to depend on the polarity of the medium. This latter methodology, when applied to probe the polarity of ionic liquids, indicated that [BMIm][BF_4], [BMIm][PF_6] and [BMIm][NTf_2] are more polar than methanol or acetonitrile [27]. Simple chemical reasoning would predict that a polar medium would dissolve polar compounds, such as, for example, carbohydrates, quite well. By this measure both the hydrophilic [BMIm][BF_4] and the hydrophobic [BMIm][PF_6] miserably fail the polarity test, because they dissolve less than 0.5 g L^{-1} of glucose at room temperature [28, 29]. [BMIm][Cl], in contrast, dissolves massive amounts of cellulose [30], and it was demonstrated that the ability of ionic liquids to dissolve complex compounds, such as sugars and proteins, mainly depends on the H-bond accepting properties of the anion [31].

The miscibility of ionic liquids with water varies widely and unpredictably. [BMIm][BF_4] and [BMIm][$MeSO_4$] are water-miscible, but [BMIm][PF_6] and

[BMIm][Tf$_2$N] are not. These ionic liquids are of similar polarity on the Reichardt scale [18], and the coordination strengths of the [BF$_4$] and [PF$_6$] anions are also comparable [25]. Neither is the partitioning of ionic liquids between water and 1-octanol a predictor of water miscibility, as the log P values of water-immiscible [BMIm][PF$_6$] and water-miscible [BMIm][AcO] and [BMIm][NO$_3$] are similar [32]. A recent measurement of the H-bond accepting properties of such ionic liquids revealed, however, that [BF$_4$] and [MeSO$_4$] are better H-bond acceptors than [PF$_6$] [30], which could perhaps explain the difference in water miscibility. Water-immiscible ionic liquids are nevertheless hygroscopic, as noted above, and readily absorb a few % of water [16], approximately corresponding with a hemihydrate. IR spectroscopic analysis has confirmed that water interacts mainly with the anion [33] via the formation of double H-bonds [34], at least when the cation is a weak hydrogen bond donor, as was the case here.

The miscibility behavior of ionic liquids and organic solvents is also rather unpredictable; dichloromethane and THF mix with, e.g., [BMIm][Tf$_2$N], whereas alkanes and ethers do not, and ethyl acetate seems to be a borderline case [35]. Supercritical carbon dioxide (scCO$_2$) does not mix with ionic liquids such as [BMIm][PF$_6$] and [OMIm][BF$_4$], but is absorbed in the ionic liquid phase in huge amounts (up to a molar fraction of 0.7) [36]. No ionic liquid dissolves in the CO$_2$ phase.

Ionic liquids are generally regarded as highly stable, and the widely used dialkylimidazolium ionic liquids are indeed thermostable up to 300 °C [4]. The propensity of the [BF$_4$] and [PF$_6$] anions to hydrolyze with liberation of HF [37], which deactivates many enzymes, has already been mentioned. The [TfO] and [Tf$_2$N] anions, in contrast, are hydrolytically stable. Dialkylimidazolium cations have a tendency to deprotonate at C-2, with ylide (heterocarbene) formation. Such ylides are strong nucleophiles and have been used as transesterification catalysts, for example [38]. These could cause enzyme deactivation as well as background transesterification when formed in small amounts from anhydrous ionic liquids and basic buffer salts, for example.

The viscosity of ionic liquids is high compared with molecular solvents and increases with the chain length. Consequently, diffusion is bound to be slow in ionic liquids. The effects on biocatalytic transformations seem to be insignificant, however, except in extreme cases, presumably because the reaction times are measured in hours rather than minutes.

In the context of green chemistry it should be pointed out that the ionic liquids that have generally been used in biocatalysis research have been designed neither for biocompatibility nor for easy biodegradability. It has recently become clear that the ecotoxicity of alkylmethylimidazolium cations is undesirably high and increases with the alkyl chain length [39, 40, 41], and the biodegradability of ionic liquids such as [BMIm][PF$_6$] and [BMIm][BF$_4$] is negligible [42]. Consequently, for future industrial application improved ionic liquids types will be required, and these are now being developed. Examples are the choline cation, which is food grade, imidazolium derivatives designed for biodegradability [43], and ionic liquids based on amino acids [44, 45]. We confidently expect that much improved and truly green ionic liquids will become available in the near future.

10.3
Enzymes in Ionic Liquids

The seminal work of Klibanov in the early 1980s [46, 47] made it clear that enzymes can be used in hydrophobic organic solvents, although at the price of a severely reduced reaction rate [48, 49]. Indeed, many lipases, as well as some proteases and acylases, are so stable that they maintain their activity even in anhydrous organic solvents. This forms the basis for their successful application in non-hydrolytic reactions, such as the (enantioselective) acylation of alcohols and amines, which now are major industrial applications [50].

The deactivation of enzymes by organic solvents, including ionic liquids, is generally attributed to removal of essential water. Many enzymes indeed require a full hydration shell to be active, but there are numerous exceptions, such as *Candida antarctica* lipase B (CaLB), which maintains its activity upon drying over phosphorus pentoxide [51], and the protease subtilisin, which requires only a few tightly bound water molecules per molecule of enzyme to be active [52]. The solvents that are tolerated well by these hydrolases – aliphatic and aromatic hydrocarbons, ethers, and alcohols (excluding methanol) – interact only weakly, and one would surmise that these are more or less like a vacuum to the enzyme. Only alcohols, which are H-bond donors and acceptors, are known to inhibit lipases. Solvents that do interact strongly with proteins, such as DMSO and DMF, even to the extent of dissolving them, also tend to cause irreversible loss of activity.

The general desire to improve the low turnover rate of enzymes in organic media was a major driving force for studying biocatalysis in ionic liquids. Low turnover rates have been ascribed to, inter alia, destabilization of the transition state as a result of the low dielectric constant of common organic media, which increases the energy of the highly polarized transition state in comparison with water [53]. Hence, ionic liquids could, on the basis of their highly polar nature, be expected to give much less transition state destabilization.

Various scenarios can be envisaged for biocatalysis in ionic liquids (Figure 10.2). The enzyme may be dissolved in a mixed aqueous-ionic liquid medium, which may be mono- or biphasic or it could be suspended or dissolved in an ionic liquid, with little or no water present. Alternatively, whole cells could be suspended in an ionic liquid, in the presence or absence of a water phase. Mixed aqueous-organic media are often used in biotransformations to increase the solubility of hydrophobic reactants and products. Similarly, mixed aqueous-ionic liquid media have been used for a variety of biotransformations, but in most cases there is no clear advantage over water-miscible organic solvents such as *tert*-butanol.

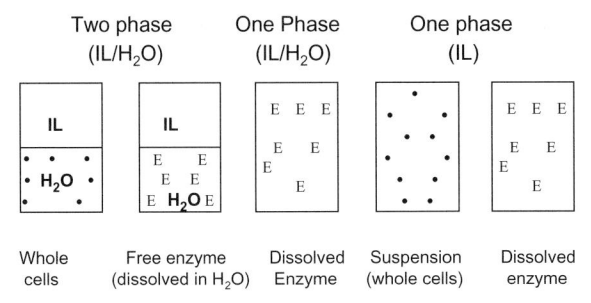

Figure 10.2 Scenarios for biocatalysis in ionic liquids.

Figure 10.3 Transesterification test reactions.

10.4
Enzymes in Nearly Anhydrous Ionic Liquids

Lipases, which are noted for their tolerance of organic solvents, were obvious candidates for biocatalysis in ionic liquids. Indeed, stable microbial lipases, such as CaLB [8, 54, 55, 56] and *Pseudomonas cepacia* lipase (PcL) [28, 55, 57] were catalytically active in the ionic liquids of the 1-alkyl-3-methylimidazolium and 1-alkylpyridinium families, in combination with anions such as [BF$_4$], [PF$_6$], [TfO] and [Tf$_2$N]. Early results were not always consistent, which may be caused by impurities that result from the preparation of the ionic liquid. Lipase-mediated transesterification reactions (Figure 10.3) in these ionic liquids proceeded with an efficiency comparable to that in *tert*-butyl alcohol [8], dioxane [57], or toluene [28]. *Candida antarctica* lipase A (CaLA), which was ten times more active in [BMPy][BF$_4$] and [BMIm][Tf$_2$N] than in diisopropyl ether (DIPE) [54] is an exception.

Other microbial lipases have also been successfully used in anhydrous ionic liquids, e.g., from *Alcaligenes* sp. (AsL) [54, 58], CaLA, *Rhizomucor miehei* lipase (RmL), and *Thermomyces lanuginosus* lipase (TlL) [54]. The lipase from pig pancreas (porcine pancreas lipase, PPL), the only mammalian lipase that has been subjected to ionic liquids, catalyzed transesterification in [BMIm][NTf$_2$] but not in [BMIm][PF$_6$]

[54, 58]. The cutinase from *Fusarium solani pisii* maintained its transesterification activity in [BMIm][BF$_4$], [OMIm][PF$_6$] and [BMIm][PF$_6$] (in order of increasing activity) at $a_w=0.2$ [59]. *Candida rugosa* lipase (CrL), which is generally much less tolerant of anhydrous media than other microbial lipases, has successfully been used in anhydrous as well as water-saturated ionic liquids [60, 61, 62, 63, 64, 65].

Esterases are much less tolerant of anhydrous media than lipases. The esterases from *Bacillus stearothermophilus* (BstE) and *Bacillus subtilis* (BsE) are exceptional, as these mediated transesterification in hexane at $a_w=0.1$ [66]. Both esterases, if immobilized on Celite 560, mediated transesterification in [BMIm][BF$_4$], [BMIm][PF$_6$], and [BMIm][Tf$_2$N] at a rate that varied from 20 to 60% of the rate in hexane or TBME.

Pencillin G acylase from *E. coli* is functionally, but not structurally, related to lipases. The enzyme would find wider use if it could be rendered tolerant of low-water media, which is the kind of problem that ionic liquids were expected to solve. It was found, however, that a covalently immobilized penicillin acylase, PGA 450, required $a_w \approx 0.8$, which also was the minimum in toluene, to stay active in the ionic liquids [BMIm][BF$_4$], [OMIm][BF$_4$], and [BMIm][PF$_6$] [67]. In a simple amine acylation test reaction (Figure 10.4), PGA 450 was somewhat less active in ionic liquids than in toluene.

Proteases have also been successfully used in ionic liquids. Papain mediated the enantioselective hydrolysis of a number of amino acid esters in an 80:20 mixture of [BMIm][BF$_4$] and water [68]. The reaction rate was approx. 50% of that in aqueous buffer and equal to that in aqueous mixtures containing 70–80% of solvents such as acetonitrile or *tert*-butyl alcohol.

α-Chymotrypsin mediated the transesterification of N-acetyl-L-aminoacid esters (Figure 10.5) in ionic liquids of the 1-alkyl-3-methylimidazolium type [69, 70, 71, 72], provided that the medium contained a small amount (~0.5%) of water. This

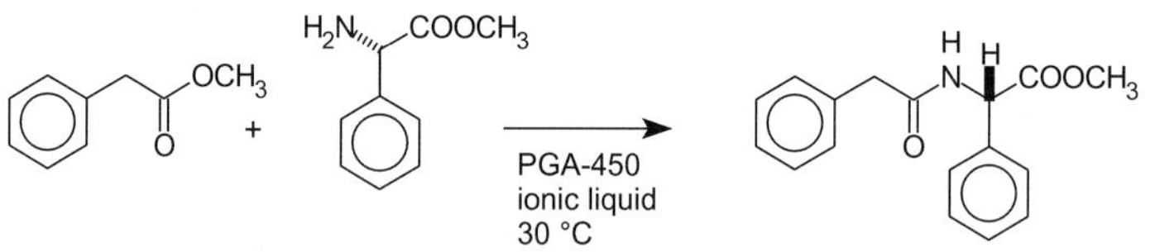

Figure 10.4 Acylation of (S)-phenylglycine methyl ester in the presence of penicillin G acylase.

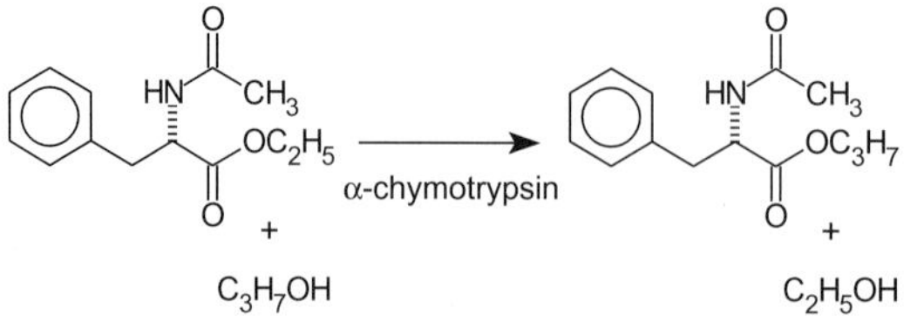

Figure 10.5 Transesterification catalyzed by α-chymotrypsin.

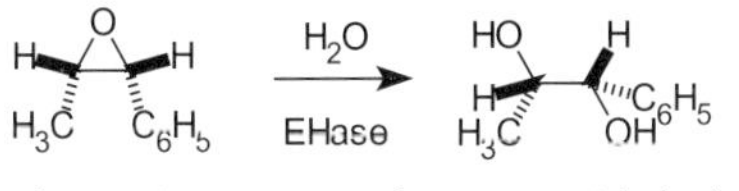

Figure 10.6 Enantioselective epoxide hydrolysis.

latter requirement was lifted when the ionic liquids were combined with scCO$_2$ [69]. The transesterification rates in [BMIm][PF$_6$] and [OMIm][PF$_6$] medium were of the same magnitude as those in isooctane or acetonitrile [69], but in [EMIm][Tf$_2$N] (under slightly different reaction conditions) the rate was nearly an order of magnitude higher [71].

Subtilisin generally shows poor transesterification activity in water-free ionic liquids, but Shah and Gupta [73] studied various approaches to maintaining its activity. The best results – up to 10 000 times increase in the initial rate N-acetylphenylalanine ethyl ester in [BMIm] [PF$_6$] – were obtained by precipitating and washing with 1-propanol.

Epoxide hydrolases (EHase) catalyze the hydrolytic ring-opening of epoxides with formation of a diol. The motivation for using an epoxide hydrolase in an ionic liquid is to improve the solubility of hydrophobic reactants. The hydrolysis of trans-β-methylstyrene oxide (Figure 10.6) in [BMIm][BF$_4$], [BMIm][PF$_6$], and [BMIm][Tf$_2$N] containing 1–10% water (depending on the enzyme preparation) was only slightly slower than the same reaction in aqueous buffer [74].

The lyophilization of enzymes from solutions containing salts or amphiphilic compounds is known to increase the activity in organic media by up to several orders of magnitude. Thus, the transesterification activity of α-chymotrypsin was increased 82-fold by co-lyophilization with pentaglyme [75]. The colyophilization of lipases and (poly ethylene)glycol (PEG) led to an enhanced transesterification activity in various ionic liquids [76, 77].

Relatively little attention has been paid to enzyme immobilization in connection with ionic liquids, and only one systematic study of this subject has appeared [78]. Itoh et al. studied the effects of the carrier material on PcL in [BMIm][PF$_6$] [79]. The activity of PsL on ceramic Toyonite carriers varied by a factor of 1000 between Toyonite 200M and Toyonite 200A. PsL adsorbed on a methacryloxypropyl-modified mesoporous silica also had a relatively high activity [78].

A novel approach to facilitating the recovery and recycling of the enzyme is by coating it with an ionic liquid having a melting point above room temperature and performing the reaction above its melting point [80, 81]. Subsequent cooling of the reaction mixture to room temperature allows for easy separation of the coated lipase as a solid, and it was shown that it retained its full activity after being reused several times.

The tolerance of lipases for anhydrous ionic liquid media discussed above was not universal. CaLB [82] was inactive in a range of ionic liquids that contained [MeSO$_4$], [NO$_3$], [AcO] or [lactate] anions [82]. All of these ionic liquids are water-miscible, and it is significant that the enzyme dissolves in these media because dissolving a protein requires the breaking of the protein-protein interactions and replacing these by stronger ones. Interestingly, a substantial fraction of the original activity was recovered from inactive solutions of CaLB in [BMIm][NO$_3$], [BMIm][lactate], [BMIm][EtSO$_4$] or [EtNH$_3$][NO$_3$] on dilution with water, suggesting that the deactivation was largely reversible [81]. The observed loss of detail in the FT-IR spectrum of CaLB in the amide I region is consistent with the notion that hydrophilic ionic liquids induce conformational changes that result in its deactivation when it dissolves [83]. Similarly, the denaturation of lysozyme on dissolution in [EtNH$_3$][NO$_3$] was observed using fluorescence spectroscopy [84].

Two strategies can be followed to avoid the loss of activity of an enzyme incumbent on its dissolution in an ionic liquid: tailor the enzyme for stability in the ionic liquid medium or tailor the ionic liquid. Bruce and coworkers have shown that ionic liquids can be designed to dissolve enzymes without denaturation [85, 86]. The key is to introduce functionalities, e.g., hydroxyl and primary amide groups, that can form hydrogen bonds with the protein to make it look more like water. Thus, morphine dehydrogenase (MDH), when dissolved in the strongly hydrogen-bonding ionic liquid [*HO*PMIm][glycolate], mediated the oxidation of codeine into codeinone (Figure 10.7) under nearly anhydrous conditions. The activity was less than in water, but better than that shown by suspensions in molecular solvents or [BMIm][PF$_6$]. As additional advantages, the reactant and the NADP cofactor dissolved in [*HO*PMIm][glycolate] and the ionic liquid also proved suitable for the subsequent chemical transformation of codeinone into oxycodone. The ionic liquid was also compatible with an enzymatic recycle system for the NADP cofactor (Figure 10.7).

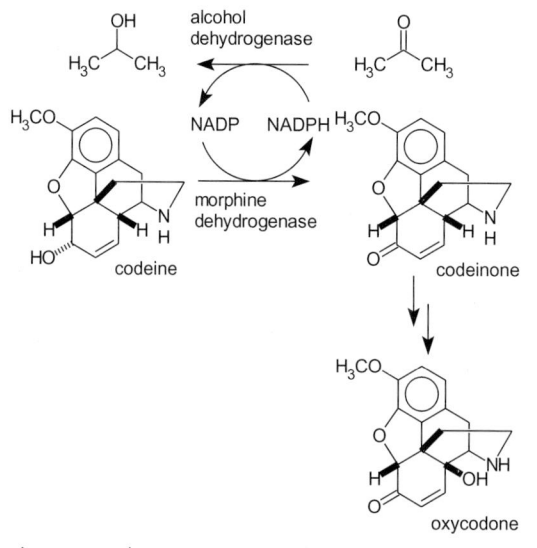

Figure 10.7 Chemoenzymatic synthesis of oxycodone and NADP recycle.

The redox protein cytochrome *c* dissolved in both [BMPrl][H$_2$PO$_4$] and [*HO*EtMe$_3$N][H$_2$PO$_4$] [87] indicating that a hydrogen-bonding capability of the cation is not required, in contrast with the above example. FT-IR measurements showed that cyt *c* maintained its native secondary structure upon dissolution in these ionic liquids.

Alternatively, the enzyme can be modified such that it dissolves in a hydrophobic ionic liquid with retention of activity. This approach was demonstrated with cyt *c*, which, when covalently modified with polyethylene glycol (PEG), dissolved in [EMIm][Tf$_2$N] with retention of activity. The best results were obtained when the molecular weight of the polymer chain was >2000 [88]. Similarly, a copolymer of PEG and maleic anhydride solubilized subtilisin in [EMIm][NTf$_2$] and a range of similar ionic liquids with good retention of activity and operational stability [89, 90].

The incompatibility of enzymes and certain anhydrous ionic liquids can be rationalized on the basis of hydrogen bonding. In particular, anions that can form strong hydrogen bonds can dissociate the hydrogen bonds that maintain the structural integrity of the α-helices and β-sheets, causing the protein to unfold wholly or partially. The lactate ion, for example, can form stable hydrogen bonds with the polypeptide backbone. It has already been noted that the deactivation of CaLB by [BMIm][NO$_3$], for example, is partially reversible. This finding is in agreement with the generally adopted kinetic model of enzyme deactivation, which involves a reversible first step and an irreversible second one. CaLB deactivation by [BMIm][dca], in contrast, was irreversible, and a small-angle neutron scattering experiment indicated the formation of aggregates [91], as is often observed upon unfolding. Presumably, hydrogen bonds also maintain the conformation of reversibly deactivated enzymes. Reconstitution requires these hydrogen bonds to be dissociated and remade into the native ones. Dilute denaturants often facilitate reconstitution, which includes denaturing ionic liquids, as discussed above, presumably by facilitating the reconstitution via the formation of transient hydrogen bonds.

10.5
Stability of Enzymes in Nearly Anhydrous Ionic Liquids

The (thermal) stability (activity over time) of enzymes is often better in organic media, in particular at low water activity, than in an aqueous medium [92], and the same is true for ionic liquids. The storage stability of CaLB at 50 °C over 4 days in water, hexane, [EMIM][Tf$_2$N], and [BuMe$_3$N][NTf$_2$] was compared by measuring the residual hydrolytic activity observed after rehydration. The deactivation was 3–4 times slower in the ionic liquids than in water or hexane [93]. Similarly, the storage stability of free (Novozym SP525) as well as carrier-adsorbed (Novozym 435) CaLB in anhydrous [BMIm][PF$_6$] at 80 °C were compared [81]. The activity of the free enzyme was found to increase in 20 h to 120% of an untreated sample, which was maintained for at least 100 h. In contrast, a linear deactivation versus time was observed in *tert*-butyl alcohol. The activity of Novozym 435 even increased to 350% in 40 h, though on continued incubation it slowly decreased to 210% after 120 h. In contrast, the incubation of a CLEC or CLEA [94] of CaLB in [BMIm][PF$_6$] at 80 °C resulted in a progressive loss of activity, comparable with that observed in *tert*-butyl alcohol. Lozano and coworkers [95] compared the stability of CaLB in the presence of 2% water in a range of ionic liquids with that in 1-butanol or hexane by preincubation and found that it was similar or better. Interestingly, the life-time of CaLB increased by three orders of magnitude when substrate was present [94]. The conformational stability of CaLB, as monitored by fluorescence and CD spectroscopy, was much greater in [EMIM][NTf$_2$] and [BuMe$_3$N][NTf$_2$] at 50 °C than in water or hexane [92]. It was also noted that the operational stability of CaLB at elevated temperatures in water-free molecular solvents is high. For example, Novozym 435 retained its full transesterification activity in refluxing *tert*-butylalcohol for 7 days [96]. CaLB was also exceptionally stable in a series of alkyltrimethylammonium [NTf$_2$] ionic liquids and in (biphasic) [BMIm][Tf$_2$N]-scCO$_2$ systems in temperatures up to 100 °C [97].

Proteases have received less attention than lipases, but in one of the earliest papers on biocatalysis in ionic liquids it was noted that the activity loss of thermolysin during preincubation proceeded much more slowly in [BMIm][PF$_6$] than in ethyl acetate [8]. The storage stability of α-chymotrypsin in the ionic liquid [EMIm][Tf$_2$N] was compared with that in water, 3 M sorbitol, and 1-propanol. The residual hydrolytic activity (after dilution with aqueous buffer) was measured vs time, and structural changes were monitored by fluorescence and CD spectroscopy as well as DSC [98]. The enzyme's life-time in [EMIm][Tf$_2$N] at 30 °C was more than twice that in 3 M sorbitol, six times as long as that in water, and 96 times as long as that in 1-propanol.

10.6
Whole-Cell Biotransformations in Ionic Liquids

Most studies of biocatalysis in ionic liquids have been concerned with the use of isolated enzymes. It should not be overlooked, however, that the first report on biocatalysis and ionic liquids involved a whole-cell preparation *Rhodococcus* R312 B in a biphasic [BMIm][PF$_6$]-water system [7]. It was shown, using a nitrile hydrolysis test reaction, that the microorganism maintained its activity better in ionic liquid than in a biphasic toluene-water system.

It was subsequently shown that baker's yeast [99], as well as *Rhodococcus* R312 and *E. coli* [100] maintain their activity in ionic liquids containing very little or no separate aqueous phase. The advantage of using biphasic aqueous-organic reaction systems for whole-cell biotransformations is that the organic phase acts as a reservoir for the hydrophobic reactants and products, thereby suppressing substrate and/or product inhibition or deactivation of the cells. The advantage of ionic liquids over common organic solvents such as toluene is that they are much less toxic to the cell membranes [7]. The cell membrane integrity of *Lactobacillus kefir* was preserved much better in biphasic aqueous/ionic liquid systems, in particular [BMIm][Tf$_2$N], than in aqueous organic systems such as decane, octanol, and TBME [101]. Similarly, the same ionic liquids were shown to be harmless to the cell viability of *E. coli* and *S. cerevisiae* [102].

We conclude that hydrophobic ionic liquids are promising and attractive replacements for molecular solvents in whole-cell biotransformations.

10.7
Biotransformations in Ionic Liquid Media

10.7.1
Lipases and Proteases

The application of lipases in synthetic biotransformations encompasses a wide range of solvolytic reactions of the carboxyl group, such as esterification, transesterification (alcoholysis), perhydrolysis, and aminolysis (amide synthesis) [103]. Transesterification and amide synthesis are preferably performed in an anhydrous medium, often in the presence of activated zeolite, to suppress unwanted hydrolytic side reactions. CaLB (which readily tolerates such conditions [104, 105]), PsL, and PcL are often used as the biocatalyst [106].

Lipase-catalyzed triglyceride modification is practiced on a large scale in the food industry [102, 107]. In a study of the CaLB-catalyzed glycerolysis of commercial oils and fats into the di- and monoglycerides, the solventless procedure was compared with reactions in *tert*-butyl alcohol and in the amphiphilic tetraalkylammonium ionic liquid [CPMA][MeSO$_4$] [108, 109]. It was found that the glycerolysis of sunflower oil was intrinsically faster in the ionic liquid as judged by V_{max}, but a high K_m of approx. 0.8 M actually caused the reaction to be slower than in *tert*-butyl alcohol [108].

Lipase-catalyzed transesterification to prepare polyesters (replacing the traditional chemical polymerization at >200 °C) has received considerable attention in recent years. CaLB was found to mediate polyester synthesis in the ionic liquids [BMIm][BF$_4$], [BMIm][PF$_6$], and [BMIm][Tf$_2$N] at 60 °C [110, 111, 112], but the molecular weight of the product was rather low compared with that in a solventless system [113], perhaps owing to the high viscosity of ionic liquid media.

Fatty acids of sugars are potentially useful and fully green nonionic surfactants, but the lipase-mediated esterification of carbohydrates is hampered by the low solubility of carbohydrates in reaction media that support lipase catalysis in general. Because the monoacylated product (Figure 10.8) is more soluble in traditional solvents than is the starting compound, the former tends to undergo further acylation into a diester. In contrast, the CaLB-catalyzed esterification of glucose with vinyl acetate in the ionic liquid [EMIm][BF$_4$] was completely selective. The reaction became much faster and somewhat less selective when conducted in

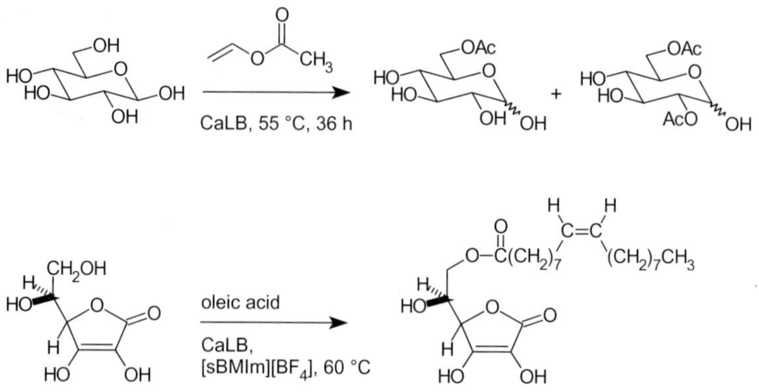

Figure 10.8 Transesterification of glucose and L-ascorbic acid.

[MOEMIm][BF$_4$], in which 5 g L^{-1} of glucose dissolves at 55 °C (100 times faster than in acetone) [28]. The disaccharide maltose was also acylated in the presence of CaLB in [MOEMIm][BF$_4$] [28].

The synthesis of long-chain fatty acid esters of carbohydrates is inherently more demanding. It was found that glucose did not react with vinyl laurate in a pure ionic liquid medium, but in biphasic *tert*-butyl alcohol/[BMIm][PF$_6$], glucose could be acylated by the vinyl esters of C$_{12}$–C$_{16}$ fatty acids. The best results were obtained with CaLB, which was twice as active as TlL, and the selectivity for acylation at C-6 was high [114]. The esterification of glucose with palmitic acid, which is, in an industrial context, to be preferred over transesterification, has recently been demonstrated in *tert*-butyl alcohol/[BMIm][PF$_6$] medium [115].

L-Ascorbic acid (Figure 10.8) proved to be less recalcitrant than glucose and could be esterified with palmitic acid in the presence of CaLB in [BMIm][BF$_4$] and similar ionic liquids [116, 117]. The equilibrium was shifted toward the product by applying a vacuum to remove the water, and undesirable precipitation of the reaction product on the biocatalyst was obviated by the addition of a hydrophobic phase such as hexane or polypropylene beads [116].

The enzymatic esterification of polysaccharides, e.g., glucomannan, a copolymer of glucose and mannose, has also been reported [118].

Acylations of carbohydrate derivatives such as alkyl glucosides and galactosides have also been successfully performed in ionic liquids [63]. Similarly, the flavonoid glycosides naringin and rutin were acylated with vinyl butyrate in ionic liquid media in the presence of a number of lipases, e.g., CaLB (Novozym 435), immobilized TlL, and RmL [119]. The products are of interest for application as strong antioxidants in hydrophobic media.

Lipase-mediated enantioselective hydrolysis of an N-unprotected aminoacid ester has been demonstrated with methyl phenylglycinate (Figure 10.9). In the presence of CaLB the *E* ratio was a rather modest 12, which improved when acetonitrile (ACN) or *tert*-butyl alcohol was added to the medium and further improved to 34 with 20% [BMIm][BF$_4$] [68, 120]. The addition of more strongly hydrogen-

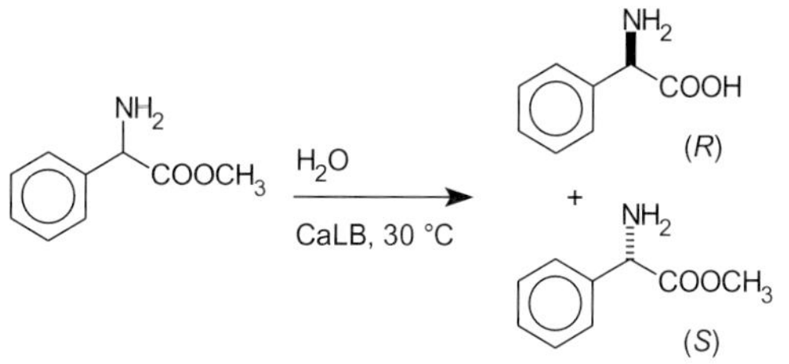

Figure 10.9 Enantioselective hydrolysis of methyl phenylglycinate.

Cosolvent	(%)	V_0 (rel, %)	E
None		100	12
[BMIm][BF$_4$]	20	107	34
[BMIm][Cl]	20	96	7
[BMIm][Br]	20	96	6
ACN	10	79	19
tert-Butyl alcohol	20	67	20

bonding ionic liquids, such as [BMIm][Cl] or [BMIm][Br], in contrast, reduced the E ratio.

The enzymatic enantioselective hydrolysis of esters of naproxen and ibuprofen has attracted considerable attention because the (S)-enantiomers of these non-steroidal anti-inflammatory drugs (NSAIDs) are the pharmacologically active isomers. These reactions have been successfully performed in a range of ionic liquids (Figure 10.10) [60, 65, 121].

Enzymatic kinetic resolution is a key step in the synthesis of the platelet aggregation inhibitor Lotrafiban (Figure 10.11). A disclosed process involves CaLB in tert-butyl alcohol/water (88:12) at 50°C; the substrate concentration was only 5 g L^{-1} owing to its low solubility in this medium [122]. By exploiting the higher solubility in 88% [BMIm][PF$_6$] and the better thermal stability of the biocatalyst in this medium, a higher rate was observed, the reaction was performed at 40 g L^{-1} at 75°C, and the biocatalyst (Novozym 435) could be recycled 10 times.

Similarly, CrL has also been employed in the resolution of a range of 2-substituted propionic acid derivatives (Figure 10.12) in ionic liquids [62].

Aminolysis with butylamine, rather than hydrolysis or transesterification, has been employed in the kinetic resolution of methyl mandelate (Figure 10.13) [123]. Enantioselectivities in the kinetic resolution of mandelic acid via transesterification are generally low. Aminolysis (or ammoniolysis) may improve the resolution, as has been shown in some cases [124, 125], presumably by a shift of the rate-determining step. Resolution with CaLB in conventional media afforded quite modest E ratios, which became near-quantitative when 10% [BMIm][BF$_4$] was added to the medium [122].

The resolution of chiral alcohols through lipase-mediated enantioselective acylation (transesterification) is one of the major industrial applications of lipases [50]. Hence, the effects of ionic liquid reaction media on the resolution of various

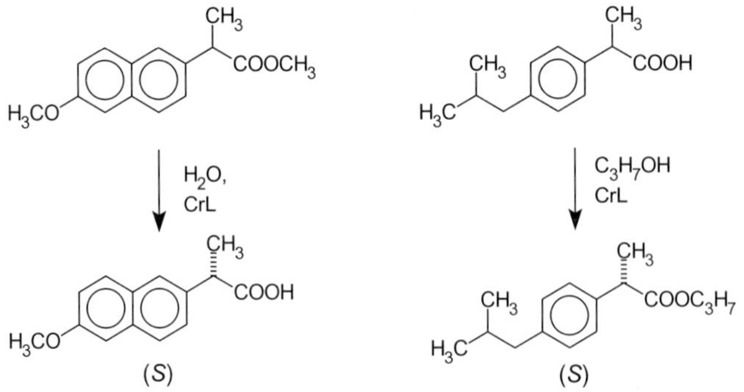

Figure 10.10 Enantioselective hydrolysis of naproxen and ibuprofen esters.

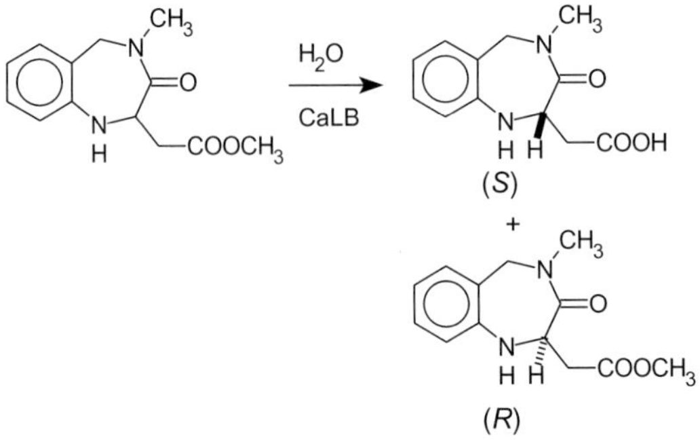

Medium	[H$_2$O] (M)	E
[BMIm][BF$_4$]	2.8	170
[HMIm][BF$_4$]	0.4[a]	> 200
[BMIm][PF$_6$]	0.39[a]	> 200
[HMIm][PF$_6$]	0.28[a]	> 200
Isooctane	n.d.[a]	88

Medium	Conv. (%)	E
[BMIm][BF$_4$]	33	6.4
[BMIm][PF$_6$]	30	24
[BMIm][MeSO$_4$]	50	1.2
Isooctane	29	13

a. Water saturated.

Figure 10.11 Enantioselective hydrolysis of a Lotrafiban precursor.

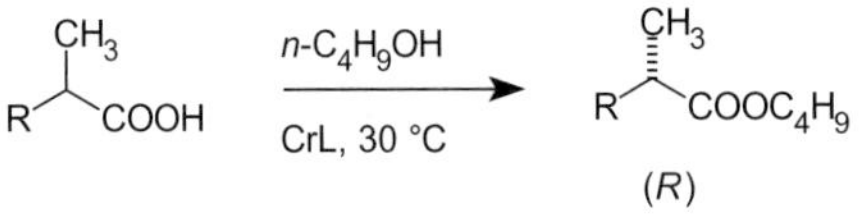

Figure 10.12 Enantioselective esterification of 2-substituted propanoic acids.

R	E	
	Hexane	[BMIm][PF$_6$]
CH$_3$O	16	25
C$_2$H$_5$O	13	21
n-C$_3$H$_7$O	7	19
i-C$_3$H$_7$O	7	14
C$_6$H$_5$O	4	10
Cl	10	20
Br	18	29

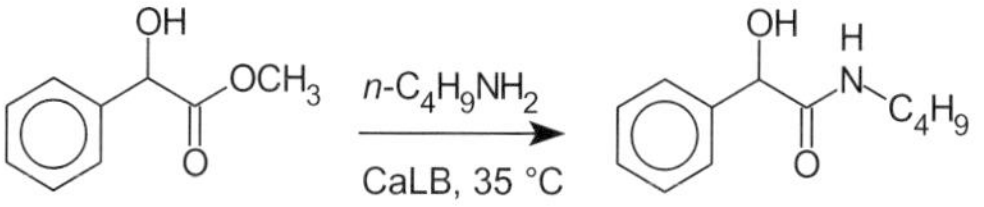

Figure 10.13 Enantioselective aminolysis of methyl mandelate.

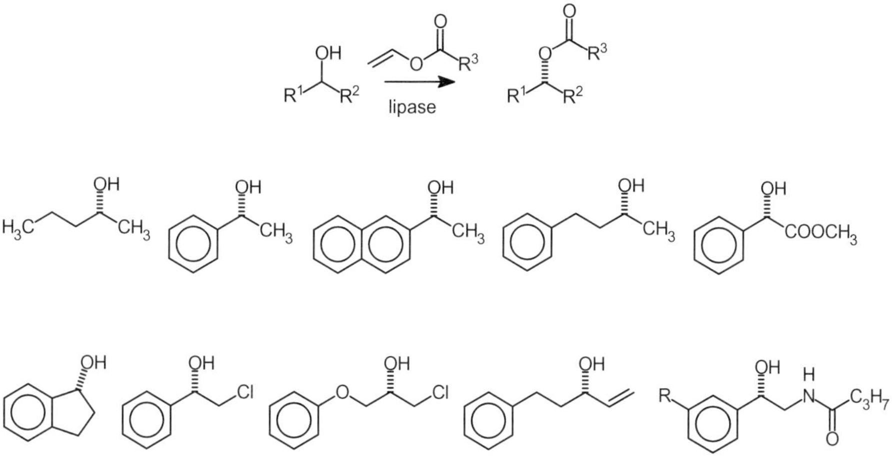

Solvent	E
t-BuOH	10 (R)
t-BuOH-[BMIm][BF$_4$] (90:10)	> 200 (R)
CHCl$_3$	22 (S)
CHCl$_3$-[BMIm][BF$_4$] (90:10)	> 200 (S)

Figure 10.14 Resolution of chiral alcohols.

arylalkanols (Figure 10.14) in the presence of, mainly, CaLB and PcL have been extensively investigated, generally using vinyl acetate as the acyl donor [28, 54, 55, 58, 126, 127]. The alcohols studied were, in general, resolved in traditional media with already good-to-excellent enantioselectivity; hence, there was not much margin for improvement. Nonetheless, the enantiomeric ratio of some of these resolutions improved considerably when the reaction was performed in an ionic liquid.

The enantiorecognition of methyl mandelate (Figure 10.14) in enantioselective acylation with vinyl acetate is often modest. Itoh et al. studied this latter reaction with immobilized PcL in [BMIm][PF$_6$] and found that the E-ratio varied from 10 to >250, depending on the carrier [128]. The best rate and enantioselectivity were obtained with PcL immobilized on a methacryloxypropyl-modified macroporous SBA-15 silica.

The high thermostability of lipases in ionic liquids has stimulated research into kinetic resolutions at elevated temperatures [129]. The PsL-mediated acylation of 1-phenylethanol by vinyl acetate in [BMIm][Tf$_2$N] remained highly

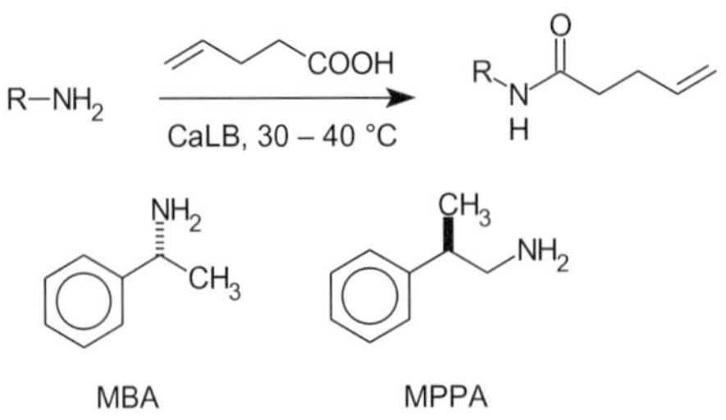

Figure 10.15 Enantioselective acylation of chiral amines mediated by CaLB.

Reaction scheme: R–NH$_2$ + CH$_2$=CH–CH$_2$–CH$_2$–COOH → R–NH–C(O)–CH$_2$–CH$_2$–CH=CH$_2$ (CaLB, 30–40 °C)

Substrates: MBA (1-phenylethylamine) and MPPA (β-methylphenethylamine).

Ionic liquid	Compound, rel. rate (%)	
	MBA	MPPA
[EMIm][TfO]	86	89
[EMIm][BF$_4$]	72	88
[BMIm][BF$_4$]	69	78
[HMIm][PF$_6$]	69	53
[OMIm][BF$_4$]	57	53
[BMIm][PF$_6$]	44	**100**
[HMIm][BF$_4$]	32	21
[OMIm][PF$_6$]	28	11

enantioselective, with E decreasing from 200 to 150, when the temperature was raised from 25 to 90 °C. In contrast, the enantioselectivity in TBME medium dropped dramatically (from $E=200$ to $E=4$) at 55 °C, which corresponds with the boiling point of TBME. In both solvents a decrease in E was observed at the boiling point of either the solvent (TBME) or vinyl acetate [128].

The resolution of chiral amines via lipase-catalyzed enantioselective acylation is now a major industrial process, but interest in adopting ionic liquid reaction media has been surprisingly scant. Interestingly, acids could be used as the acyl donor (Figure 10.15) rather than the usual activated ester in a range of ionic liquids. CaLB was employed as the biocatalyst, and water was removed to shift the equilibrium toward the product [130, 131]. The highest rates were found in [BMMIm][TfO], [EMIm][TfO], and [EMIm][BF$_4$].

Summarizing, it has become clear that the medium, either ionic liquid or traditional, has to be fine-tuned to the reactant and biocatalyst for optimum enantio-recognition. In other words, there is no optimum ionic liquid for performing a kinetic resolution, just as there is no optimum organic solvent, and the theoretical basis for selecting one is still embryonic [132]. With the advent of ionic liquids, the choice of solvents and thus the chance to find one that is satisfactory has increased enormously.

Proteases have been much less studied than lipases in ionic liquid media and generally require the presence of water for activity. We note that the thermolysin-catalyzed amide coupling of benzoxycarbonyl-L-aspartate and L-phenylalanine methyl ester into Z-aspartame in [BMIm][PF$_6$] was an early example of an enzymatic reaction in an ionic liquid medium [8].

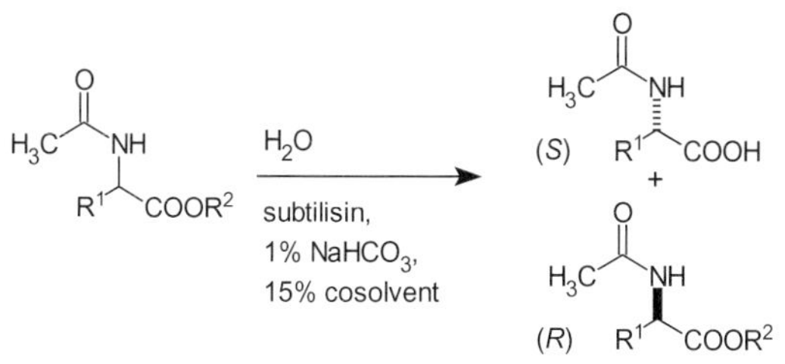

Figure 10.16 Subtilisin-catalyzed hydrolysis of N-acylamino acid esters.

R^1	R^2	Cosolvent, ee_{acid} (% S)	
		ACN	[EPy][TFA]
CH_3	C_2H_5	63	86
$HOCH_2$	CH_3	NA	90
$CH_3CH(OH)$	CH_3	92	97
$CH_3S(CH_2)_2$	CH_3	83	89
$C_6H_5(CH_2)_2$	C_2H_5	95	93
$p\text{-Cl-}C_6H_4CH_2$	C_2H_5	NA	96
$CH_3(CH_2)_3$	CH_3	18	88

NA: no activity

Subtilisin is an endoprotease that has been used in the enantioselective hydrolysis of N-acylamino acid esters (Figure 10.16) into the corresponding (S)-amino acid derivatives. An organic solvent, such as acetonitrile, is often added to improve the solubility of the amino acid derivative, and this function can also be performed by an ionic liquid mixture [133, 134, 135].

10.7.2
Dynamic Kinetic Resolution of Chiral Alcohols

Kinetic resolutions, such as the ones discussed above, are limited to a 50% yield. Consequently, the undesired enantiomer needs to be recovered, racemized, and recycled, which makes the process more complex and leads to an increased solvent use. The obvious solution is to racemize the slow-reacting enantiomer *in situ*. With chiral alcohols, the racemization catalysts of choice are based on ruthenium (Figure 10.17).

Racemization of (S)-1-phenylethanol in the presence of an Ru p-cymene binuclear complex and triethylamine was much faster in [BMIm][BF$_4$] or [BMIm][PF$_6$] than in toluene [136]. A range of chiral alcohols (Figure 10.17) were resolved in the presence of this complex and immobilized PsL. The reactions were performed in [BMIm][PF$_6$] with the activated ester 2,2,2-trifluoroethyl acetate as the acyl donor (Figure 10.17). A hydrogen donor was required to prevent the formation of partially oxidized byproducts. Enantiomerically pure acetates were isolated in high yield (>85%).

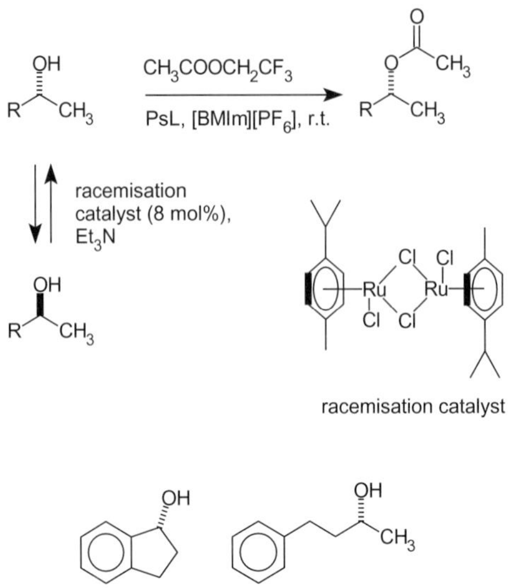

Figure 10.17 Dynamic kinetic resolution of chiral alcohols.

The enantiopreference of the protease subtilisin in the acylation of chiral alcohols is known to be opposite to that observed with lipases, providing for access to both enantiomers with DKR, depending on the enzyme used [137, 138, 139]. Acylation using 2,2,2-trifluoroethyl butyrate as the acyl donor was combined with in situ racemization, affording the corresponding esters in high yield and ee [135].

10.7.3
Glycosidases

In their natural role, glycosidases hydrolyze glycosidic bonds, but they are also widely used as biocatalysts for carbohydrate synthesis in vitro. Two methodologies are applied: condensation (reversed hydrolysis) and transglycosylation. Such reactions are commonly carried out in aqueous-organic mixtures. The condensation of two monosaccharides, such as galactose and glucose (Figure 10.18), is thermodynamically controlled. Hence, the product yield cannot exceed equilibrium, which depends on the reactant and product concentrations, in particular that of water. The β-galactosidase from B. circulans was active in [MMIm][MeSO$_4$] containing only 0.6% water, affording lactose in 18% yield [140]. It should be noted that the reaction was not monitored over time, and the equilibrium conversion may be higher.

Tranglycosylations have also been performed in [MMIm][MeSO$_4$]/water (25:75 v/v) with increased yields compared to those obtained in aqueous buffer [141, 142].

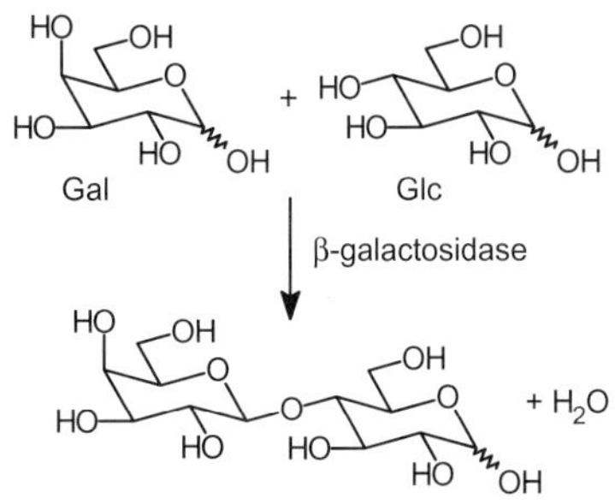

Figure 10.18 Condensation of galactose (Gal) and glucose (Glc) to form lactose.

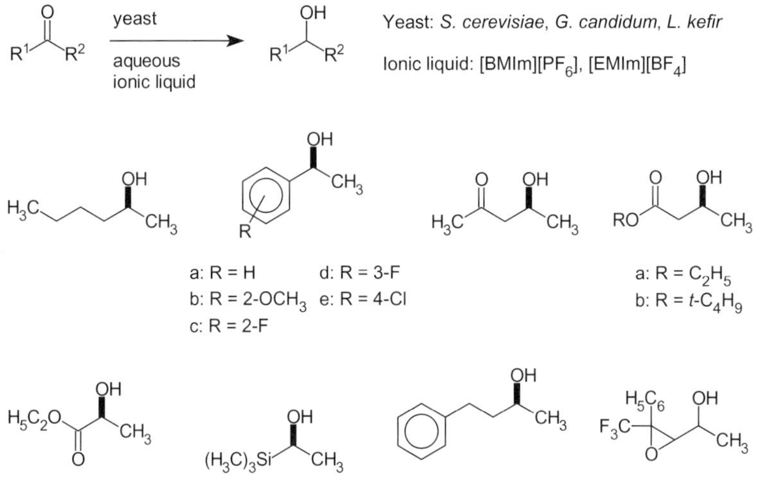

Figure 10.19 Enantioselective ketone reduction with baker's yeast in an ionic liquid.

10.7.4
Oxidoreductases

Biocatalytic redox reactions are often carried out using whole-cell biocatalysts, because of the necessity of recycling the redox cofactor. It was shown that the organic phase, which is often used to store the sparingly soluble reactants and products, can be replaced by an ionic liquid, which seems less harmful to the cell membranes [99, 100]. Thus, a range of ketones was enantioselectively reduced to the corresponding (S)-alcohols by an immobilized yeast in [BMIm][PF$_6$]/water (91:9) biphasic medium (Figure 10.19) [98]. Results were comparable with those obtained in a conventional aqueous-organic medium.

The reduction of 4-chloroacetophenone mediated by *L. kefir* stopped at 46% conversion owing to the reactant and/or the product being toxic to the cell membrane [100]. The addition of TBME (20%, v/v) made the situation worse because

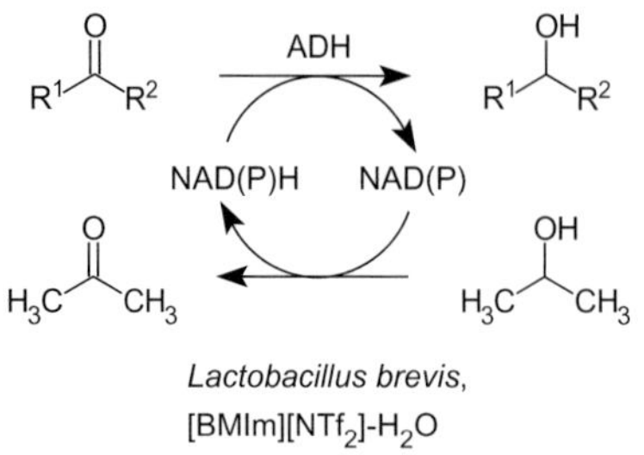

Figure 10.20 Enzymatic ketone reduction with regeneration of NAD(P)H.

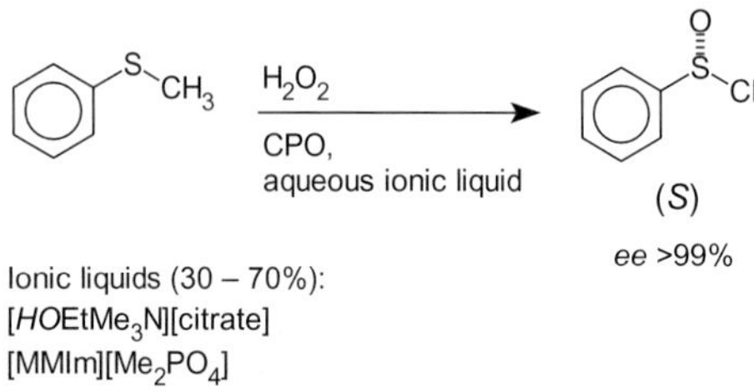

Figure 10.21 Enantioselective sulfoxidation catalyzed by CPO in aqueous-ionic liquid mixtures.

of the toxicity of this solvent to the membrane, and the product yield dropped to 4%. With ionic liquids, in particular [BMIm][NTf$_2$], as the organic phase, the conversion increased and the product *ee*, which was always high, even improved to >99% [100].

Enantioselective reduction of a range of ketones mediated by resting cells of *G. candidum* [143], employing isopropyl alcohol to regenerate the NADPH cofactor (Figure 10.20), was performed in biphasic [BMIm][PF$_6$]/water medium and also in monophasic [EMIm][BF$_4$]/water (67:33) provided that the cells were protected by a water-absorbing polymer, and the enantiomerically pure (S)-alcohols were obtained [95]. A large excess of isopropyl alcohol is usually required to push the reduction toward complete conversion. A cosolvent in which the acetone coproduct selectively partitions would further tilt the system toward reduction of the reactant. [BMIm][NTf$_2$] was found to meet this latter requirement, and accordingly the reduction rate of 2-octanone into (S)-2-octanol, catalyzed by *Lactobacillus brevis* alcohol dehydrogenase, was much higher in a biphasic buffer/[BMIm][NTf$_2$] system than in buffer/TBME or pure buffer [144].

Rather less attention has been paid to enzymatic oxidations in ionic liquids. Chloroperoxidase (CPO) catalyzed the chemo- and enantioselective sulfoxidation of thioanisole (Figure 10.21) in aqueous mixtures containing up to 50% [*HOEtMe$_3$N*] [citrate] or [MMIm][Me$_2$PO$_4$] [145].

Enantioselective sulfoxidations were performed with the peroxidase from *Coprinus cinereus* (CiPx) by employing glucose oxidase-catalyzed oxidation of glucose for the *in situ* generation of hydrogen peroxide in [BMIm][PF$_6$]/buffer (90:10, optimum) [146]. Conversions were rather modest (<32%) and enantioselectivities <70% to >90% ee.

10.8
Downstream Processing

After performing the bioconversion in an ionic liquid, the product needs to be recovered and the biocatalyst and the ionic liquid recycled. Relatively volatile products can be removed by evaporation. Alternatively, immiscible organic solvents can be used to extract the product, and the biocatalyst can be recycled as a suspension in the ionic liquid phase [58]. A more elegant, green method, which avoids the use of volatile organic solvents altogether, involves the use of supercritical carbon dioxide as the extractive phase [96, 147, 148].

The principle has been demonstrated with CaLB in simple model transesterifications as well as in the enantioselective acylation of 1-phenylethanol, in batchwise and continuous procedures. The high operational stability of CaLB, which contrasts with the generally rapid deactivation in pure $scCO_2$, is one of the attractive aspects of this approach. The reaction rate was approximately eight times better than that in pure $scCO_2$ under otherwise identical conditions [96].

The continuous reaction system could be combined with solid acid-catalyzed *in situ* racemization of the slow-reacting alcohol enantiomer [149]. The racemization catalyst and the lipase (Novozym 435) were coated with ionic liquid and kept physically separate in the reaction vessel. Another variation on this theme, which has yet to be used in combination with biocatalysis, involves the use of $scCO_2$ as an anti-solvent in a pressure-dependent miscibility switch [150].

Another emerging technique which deserves mention is the use of a supported ionic liquid membrane. This involves two liquid phases that both contain an enzyme and are separated by the membrane. Lipase-catalyzed esterification takes place in the feed phase to afford a mixture of the (R)-acid and the (S)-ester (Figure 10.22). The latter diffuses through the membrane and is hydrolyzed in the receiving phase to afford the (S)-acid [151, 152]. The methodology has been applied, for example, to the resolution of ibuprofen [151].

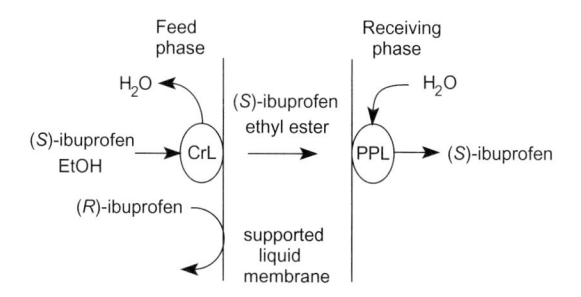

Figure 10.22 Supported ionic liquid membrane in the enzymatic resolution of ibuprofen.

10.9
Conclusions

A large variety of enzymes appear to tolerate aqueous-ionic liquid mixtures as the reaction medium, generally to higher concentrations than those tolerated in water-miscible molecular solvents. Many hydrolases, particularly those that tolerate conventional organic solvents, are eminently capable of performing non-hydrolytic reactions in ionic liquids. Activities are generally comparable with or higher than those observed in conventional organic solvents. Furthermore, enhanced thermal and operational stabilities and regio- or enantioselectivities have been observed in many cases.

Ionic liquids have obvious potential as reaction media for biotransformations which cannot be performed in water owing to equilibrium limitations, i.e. those of highly polar substrates such as (poly)saccharides. The development of less expensive, non-toxic and biodegradable ionic liquids will also further stimulate their use in industrial biotransformations. We confidently expect that green and biocompatible ionic liquids will become available soon, as is absolutely required for ionic liquids to contribute to a greener chemical industry. Another fundamental contribution to the greening of industrial biocatalysis is to be expected from innovative reaction methodologies, downstream processing, and biocatalyst recycling, based on the unique solvent properties of ionic liquids and combinations of ionic liquids with supercritical carbon dioxide. Furthermore, it is to be expected that ionic liquid-based solvent systems will have enormous potential in multicatalyst (chemo)enzymatic transformations. Work toward these ends has only just started. In short, we believe that biocatalysis in ionic liquids holds much promise for the future.

Abbreviations

ACN	acetonitrile
AsL	*Alcaligenes* sp. lipase
BMIm	1-butyl-3-methyl imidazolium
BMMIm	1-butyl-2,3-dimethyl imidazolium
BMPrl	butylmethyl pyrrolidinium
BMPy	1-butyl-4-methyl pyridinium
BsE	*Bacillus subtilis* lipase
BstE	*Bacillus stearothermophilus* lipase
BuMe$_3$N	butyltrimethylammonium
CaLA	*Candida Antarctica* lipase A
CaLB	*Candida Antarctica* lipase B
CPO	chloroperoxidase
CrL	*Candida rugosa* lipase
EMIm	1-ethyl-3-methyl imidazolium
HOEtMe$_3$N	choline

HOPMIm	1(3-hydroxypropyl)-3-methyl imidazolium
MOEMIm	1-methoxyethyl-3-methyl imidazolium
OMIm	1-octyl-3-methyl imidazolium
PcL	*Pseudomonas cepacia* lipase
PsL	*Pseudomonas* sp. lipase
PPL	porcine pancreatic lipase
RmL	*Rhizomucor miehei* lipase
Tf$_2$N	bistriflicimide
TfO	triflate
TlL	*Thermomyces lanuginosus* lipase

References

1. D. R. Macfarlane, K. R. Seddon, *Aust. J. Chem.* 2007, **60**, 3.
2. M. Deetlefs, K. R. Seddon, *Chim. Oggi* 2006, **24**(2), 16.
3. M. J. Earle, J. M. S. S. Esperança, M. A. Gilea, J. N. Canongia Lopes, L. P. N. Rebelo, J. W. Magee, K. R. Seddon, J. A. Widegren, *Nature* 2006, **439**, 831.
4. M. Kosmulski, J. Gustafsson, J. B. Rosenholm, *Thermochim. Acta* 2004, **412**, 47.
5. *Ionic Liquids in Synthesis*; P. Wasserscheid, T. Welton, Eds.; Wiley-VCH: Weinheim, 2003.
6. Some commercial suppliers: Acros (www.acros.be), Covalent Associates (www.covalentassociates.com), IOLITEC (www.iolitec.de), Merck (www.ionicliquids-merck.de), Sachem (www.sacheminc.com), Sigma-Aldrich (www.sigmaaldrich.com), Solvent Innovation (www.solvent-innovation.de) and TCI (www.tciamerica.com).
7. S. G. Cull, J. D. Holbrey, V. Vargas-Mora, K. R. Seddon, G. J. Lye, *Biotechnol. Bioeng.* 2000, **69**, 227.
8. M. Erbeldinger, A. J. Mesiano, A. J. Russell, *Biotechnol. Progr.* 2000, **16**, 1129.
9. R. Madeira Lau, F. van Rantwijk, K. R. Seddon, R. A. Sheldon, *Org. Lett.* 2000, **2**, 4189.
10. U. Kragl, M. Eckstein, N. Kaftzik, *Curr. Opin. Biotechnol.* 2002, **13**, 565.
11. F. van Rantwijk, R. Madeira Lau, R. A. Sheldon, *Trends Biotechnol.* 2003, **21**, 131.
12. S. Park, R. J. Kazlauskas, *Curr. Opin. Biotechnol.* 2003, **14**, 432.
13. Z. Yang, W. Pan, *Enzyme Miocrob. Technol.* 2005, **37**, 19.
14. Y. H. Moon, S. M. Lee, S. H. Ha, Y.-M. Koo, *Korean J. Chem. Eng.* 2006, **23**, 247.
15. P. J. Scammels, J. L. Scott, R. D. Singer, *Aust. J. Chem.* 2005, **58**, 155.
16. K. R. Seddon, A. Stark, M. J. Torres, *Pure Appl. Chem.* 2000, **72**, 2275.
17. S. H. Lee, S. H. Ha, S. B. Lee, Y.-M. Koo, *Biotechnol. Lett.* 2006, **28**, 1335.
18. C. Reichardt, *Green Chem.* 2005, **7**, 339.
19. For a recent review on ionic liquid solvent properties see: C. Chiappe, D. Pieraccini, *J. Phys. Org. Chem.* 2005, **18**, 275.
20. J. F. Deye, T. A. Berger, A. G. Anderson, *Anal. Chem.* 1990, **62**, 615.
21. Solvatochromic dyes are compounds with a visible absorption maximum that depends on the polarity of the solvent.
22. T. Soujanaya, T. S. R. Krishna, A. Samanta, *J. Phys. Chem.* 1992, **96**, 8544.
23. A. J. Carmichael, K. R. Seddon, *J. Phys. Org. Chem.* 2000, **13**, 591.
24. S. N. V. K. Aki, J. F. Brennecke, A. Samanta, *Chem. Commun.* 2001, 413.
25. M. J. Muldoon, C. M. Gordon, I. R. Dunkin, *Chem. Soc., Perkin Trans. 2* 2001, 433.
26. C. Reichardt, *Chem. Rev.* 1994, **94**, 2319.
27. M. J. Earley, B. S. Engel, K. R. Seddon, *Aust J. Chem.* 2004, **57**, 149.
28. S. Park, R. Kazlauskas, *J. Org. Chem.* 2001, **66**, 8395.

29 Q. Liu, M. H. A. Janssen, F. van Rantwijk, R. A. Sheldon, *Green Chem.* 2005, **7**, 39.
30 R. P. Swatloski, S. K. Spear, J. D. Holbrey, R. D. Rogers, *J. Am. Chem. Soc.* 2002, **124**, 4974.
31 J. L. Anderson, J. Ding, T. Welton, D. W. Armstrong, *J. Am. Chem. Soc.* 2002, **124**, 14247.
32 J. L. Kaar, A. M. Jesionowski, J. A. Berberich, R. Moulton, A. J. Russell, *J. Am. Chem. Soc.* 2003, **125**, 4125.
33 L. Cammarata, S. G. Kazarian, P. A. Salter, T. Welton, *Phys. Chem. Chem. Phys.* 2001, **3**, 5192.
34 T. Köddermann, C. Wertz, A. Heintz, R. Ludwig, *Angew. Chem. Int. Ed.* 2006, **45**, 3697.
35 P. Bonhôte, A.-P. Dias, N. Papageorgiou, K. Kalyanasundaram, M. Grätzel, *Inorg. Chem.* 1996, **35**, 1168.
36 L. A. Blanchard, Z. Gu, J. F. Brennecke, *J. Phys. Chem. B* 2001, **105**, 2437.
37 R. P. Swatloski, J. D. Holbrey, R. D. Rogers, *Green Chem.* 2003, **5**, 361.
38 R. Singh, R. M. Kissling, M.-A. Letellier, S. P. Nolan, *J. Org. Chem.* 2004, **69**, 209.
39 K. M. Docherty, C. F. Kulpa, Jr., *Green Chem.* 2005, **7**, 185.
40 A. S. Wells, V. T. Coombe, *Org. Proc. Res. Dev.* 2006, **10**, 794.
41 S. Stolte, J. Arning, U. Bottin-Weber, M. Matzke, F. Stock, K. Thiele, M. Uerdingen, U. Welz-Biermann, B. Jastorff, J. Ranke, *Green Chem.* 2006, **8**, 621.
42 N. Gathergood, M. T. Garcia, P. J. Scammels, *Green Chem.* 2004, **6**, 166.
43 N. Gathergood, P. J. Scammels, M. T. Garcia, *Green Chem.* 2006, **8**, 156.
44 K. Fukumotu, M. Yoshizawa, H. Ohno, *J. Am. Chem. Soc.* 2005, **127**, 2398.
45 G. Tao, L. He, W. Liu, L. Xu, W. Xiong, T. Wang, Y. Kou, *Green Chem.* 2006, **8**, 639.
46 A. Zaks, A. M. Klibanov, *Proc. Natl. Acad. Sci. USA* 1985, **82**, 3192.
47 A. M. Klibanov, *CHEMTECH* 1986, **16**, 354.
48 A. M. Klibanov, *Trends Biotechnol.* 1997, **15**, 97.
49 Z. Yang, A. J. Russell, *Fundamentals of non-aqueous enzymology.* In *Enzymatic Reactions in Organic Media*, A. M. P. Koskinen, A. M. Klibanov, Eds., Blackie Academic and Professional: London, 1996, p 43.
50 A. Schmidt, J. S. Dordick, B. Hauer, A. Kiener, M. Wubbolts, B. Witholt, *Nature* 2001, **409**, 258.
51 A. T. J. W. de Goede, W. Benckhuijsen, F. van Rantwijk, L. Maat, H. Van Bekkum, *Recl. Trav. Chim. Pays-Bas* 1993, **112**, 567.
52 M. Dolman, P. J. Halling, B. D. Moore, S. Waldron, *Biopolymers* 1997, **41**, 313.
53 D. S. Clark, *Philos. Trans. R. Soc. London, Ser. B*, 2004, **359**, 1299.
54 S. H. Schöfer, N. Kaftzik, P. Wasserscheid, U. Kragl, *Chem. Commun.* 2001, 425.
55 K.-W. Kim, B. Song, M.-Y. Choi, M.-J. Kim, *Org. Lett.* 2001, **3**, 1507.
56 P. Lozano, T. De Diego, D. Carrié, M. Vaultier, J. L. Iborra, *J. Mol. Catal. B: Enzym.* 2003, **21**, 9.
57 S. J. Nara, J. R. Harjani, M. M. Salunkhe, *Tetrahedron Lett.* 2002, **43**, 2979.
58 T. Itoh, E. Akasaki, K. Kudo, S. Shirakami, *Chem. Lett.* 2001, 262.
59 S. Garcia, N. M. T. Lourenço, D. Lousa, A. F. Sequiera, P. Mimoso, J. M. S. Cabral, C. A. M. Afonso, S. Barreiros, *Green Chem.* 2004, **6**, 466.
60 J.-Y. Xin, Y.-J. Zhao, G.-L. Zhao, Y. Zheng, X.-S. Ma, C.-G. Xia, S.-B. Li, *Biocatal. Biotransform.* 2005, **23**, 353.
61 L. Gubicza, N. Nemestóthy, T. Fráter, K. Bélafi-Bakó, *Green Chem.* 2003, **5**, 236.
62 O. Ulbert, T. Fráter, K Bélafi-Bakó, L. Gubicza, *J. Mol. Catal. B: Enzym.* 2004, **31**, 39.
63 M.-J. Kim, M. Y. Choi, K. L. Lee, Y. Ahn, *J. Mol. Catal. B: Enzym.* 2003, **26**, 115.
64 O. Ulbert, K. Bélafi-Bakó, K. Tonova, L. Gubicza, *Biocatal. Biotransform.* 2005, **23**, 177.
65 H. Yu, J. Wu, C. B. Ching, *Chirality* 2005, **17**, 16.
66 M. Persson, U. T. Bornscheuer, *J. Mol. Catal. B: Enzym.* 2003, **22**, 21.
67 A. Basso, S. Cantone, P. Linda, C. Ebert, *Green Chem.* 2005, **7**, 671.

68 Y.-Y. Liu, W.-Y. Lou, M.-H. Zong, R. Xu, X. Hong, H. Wu, *Biocatal. Biotransform.* 2005, **23**, 89.

69 J. A. Laszlo, D. L. Compton, *Biotechnol. Bioeng.* 2001, **75**, 181.

70 J. A. Laszlo, D. L. Compton, *Chymotrypsin-catalyzed transestrification in ionic liquids and ionic liquid/supercritical carbon dioxide*. In *Ionic Liquids*, R. D. Rogers, K. R. Seddon, Eds., ACS Symposium Series Vol. 818, ACS: Washington D.C., 2002, p 387.

71 P. Lozano, T. De Diego, J.-P. Guegan, M. Vaultier, J. L. Iborra, *Biotechnol. Bioeng.* 2001, **75**, 563.

72 M. Eckstein, M. Sesing, U. Kragl, P. Adlercreutz, *Biotechnol. Lett.* 2002, **24**, 867.

73 S. Shah, M. N. Gupta, *Biochim. Biophys. Acta*, 2007, **1770**, 94.

74 C. Chiappe, E. Leandri, S. Lucchesi, D. Pieraccini, B. D. Hammock, C. Morisseau, *J. Mol. Catal. B: Enzym.* 2004, **27**, 243.

75 J. Broos, J. F. J. Engbertsen, I. K. Sakodinskaya, W. Verboom, D. N. Reinhoudt, *J. Chem. Soc., Perkin Trans 1* 1995, 2899.

76 T. Maruyama, S. Nagasawa, M. Goto, *Biotechnol. Lett.* 2002, **24**, 1341.

77 T. Maruyama, H. Yamamura, T. Kotani, N. Kamiya, M. Goto, *Org. Biomol. Chem.* 2004, **2**, 1239.

78 P. Lozano, T. De Diego, J. L. Iborra, *Immobilization of enzymes for use in ionic liquids*. In *Immobilization of Enzymes and Cells*, J. M. Guisán, Eds., Methods in Biotechnology Vol. 22, Humana Press: Totowa NJ, 2nd Eds., 2006, p 257.

79 T. Itoh, Y. Nishimura, M. Kashiwagi, M. Onaka, *Efficient lipase-catalyzed enantioselective acylation in an ionic liquid solvent system*. In *Ionic Liquids as Green Solvents*, R. D. Rogers, K. R. Seddon, Eds., ACS Symposium Series Vol. 856, ACS: Washington D.C., 2003, p 251.

80 J. K. Lee, M.-J. Kim, *J. Org. Chem.* 2002, **67**, 6845.

81 T. Itoh, S. Han, Y. Matsushita, S. Hayase, *Green Chem.* 2004, **6**, 437.

82 R. A. Sheldon, R. Madeira Lau, M. J. Sorgedrager, F. van Rantwijk, K. R. Seddon, *Green Chem.* 2002, **4**, 147.

83 R. Madeira Lau, M. J. Sorgedrager, G. Carrea, F. van Rantwijk, F. Secundo, R. A. Sheldon, *Green Chem.* 2004, **6**, 483.

84 C. A. Summers, R. A. Flowers II, *Protein Sci.* 2000, **9**, 2001.

85 A. J. Walker, N. C. Bruce, *Tetrahedron* 2004, **60**, 561–568.

86 A. J. Walker, N. C. Bruce, *Chem. Commun.* 2004, 2570.

87 K. Fujita, D. R. MacFarlane, M. Forsyth, *Chem. Commun.* 2005, 4804.

88 H. Ohno, C. Suzuki, K. Fukumoto, M. Yoshizawa, K. Fujita, *Chem. Lett.* 2003, **32**, 450.

89 K. Nakashima, T. Maruyama, N. Kamiya, M. Goto, *Chem. Commun.* 2005, 4297.

90 K. Nakashima, T. Maryama, N. Kamiya, M. Goto, *Org. Biomol. Chem.* 2006, **4**, 3462.

91 D. Sate, M. H. A. Janssen, R. A. Sheldon, J. Lu, paper in preparation.

92 A. Zaks, A. M. Klibanov, *Science* 1984, **224**, 1249.

93 T. De Diego, P. Lozano, S. Gmouh, M. Vaultier, J. L. Iborra, *Biomacromolecules* 2005, **6**, 1457.

94 P. López-Serrano, L. Cao, F. van Rantwijk, R. A. Sheldon, *Biotechnol. Lett.* 2002, **24**, 1379.

95 P. Lozano, T. De Diego, D. Carrié, M. Vaultier, J. L. Iborra, *Biotechnol. Lett.* 2001, **23**, 1529.

96 M. Woudenberg-van Oosterom, F. Van Rantwijk, R. A. Sheldon, *Biotechnol. Bioeng.* 1996, **49**, 328.

97 P. Lozano, T. De Diego, D. Carrié, M. Vaultier, J. L. Iborra, *Chem. Commun.* 2002, 692.

98 T. De Diego, P. Lozano, S. Gmouh, M. Vaultier, J. L. Iborra, *Biotechnol. Bioeng.* 2004, **88**, 916.

99 J. Howarth, P. James, J. Dai, *Tetrahedron Lett.* 2001, **42**, 7517.

100 N. J. Roberts, A. Seago, G. J. Lye, *Biocatalytic routes to the efficient synthesis of pharmaceuticals in ionic liquids*. In *Book of Abstracts, International Congress on Biocatalysis*, Hamburg, July 28-31, 2002, p 117.

101 H. Pfruender, M. Amidjojo, U. Kragl, D. Weuster-Botz, *Angew. Chem. Int. Ed.* 2004, **43**, 4529.
102 H. Pfruender, R. Jones, D. Weuster-Botz, *J. Biotechnol.* 2006, **124**, 182.
103 R. D. Schmidt, R. Verger, *Chem. Int. Ed.* 1998, **37**, 1608.
104 E. M. Anderson, K. M. Larsson, O. Kirk, *Biocatal. Biotransform.* 1998, **16**, 181.
105 O. Kirk, M. W. Christensen, *Org. Proc. Res. Dev.* 2002, **6**, 446.
106 U. T. Bornscheuer, R. J. Kazlauskas, *Hydrolases in Organic Synthesis: Regio- or Stereoselective Transformations*, Wiley-VCH: Weinheim, 1999.
107 Sheldon R. A. Large-scale enzymatic conversions in non-aqueous media. In *Enzymatic Reactions in Organic Media*; A. M. P. Koskinen, A. M. Klibanov, Eds.; Blackie: London 1996; p 266.
108 Z. Guo, X. Xu, *Org. Biomol. Chem.* 2005, **3**, 2615.
109 Z. Guo, X. Xu, *Green Chem.* 2006, **8**, 54; see also X. Guo, B. Chen, R. L. Murillo, T. Tan, X. Xu, *Org. Biomol. Chem.* 2006, **4**, 2772.
110 H. Uyama, T. Takamoto, S. Kobayashi, *Polymer J.* 2002, **34**, 94.
111 S. J. Nara, J. R. Harjani, M. M. Salunkhe, A. T. Mane, P. P. Wadgaonkar, *Tetrahedron Lett.* 2003, **44**, 1371.
112 R. Marcilla, M. De Geus, D. Mecerreyes, C. J. Duxbury, C. E. Koning, A. Heise, *Eur. Polymer J.* 2006, **42**, 1215.
113 F. Binns, P. Harffey, S. M. Roberts, A. Taylor, *J. Chem. Soc., Perkin Trans 1* 1999, 2671.
114 F. Ganske, U. T. Bornscheuer, *J. Mol. Catal. B: Enzymatic* 2005, **36**, 40.
115 F. Ganske, U. T. Bornscheuer, *Org. Lett.* 2005, **7**, 3097.
116 S. Park, F. Viklund, K. Hult, R. J. Kazlauskas, in *Ionic liquids as Green Solvents*, R. D. Rogers, K. R. Seddon, Eds., ACS Symposium Series Vol 856, ACS: Washington D.C., 2003, p 225.
117 S. Park, F. Viklund, K. Hult, R. J. Kazlauskas, *Green Chem.* 2003, **5**, 715.
118 Z.-G. Chen, M.-H. Zong, G.-J. Li, *Abstracts of Papers, 231st ACS National Meeting*, Atlanta, GA, United States, March 26–30, 2006, POLY-164; see also: Z.-G. Chen, M.-H. Zong, G.-J. Li, *Proc. Biochem.* 2006, **41**, 1514.
119 M. H. Katsoura, A. C. Polydera, L. Tsironis, A. D. Tselepis, H. Stamatis, *J. Mol. Catal. B: Enzymatic* 2006, **123**, 491.
120 W.-Y. Lou, M.-H. Zong, Y.-Y. Liu, J.-F. Wang, *J. Biotechnol.* 2006, **125**, 64.
121 F. J. Contesini, P. O. Carvalho, *Tetrahedron: Asymmetry* 2006, **17**, 2069.
122 N. J. Roberts, A. Seago, J. S. Carey, R. Freer, C. Preston, G. L. Lye, *Green Chem.* 2004, **6**, 475.
123 C. Pilisão, M. G. Nascimento, *Tetrahedron: Asymmetry* 2006, **17**, 428.
124 M. C. de Zoete, A. C. Kock-van Dalen, F. van Rantwijk, R. A. Sheldon, *Biocatalysis*, 1994, **10**, 307.
125 W. A. J. Starmans, R. G. Doppen, L. Thijs, B. Zwanenburg, *Tetrahedron: Asymmetry* 1998, **9**, 429.
126 T. Itoh, E. Akasaki, Y. Nishimura, *Chem. Lett.* 2002, 154.
127 M. Noël, P. Lozano, M. Vaultier, J. L. Iborra, *Biotechnol. Lett.* 2004, **26**, 301.
128 T. Itoh, Y. Nishimura, M. Kashiwagi, M. Onaka, in *Ionic Liquids as Green Solvents*, R. D. Rogers, K. R. Seddon, Eds., ACS Symposium Series, Vol 856, ACS, Washington D.C., 2003, p.251.
129 M. Eckstein, P. Wasserscheid, U. Kragl, *Biotechnol. Lett.* 2002, **24**, 763.
130 R. Irimescu, K. Kato, *Tetrahedron Lett.* 2004, **45**, 523.
131 R. Irimescu, K. Kato, *J. Mol. Catal. B: Enzymatic* 2004, **30**, 189.
132 T. Ke, C. R. Wescott, A. M. Klibanov, *J. Am. Chem. Soc.* 1996, **118**, 3366.
133 H. Zhao, S. V. Malhotra, *Biotechnol. Lett.* 2002, **24**, 1257.
134 H. Zhao, S. M. Campbell, L. Jackson, Z. Song, O. Olubajo, *Tetrahedron: Asymmetry* 2006, **17**, 377.
135 The pH could have drifted in the course of the reactions as the buffer concentration was rather low.
136 M.-J. Kim, H. M. Kim, D. Kim, Y. Ahn, J. Park, *Green Chem.* 2004, **6**, 471.
137 P. A. Fitzpatrick, A. M. Klibanov, *J. Am. Chem. Soc.* 1991, **113**, 3166.
138 R. J. Kazlauzkas, A. N. E. Weissfloch, *J. Mol. Catal. B: Enzymatic* 1997, **3**, 65.

139 R. C. Lloyd, M. Dickman, J. B. Jones, *Tetrahedron: Asymmetry* 1998, **9**, 551.

140 N. Kaftzik, S. Neumann, M. R. Kula, U. Kragl, in *Ionic Liquids as Green Solvents*, R. D. Rogers, K. R. Seddon, Eds., ACS Symp. Ser., Vol. 856, ACS, Washington D.C., 2003, p. 206.

141 N. Kaftzik, P. Wasserscheid, U. Kragl, *Proc. Res. Dev.* 2002, **6**, 553.

142 M. Lang, T. Kamrat, B. Nidetzky, *Biotechnol. Bioeng.* 2006, **95**, 1093.

143 Y. Matsuda, Y. Yamagishi, S. Koguchi, N. Iwai, T. Kitazume, *Tetrahedron Lett.* 2006, **47**, 4619.

144 M. Eckstein, M. Villela Filho, A. Liese, U. Kragl, *Chem. Commun.* 2004, 1084.

145 C. Chiappe, L. Neri, D. Pieracini, *Tetrahedron Lett.* 2006, **47**, 5089.

146 K. Okrasa, E. Guibé-Jampel, M. Therisod, *Tetrahedron: Asymmetry* 2003, **14**, 2487.

147 M. T. Reetz, W. Wiesenhöfer, G. Franciò, W. Leitner, *Chem. Commun.* 2002, 992.

148 M. T. Reetz, W. Wiesenhöfer, G. Franciò, W. Leitner, *Adv. Synth. Catal.* 2003, **345**, 1221.

149 P. Lozano, T. De Diego, M. Larnicol, M. Vaultier, J. L. Iborra, *Biotechnol. Lett.* 2006, **28**, 1559.

150 M. C. Kroon, J. van Spronsen, C. J. Peters, R. A. Sheldon, G. J. Witkamp, *Green Chem.* 2006, **8**, 246.

151 E. Miyako, T. Maruyama, N. Kamiya, M. Goto, *Biotechnol. Lett.* 2003, **25**, 805.

152 E. Miyako, T. Maruyama, N. Kamiya, M. Goto, *Chem. Commun.* 2003, 2926.

11
Solid/Gas Biocatalysis

Isabelle Goubet, Marianne Graber, Sylvain Lamare, Thierry Maugard and Marie-Dominique Legoy

11.1
Introduction

Solid/gas biocatalysis consists in the use of a biocatalyst as a solid phase acting on gaseous substrates. Solid/gas bioreactors offer the ability to control precisely all the thermodynamic parameters influencing not only the kinetics of the reactions performed but also the stability of the biocatalysts when working with biological catalyst at elevated temperatures.

From a technological point of view, solid/gas systems offer very high production rates for minimal plant sizes, significant reduction of treated volumes, and simplified downstream processes.

Although these systems present many interesting features, the field of application of solid/gas technology remains limited compared to conventional liquid systems, since it is based on the volatile character of the substrates and products of the reaction.

Physical and chemical properties of substrates and products of a reaction are of crucial importance, since they condition the efficiency of the technology.

This is probably why studies of biocatalysts suspended in mixtures of substrates and water vapor remain relatively few in number in contrast to those where enzymes are placed directly in aqueous or non-aqueous solvents.

Nevertheless, because it offers total thermodynamic control of the reaction environment (unlike liquid systems), it is possible to modulate and to study precisely the effect of each component present in the microenvironment of a catalyst on its stability, its activity, or its specificity.

Finally, since solid/gas systems avoid the use of any solvent, they can be seen as a new clean technology fulfilling many of the green chemistry requirements.

There is only one example in nature (reported by Yagi and collaborators [1]) of an enzyme acting on gaseous substrates. Hydrogenase is a unique enzyme, able to bind in the dry state the gaseous hydrogen molecule, rendering it activated, and resulting in parahydrogen-orthohydrogen conversion. Subsequently, in a following study, it was shown that, using purified hydrogenase, the reversible

Organic Synthesis with Enzymes in Non-Aqueous Media. Edited by Giacomo Carrea and Sergio Riva
Copyright © 2008 WILEY-VCH Verlag GmbH & Co. KGaA, Weinheim
ISBN: 978-3-527-31846-9

oxido-reduction of the electron carrier cytochrome c3 with gaseous hydrogen was possible [2]. These pieces of work constitute the beginning of the development of solid/gas biocatalysis.

Following these pioneer papers, about twenty years ago, examples of gas/solid systems using either entire cells or isolated enzymes were reported in the literature. These involved biocatalysts traditionally acting on liquid or soluble substrates but able to catalyze reactions at the solid/gas interface once the vaporization of substrates and products was possible.

Different purified or partially purified enzymes were tested successfully, such as horse liver dehydrogenase [3], *Sulfolobus solfataricus* dehydrogenase [4], *Pischia pastoris* alcohol oxidase [5, 6], the baker's yeast alcohol dehydrogenase [7], and finally lipolytic enzymes, which probably constitute the major part of the work devoted to the use of enzymes working at the solid/gas interface, as summarized in a recent publication [8].

From the beginning of the 1990s, the use of entire cells as catalysts in solid/gas systems was also reported, showing that the use of purified enzymes can be avoided, and thus simplifying access to cheaper catalysts or other catalytic activities.

During this period, work was reported concerning the use of *Methylocystis sp* [9], *Pseudomonas putida* [10], *Sacchuromyces cerevisiae* [11-13], and *Rhodococcus erythropolis* cells [14-16] for applications concerning mainly the bio remediation of VOC-containing gaseous pollutants.

Table 11.1 below gives a non-exhaustive view of the catalyzed reactions tested at the solid / gas interface, using purified enzymes, cell extracts, or whole cells.

As well as demonstrating that solid/gas biocatalysis was possible with enzymes that usually act on liquid substrates, this system (compared to other non-conventional methods developed for overcoming problems in enzymatic catalysis such as substrate or product solubility or to permit modification of thermodynamic constraints) allows very precise control and independent variation of the thermodynamic activity of any substrate or other added component in the gaseous phase.

This is a key feature of the system for anyone who wants to understand and rationalize the effects of the microenvironment of a biocatalyst on its activity, its stability, or its specificity. Since for many years the use of thermodynamic activity was recommended for quantifying substrate availability in non-conventional media [17, 18], the replacement of concentrations of species by their thermodynamic activities in liquid non-conventional media requires a knowledge of their activity coefficients (γ values). And this point is still far from being straightforward, as (a) γ values depend on molar ratios of other species present in the medium, and (b) methods used to estimate these values, such as UNIFAC group contribution method [19], are often called into question, and claimed to be sources of inaccuracy [20, 21].

In solid/gas technology, however, the thermodynamic activity of each compound can be precisely controlled by fixing (or measuring) the partial pressure of the compound in the gas, the thermodynamic activity being calculated from the ratio

Table 11.1 Examples of biocatalysts and reactions tested in solid/gas systems.

Biocatalyst	Reaction, application field	References
	Purified enzymes or cell extracts	
Hydrogenase	Otho $H_2 \leftrightarrow$ para H_2 conversion, cytochrome c3 oxidoreduction	Yagi et al., 1969 [1] Kimura et al.,1979 [2]
Alcohol dehydrogenases (NAD or NADP dependent).	Synthesis of aldehydes and alcohols by coupled oxidoreduction with *in situ* cofactor regeneration Enantioselective reduction with *in situ* cofactor regeneration	Ferloni et al, 2004 [44] Grizon at al., 2004 [13] Pulvin et al., 1986, 1988 [3, 4] Trivedi et al, 2006a, 2006b [45, 46] Yang and Russell, 1996a, 1996b [7, 43]
Alcohol oxidase from *Pischia pastoris*	Ethanol oxidation	Kim and Rhee, 1992 [60] Barzana et al., 1989; 1987 [5, 6] Hwang et al., 1993 [56]
Haloalkane dehalogenase from *Rhodococcus rhodocrous*	Dehalogenation reaction	Dravis et al., 2000 [47]
Lipases	Esterification, alcoholysis, hydrolysis, and transesterification reactions Resolution of racemic alcohols	Bousquet-Dubouch et al, 2001 [33] Cameron et al., 2002 [57] Debeche et al., 2005 [58] Graber et al., 2003a, 2003b [40, 41] Lamare and Legoy,1999 [59] Léonard et al., 2004, 2007 [24, 32] Letisse et al., 2003 [42] Robert et al., 1992 [37]
Cutinase from *Fusarium solani*	Transesterification reactions	Parvaresh et al., 1992 [61] Lamare and Legoy,1995, 1997 [39, 62]
	Whole cells	
Methylosinus sp.	Propylene oxide production	Hou, 1984 [9]
Hansenula polymorpha	Acetaldehyde production	Kim and Rhee, 1992 [60]
Saccharomyces cerevisiae	Aldehydes and alcohols production	Maugard et al., 2001 [11] Goubet et al., 2002 [12] Grizon et al., 2004 [13]
Methylocystis sp.	Trichloroethylene degradation, depollution	Uchiyama et al., 1992 [63]
Pseudomonas putida	Phenol degradation, depollution	Zilli et al., 1992 [10]
Rhodococcus erythropolis	Halogenated compounds degradation, depollution	Erable et al., 2004, 2006 [14, 15]
Xanthobacter autotrophicus	Halogenated compounds degradation, depollution	Erable et al., 2005, 2006 [15, 16]

of the compound's partial pressure to its saturation vapor pressure at the operating temperature.

In solid/gas reactors, the molar fraction of each different species in the inlet gas can be chosen independently. As a result, thermodynamic activities of the different species can be fixed independently very easily. This constitutes a significant advantage over the solid-liquid system, in which reaction species and solvent molar fractions are linked together and for which liquid solvents totally inert towards solvation of substrates, products, and catalysts do not exist.

11.2
Operating Solid/Gas Systems

Except for a few cases among all the systems described in the literature, all the solid/gas bioreactors are continuous (packed-bed type) and can be divided in two main categories based on the technique used for controlling the thermodynamic activity of the compounds present in the inlet gas.

To be able to control the thermodynamic activity of each compound, the inlet gas must be characterized by its molar composition, working temperature, and the total pressure of the system. For inert gas having a low content of organics, assuming that it can be considered as an ideal gas, these few parameters allow a complete definition of the thermodynamic parameters of the gas entering the reactor.

While the molar composition of the inlet gas and the total pressure are known, it is possible to calculate easily the partial pressure of each compound (Pp_x), according to Equation 1, assuming that n_x is equal to the number of moles of X in a finite gas volume containing a total of n_{total} moles at the absolute pressure of Pa.

$$Pp_x = \frac{n_x}{n_{total}} Pa = \frac{\dot{Q}_x}{\dot{Q}_{total}} Pa \text{(atm)} \tag{1}$$

For continuous systems, molar flow rates $\dot{Q}$ can be used instead of n. The thermodynamic activity (a_x) can be calculated according to Equation 2, but requires knowledge of the saturation pressure of the pure compound ($Ppsat_x$). This data can be obtained from the saturation curves (vapor-liquid equilibrium curves) and is taken at the working temperature of the gas stream. The thermodynamic activity is then calculated using the following equation:

$$a_x = \frac{Pp_x}{Ppsat_x} \text{(dimensionless)} \tag{2}$$

The precise control thus results from the technique used for producing the inlet gas.

11.2 Operating Solid/Gas Systems

For systems working at atmospheric pressure or above, saturation of inert gases is the most employed technique, and consists in mixing different flow rates of inert gases that have previously been separately saturated by organic molecules.

Figure 11.1 gives a schematic diagram of a continuous packed-bed reactor working at atmospheric pressure, with control of the inlet gas composition based on the mixing of saturated gases.

In the scheme depicted in Figure 11.1, separate lines of an inert carrier gas (N_2) are saturated by bubbling through pure substrate solutions (C1, C2, and C3) maintained in separate saturation vessels (SV1, SV2, and SV3) at controlled temperature.

Since a sufficient contact time can be achieved, the gaseous phase above the liquid reaches equilibrium with the liquid phase, so that the partial pressure of the substrate in the gas leaving the saturation apparatus is equal to the vapor pressure corresponding to the saturation pressure above the pure compound at the temperature of the liquid.

SV01/02/03: carrier gas saturation vessel
MC01: gas mixing chamber
R01: packed bed bioreactor
V01: phase separator

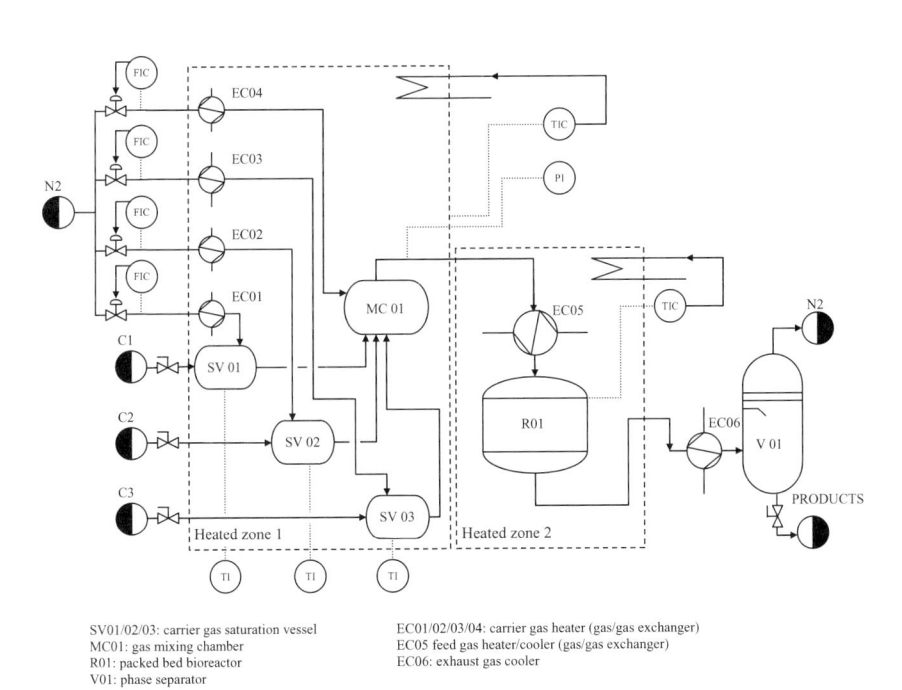

EC01/02/03/04: carrier gas heater (gas/gas exchanger)
EC05 feed gas heater/cooler (gas/gas exchanger)
EC06: exhaust gas cooler

Figure 11.1 Schematic diagram of a packed-bed solid / gas bioreactor working at atmospheric pressure based on saturation of a carrier by organic compounds in separate lines. The different lines are then mixed and partial pressures of the final gas may be adjusted by a fourth line of pure carrier gas.

In order to calculate the composition of the gas entering the reactor (R01), the different molar flow rates for each compound (C1 + C2 + C3) have to be known.

Molar carrier flow rate in each line $\dot{Q}_{N_2}^n$ is calculated from Equation 3, knowing the volumetric normalized flow rates $\dot{Q}v_{N_2}^n$ normalized (L h^{-1} at 273.15 K, 1 atm) of carrier gas used for saturation, usually obtained by mass flow meters:

$$\dot{Q}_{N_2}^n = \frac{\dot{Q}v_{N_2}^n \, normalized}{R.T} \text{ (mol h}^{-1}\text{ for T} = 273.15 \text{K)} \tag{3}$$

Then, the different molar flow rates leaving the saturation vessels can be determined by using the saturation pressure $Ppsat_X^n$ determined at the temperature of saturation, according to Equation 4.

$$\dot{Q}_X^n = \dot{Q}_{N_2}^n \frac{Ppsat_X^n}{(Pa - Ppsat_X^n)} \text{ (mol h}^{-1}\text{)} \tag{4}$$

After mixing the different lines in the mixing chamber (MC01), the partial pressure of each compound Pp_X^n in the gas entering the bioreactor is determined using Equation 5.

$$Pp_X^n = \frac{\dot{Q}_X^n}{\sum_{1}^{n}(\dot{Q}_{N_2}^n + \dot{Q}_X^n)} Pa \text{(atm)} \tag{5}$$

The activity of each compound a_x in the reactor stage (R01) is calculated according to Equation 6 with the respective saturation pressure $Ppsat_X^n$ determined at the temperature of the bioreactor.

$$a_x = \frac{Pp_X^n}{Ppsat_X^n} \text{(dimensionless)} \tag{6}$$

Finally, the total volumetric flow rate $\dot{Q}v_{total}$ in the bioreactor can be estimated, using Equation 7, by applying the ideal gas law PV = nRT.

$$\dot{Q}v_{total} = \frac{RT\sum_{1}^{n}(\dot{Q}_{N_2}^n + \dot{Q}_X^n)}{Pa} \text{ (L h}^{-1}\text{ at the temperature of the bioreactor in K)} \tag{7}$$

This simple method for creating a gas with controlled partial pressure for each compound may present limitations in some cases. By this technique, the realization of a final gas with a partial pressure for each component close to its saturation pressure (i.e. presenting high thermodynamic activity values) is quite impossible, since it would require a saturation temperature close to the boiling point for each compound in order to greatly reduce the inert gas flow rate.

Such operating conditions unavoidably lead to serious inaccuracy due to the significant variation of Pp_{sat} values observed for small temperature variations.

The unique solution consists in changing the reaction temperature so as to be able to study the effect of high thermodynamic values. By performing the reaction at a lower temperature, it is possible to increase the thermodynamic activity of each component without changing the composition of the gas, simply by lowering the Pp_{sat} values at the reaction stage.

To achieve this, and to facilitate the control of the operating parameters, some authors recommended using a technique based on the flash vaporization of liquid substrates in an inert gas. This technique also increases the range of operating pressure, thus allowing the realization of reactions under reduced pressure. Figure 11.2 shows a solid/gas setup in which liquid substrates are injected in a flash evaporator and mixed with an inert gas to produce the required gas mixture.

In such a system, known quantities of the different compounds required to produce the final gas mixture are easily controlled (either by using high-precision metering pumps or mass flow meters) and injected into a flash evaporator (EV01) working at high temperature. Once the vaporization is performed, the gas is

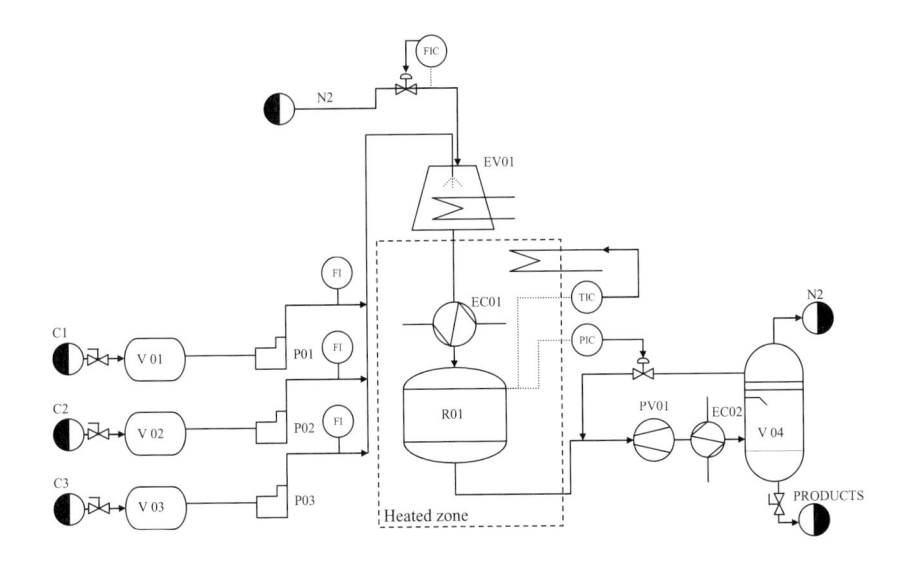

V01/02/03: substrates storage drums   P01/02/03: feeding metering pumps   EV01: evaporator (Flashing unit)
EC01: feed gas cooler (gas/gas exchanger)   R01: packed bed bioreactor   PV01: vacuum pump
EC02: exhaust gas cooler   V04: phase separator

Figure 11.2 Schematic diagram of a packed-bed solid gas bioreactor working at reduced pressure based on the flash vaporization of liquid substrates in an inert carrier gas.

simply cooled or heated to the working temperature by passing over a heat exchanger (EC01), and it can then be injected into the bioreactor (R01) to perform the reaction. In this system, EV01, EC01, and R01 can be maintained at a given pressure, giving the opportunity to modulate the thermodynamic activity simply by changing the absolute working pressure.

Thermodynamic control is achieved by calculating the different partial pressures based on the different molar flow rates. When $\dot{Q}_{N_2}$ and $\dot{Q}_x^n$ are known, then total molar flow rate $\dot{Q}_{total}$ can be determined using Equation 8, and resulting partial pressures can be easily calculated according to Equation 9.

$$\dot{Q}_{total} = \dot{Q}_{N_2} + \sum_1^n \dot{Q}_X^n (\text{mol h}^{-1}) \tag{8}$$

$$Pp_X^n = \frac{\dot{Q}_X^n}{\dot{Q}_{total}} Pa(\text{atm}) \tag{9}$$

Thermodynamic activities are then calculated according to Equation 6.

In this last system, the possibility to operate under reduced pressure offers multiple advantages, e.g., the possible use of higher boiling point compounds or improvement of production rate. The use of reduced pressure allows an important enrichment of the gaseous phase in reactants while much reducing the amount of carrier gas needed, as shown in Figure 11.3. This theoretical example shows that for an identical total number of moles, by simply working at an absolute pressure of 0.25 atm instead of 1 atm, the enrichment of the gaseous phase in substrates A and B is multiplied by a factor of 4, and the requirement for inert carrier gas is divided by a factor of 4, while maintaining the same partial

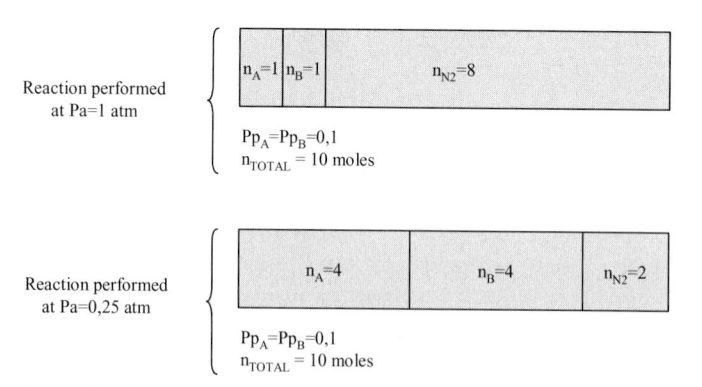

Figure 11.3 Comparison of the composition of a gas involving two reactants (A and B) in nitrogen for two different absolute pressures. While the compositions are different, the thermodynamic activities of A and B are maintained under both pressure conditions.

pressures and thermodynamic activities of substrates A and B in both conditions.

11.2.1
Examples of the Use of Solid/Gas Systems

11.2.1.1 Example 1: Synthesis of Chiral Compounds Using Enzymes in Solid/Gas Reactors

Among organic products of interest are enantiopure compounds, the demand for which has been dramatically growing for some years. The pharmaceutical industry is the main contributor to this increasing demand and the driving force behind it. Indeed, a number of new drugs under development are chiral, as enantiomers often have different and sometimes opposite pharmacological effects. However, applications of chiral organic compounds are not limited to the pharmaceutical industry: they also extend to agrochemicals, cosmetics, and fine chemicals.

Enzymes are widely recognized as valuable tools for the synthesis of optically active compounds [22]. Thus, lipase-catalyzed acylation or deacylation is one of the most efficient methods for the preparation of optically active alcohols, acids, and esters. Because lipases retain activity and selectivity in non-conventional media such as organic liquids, their use as biocatalysts in enantioselective synthetic reactions has considerably increased.

Candida antarctica lipase B (CALB) is one of the most widely used biocatalysts in organic synthesis on both the laboratory and the commercial scale. It has a broad range of applications, including resolutions of alcohols and amines and desymmetrization of complex drug intermediates. This enzyme is highly enantioselective for enantiomers of secondary alcohols. From our knowledge of enzyme structure, the origin of this enantioselectivity can be attributed to the physical restriction of the active site, the existence of a stereospecific pocket, the presence of a long, hydrophobic tunnel at the entrance of the active site, and specific bindings stabilizing the tetrahedral intermediate [23].

11.2.1.1.1 Enantioselective Reactions in the Solid/Gas Reactor versus Liquid Systems

To test the feasibility of enzyme-catalyzed enantiosective reactions in solid/gas reactors and to evaluate the efficiency of the resolution obtained in the gas phase compared to liquid systems, resolution of racemic 2-pentanol, catalyzed by CALB, through alcoholysis with methyl propanoate as acyl donor has been investigated in both liquid media and the gas phase [24]. As CALB has an enantiopreference for R enantiomers of secondary alcohols, this last reaction leads to S-2-Pentanol. This compound is a chiral intermediate in the synthesis of several potential anti-Alzheimer's drugs that inhibit β-amyloid peptide release and/or its synthesis [25]. The degree of enantioselectivity was measured by using the enantiomeric ratio E, which is defined as the ratio of the specificity constants k_{cat}/K_M for the enantiomers (R/S in this case). E can be determined from the enantiomeric excess *ee* of either the substrate or the product, defined as: $ee = \frac{[R]-[S]}{[R]+[S]}$, where [R] is the concentration of the (R)-enantiomer and [S] the concentration of the (S)-enantiomer [26]. The enantiomeric ratio measured in the solid/gas reactor when supplied with the two substrates alone (2-pentanol and methyl propanoate), with nitrogen as carrier gas, was equal to 176 ± 13. Measurement of E was performed in different organic liquid media containing the two substrates and an organic solvent. Results obtained for E were equal to 192 ± 21 with 4-dioxane as solvent, 196 ± 12 with 2-methyl-2-butanol, and 186 ± 19 with hexane [24]. It should be noted that these experimental results were obtained at the same temperature and in the absence of water for both systems, as any change in these reaction conditions are known to change enzyme enantioselectivity [27, 28].

It thus appears that the enantioselectivity of CALB for this reaction in the solid/gas bioreactor is similar to that in an organic liquid medium. Solid/gas biocatalysis therefore offers important potential for production of enantiomerically pure compounds, provided that these transformations involve components having a degree of volatility. Furthermore, as the addition of solvents is avoided in this system, separation and purification during downstream processing are simplified, and side reactions are suppressed.

11.2.1.1.2 Using solid/gas reactors to improve enzyme enantioselectivity by solvent engineering and changing reaction conditions

It is known that enantioselectivity of enzymes depends on many different parameters such as temperature, substrate structure, reaction medium, and presence of water. Enantiopreference of enzymes can be greatly affected, even reversed, by changing the reaction solvent. Such an example was reported by Ueji et al. in 1992 for *Candida cylindracea* lipase-catalyzed esterification of (±)-2-phenoxy propionic acid with 1-butanol [29].

Nevertheless, it appears from the literature that no general rules for the effect of solvent on enantioselectivity of enzymes can be established, and that the effect of other parameters having an influence on enantioselectivity (temperature, type of substrates, presence of water, etc.) are linked together and may modify the effect of solvents.

Therefore, rational improvement of enzyme enantioselectivity, e.g., by such means as solvent engineering, is complicated and mostly a matter of trial and error. To improve the situation, a more detailed understanding of the solvent-substrate-enzyme interactions needs to be obtained. From this point of view, solid/gas technology presents very interesting features, as it permits total thermodynamic control of all the constituents present in the microenvironment of the biocatalyst. It therefore allows us to rigorously follow recommendations for quantifying substrate, water, and solvent availability in non-conventional media by using their thermodynamic activity, which has been established and largely admitted for many years [17].

Indeed, in solid/gas reactors, the enzyme is permeated by a carrier gas, which simultaneously carries gaseous substrates to the enzyme and removes gaseous products. The thermodynamic activities of all gaseous components present in the reaction medium, whether they are reactants or not, can be independently controlled and adjusted by varying the partial pressure of each compound in the carrier gas as explained in the introduction part. A complete description of solid/gas reactors and an explanation of their advantages as tools for studying the enzyme microenvironment was given by Lamare et al. in 2004 [8].

In the following example, solid/gas technology was used to study the influence of the addition of different non-reactant organic components on CALB enantioselectivity [24]. The resolution of pentan-2-ol catalyzed by CALB, was studied with the addition of five different organic compounds to the solid/gas bioreactor (2-methyl-2-butanol (2M2B), hexane, 1,4-dioxane, acetone and cyclopentane), each of them being successively added at a fixed thermodynamic activity, equal to 0.3. These components were chosen because they usually serve as solvents in enzymatic reactions performed in liquid organic media, and because they cover a large range of (a) hydrophobicity, as this parameter was shown to have an influence on enzyme enantioselectivity [28], and (b) size, measured as the van der Waals volume of the solvent molecules, since Ottosson et al. in 2002 obtained a correlation between this parameter and CALB enantioselectivity [30].

Results showed that the enantioselectivity of CALB towards 2-pentanol, measured as E values, did not change significantly when adding one of the above-mentioned organic components to the gaseous reaction medium. To study whether this result could be applied more generally to other non-conventional media, CALB enantioselectivity was also measured in organic liquid media in the presence of 2M2B, hexane, and 1,4-dioxane successively. The liquid reaction media were designed to give a thermodynamic activity of 0.3 (estimated by using the UNIFAC group contribution method), so that they corresponded to the same level of availability as that in previously mentioned experiments in the gas phase. Results obtained showed that the E values obtained did not change when using different organic solvents (Figure 11.4).

The influence of the level of availability of the different organic components was also studied in a solid/gas reactor by varying their thermodynamic activity, and the result was that E remained constant.

In this example, the solid/gas reactor did not give enhancement of enantioselectivity of CALB towards 2-pentanol through alcoholysis with methyl propanoate as acyl donor. Nevertheless, it was clearly shown that this selectivity did not change when adding one of the above mentioned organic component in the medium, either gaseous or liquid, all other reaction conditions being equal. This result cannot be generalized either to other secondary alcohols or to other acyl donors. Indeed, solvent molecules can be present in the active site and can modify bindings of bigger substrates and therefore enantioselectivity of enzyme [31].

In the next example, the effect of water availability (measured as water thermodynamic activity : a_W) on CALB enantioselectivity was studied in a solid/gas reactor, using the same reaction model as that mentioned above: esterification of 2-pentanol with methyl propanoate [32]. In this case, experimental data showed a pronounced effect of a_W on enantioselectivity: E value was multiplied by 3 for a_W

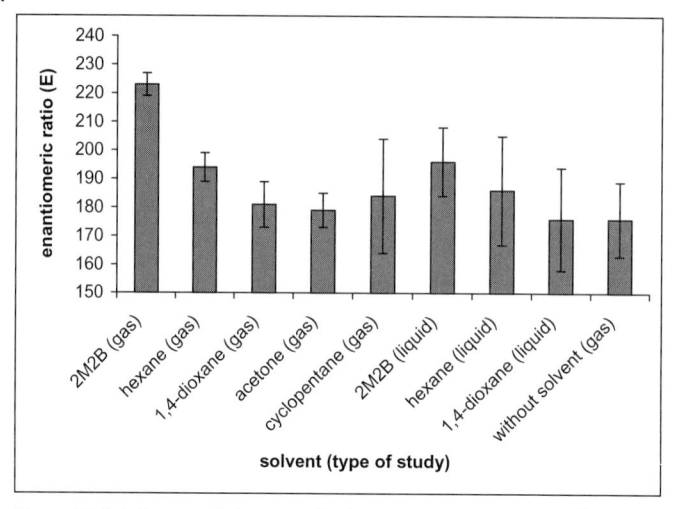

Figure 11.4 Influence of the type of solvent on the enantiomeric ratio for a thermodynamic activity of solvent equal to 0.3. For the reaction in the solid/gas bioreactor: $a_{pentan\ 2\ ol} = 0.05$ and $a_{methyl\ propionate} = 0.1$. For the reaction in liquid medium: $a_{pentan-2-ol} = 0.04$ (0.25 mol L^{-1}) and $a_{methyl\ propionate}$ = 0.8 (7.74 mol L^{-1}) for 2-methyl-2-butanol (2.06 mol L^{-1}); $a_{pentan-2-ol}$ = 0.04 (0.16 mol L^{-1}) and $a_{methyl\ propionate}$ = 0.8 (7.67 mol L^{-1}) for hexane (1.84 mol L^{-1}); $a_{pentan-2-ol}$ = 0.03 (0.17 mol L^{-1}) and $a_{methyl\ propionate}$ = 0.68 (7.29 mol L^{-1}) for 1,4-dioxane (3.26 mol L^{-1}).

increasing from 0 to 0.2, and E regularly decreased for higher a_W. By means of molecular modeling, we were able to show the possibility of a water molecule binding in the stereospecificity pocket of CALB. Being located in this pocket, the water affects the two substrate enantiomers differently, obstructing the binding of the slow reacting enantiomer (S)-2-pentanol.

This study permitted us not only to determine the optimal a_W value for enantioselectivity, but also allowed us to perform a detailed study on the interaction between water and the active site of CALB. Measurements of enantioselectivity at different a_W and temperature showed that the water molecule located in the selectivity pocket had a high affinity for this particular part of the active site, with a binding energy of 9 kJ mol^{-1}, and lost all its degrees of rotation, corresponding to an entropic energy of 37 J mol^{-1} K^{-1}.

The solid/gas reactor also allowed us to investigate the effect of a_W on CALB activity (Figure 11.5). At low a_W (up to 0.02), the alcoholytic activity increased because of water acting as a lubricant, facilitating the reaction. As the a_W increased, water started to act as a competitive nucleophile substrate, contributing to a hydrolytic activity. In parallel to water acting as a competitive substrate, it was also acting as a competitive inhibitor of methyl propanoate, causing a decrease in the total activity [33].

In these two examples it is shown how the solid/gas technology provides an accurate thermodynamic approach for studying the effect of the microenvironment on enzymatic activity and enantiospecificity.

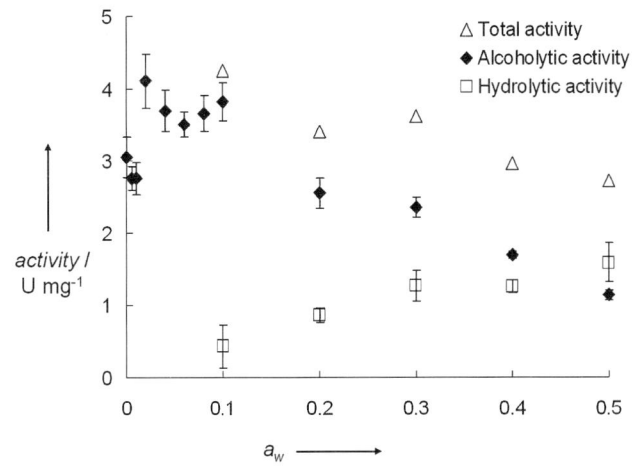

Figure 11.5 Influence of thermodynamic water activity, a_w, on the hydrolytic (□), alcoholytic (◆) and total (△) activity for the *Candida antarctica* lipase B-catalyzed acylation of 2-pentanol with methyl propanoate as acyl donor.

11.2.1.2 Example 2: Conversion of Gaseous Halogenated Compounds by Dehydrated Bacteria

Volatile Organic Compounds (VOCs) are widely produced by mobile sources, industrial, and even domestic activities. Transportation, manufacturing of petroleum products, and the handling of solvents represent a major part of these emissions [34, 35]. The induced pollution includes a very large range of chemicals with direct or indirect impact on health. Depending on the compound considered, VOCs can be irritants, especially for the eyes and the respiratory tracts, but some have been recognized as teratogenic, mutagenic, or carcinogenic. Moreover, their accumulation in the troposphere induces the increase of ozone formation and also indirectly contributes to the greenhouse effect [36].

Of the VOCs, the halogenated ones are among the most recalcitrant, and several of them can be harmful at low or very low concentrations. As a consequence, these have been designed as priority pollutants. Unfortunately, substitutes for halogenated compounds cannot always be found in industrial processes.

Several treatments have been developed for the removal of VOCs, and these can sometimes be applied to the halogenated compounds. Thermal oxidation is efficient but expensive and can therefore not be used for diluted effluents. Activated carbon can be applied to rather simple mixtures (preferentially no more than three pollutants for recycling). Biological treatments such as biofilters can be an interesting alternative for the treatment of dilute effluents or complex mixtures, but they are limited by the toxicity of the pollutant toward microorganisms. This is especially the case with halogenated compounds. Moreover, the efficiency of this kind of treatment would be affected by acidification of the biofilter during bioremediation, and dehalogenation induces the release of HCl or HBr. Nowadays there is also no example of a biofilter applied to the removal of halogenated compounds. However, solubility of halogenated VOCs is usually low and is a third important limiting factor for biological treatment of halogenated gaseous effluents.

Consequently, degradation of halogenated compounds in gaseous effluents, especially in diluted effluents, is often problematic, dehalogenation being the limiting step. In order to overcome this problem, a new process, based on solid/gas catalysis with whole dehydrated cells as the catalyst, has been proposed [14]. The aim is to convert halogenated compounds into alcohols.

First studies and applications of solid/gas catalysis have been conducted with isolated enzymes [3, 4]. Knowledge concerning the effect of different parameters on the activity and stability of enzymes has been gained from studies on lipases [24, 37–42], alcohol dehalogenases [43–46], and haloalkane dehalogenases [47, 48], and is currently under investigaton. It has been shown that whole dehydrated cells could perform the interconversion of alcohols and aldehydes at the solid/gas interface [11–13, 49]. The use of dehydrated cells was next extended to the conversion of halogenated compounds by dehydrated bacteria. *Rhodococcus erythropolis* NCIMB 13064 and *Xanthobacter autotrophicus* GJ10 were selected for this study since they produce monomeric haloalkane dehalogenases (respectively the DhaA and the DhlA) able to hydrolyze halogenated compounds without the necessity for any cofactor. The studies were mainly conduced with 1-chlorobutane as a model compound. We showed that whole dehydrated cells of *R. erythropolis* or *X. autotrophicus* were able to hydrolyze 1-chlorobutane at the solid/gas interface [14, 15]. We checked that recombinant strains of *Escherichia coli* expressing the DhaA and the DhlA were also able to perform this hydrolysis.

At 30 °C and fed at a total flow rate of $500\,\mu mol\,min^{-1}$ with 0.06 and 0.6 respective thermodynamic 1-chlorobutane and water activities, similar rates of hydrolysis of 1-chlorobutane were observed with strains producing either the DhaA (*R. erythropolis* and the corresponding *E. coli*) or those expressing the DhlA (*X. autotrophicus* and *E. coli* DhlA-producing). Strains producing the DhlA were less efficient for 1-chlorobutane degradation. Indeed their activity represented half that of the two strains producing the DhaA. This result could logically been expected, since the active site of the DhlA is optimized for the binding of short molecules like 1,2-dichloroethane [50].

These experiments also showed that whole dehydrated bacteria can be used as a catalyst for the conversion of halogenated compounds into less toxic and much more water-soluble alcohols. Interestingly, there is no need to isolate the enzyme, and the time required for production of the catalyst can be lowered using recombinant strains.

Since bacteria are able to retain their dehalogenase activity after dehydration, this new process could allow direct continuous treatment of gaseous effluents. The two main points are that there is no need to transfer the pollutant in an aqueous phase and there is also no longer limitation by solubility, and secondly that microorganisms are no longer growing. If we consider that transfers in the gas phase are much more efficient than those in the liquid phase, this also means that the rate of degradation should be far less limited by transfer and diffusion rate of the pollutant. Moreover, there is no need to provide nutrient elements for microorganisms, and no clogging of the biofilter will occur.

Solid/gas catalysis could also be a tool to overcome the limiting step of dehalogenation. Nevertheless, working with whole dehydrated cells as catalyst requires the control of several operational parameters.

11.3
Influence of Operational Parameters on the Dehalogenase Activity of Whole Dehydrated Cells

Both activity and stability of whole dehydrated bacteria are dependent on temperature, water activity, and the pH of the buffer used prior to dehydration.

11.3.1
Effect of Temperature

The effect of temperature has only been reported for R. erythropolis cells [Erable et al., 2004]. In this case, dehalogenase activity increases with temperature, in agreement with results already reported for isolated enzyme or whole dehydrated cells of Saccharomyces cerevisiae in a solid/gas reactor [12]. Over the range of temperatures tested, the highest activity was obtained at 60°C, whereas the optimal temperature for resting cells in the aqueous phase was 45°C [16]. For R. erythropolis cells, solid/gas catalysis also allows the use of higher temperatures than conventional medium, in agreement with what has already been shown with lipolytic enzyme [37, 39]. Nevertheless, the increase in activity was concomitant with the decrease in stability. Indeed at 40°C, the half-life of the catalyst was 80% of that measured at 30°C, and it dropped to less than 30% at 50°C. Solid/gas catalysis protects the catalyst from thermal denaturation to a certain extent, but this protection is insufficient to allow use for a long period of time at high temperatures, in contrast to what has been shown for isolated lipase exposed to a water activity of 0.1 [51]. This could be related to the combined action of temperature and the relatively high water activity (0.6 to 0.8) used with whole cells.

For practical application, a compromise between activity and stability has to be found. For air treatment, the use of ambient temperature should not be a limitation, since such processes have to be as low-energy as possible.

11.3.2
Effect of Water Activity

Increase in water activity has an opposite effects on activity and stability of whole cells. Both for R. erythropolis and X. autotrophicus an increase in dehalogenation was noticed on raising water activity [14, 15]. Interestingly, a minimum water activity of 0.4 was required in both cases to observe hydrolysis of 1-chlorobutane, while Dravis et al. [47] showed that the isolated haloalkane dehalogenase of R.

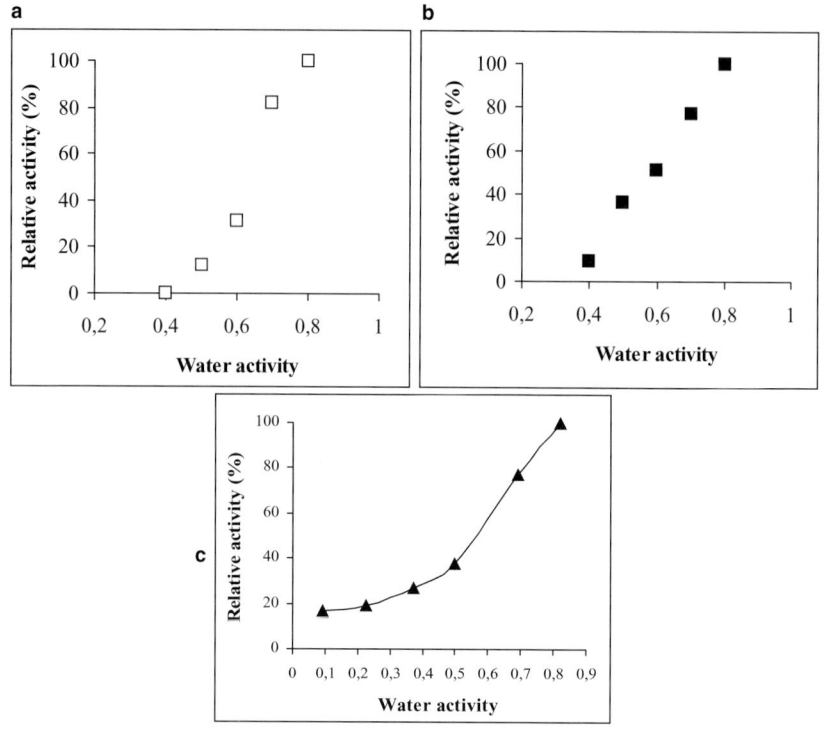

Figure 11.6 (a) Effect of water activity on the relative dehalogenase activity of dehydrated *R. erythropolis* cells maintained at 40 °C (□) or (b) *X. autotrophicus* cells maintained at 30 °C (■) in a solid/gas reactor. The maximal dehalogenase activity obtained with 0.8 water activity was taken as a reference in each case. The total flow through the reactor was 500 µmol min^{-1} and 1-chlorobutane was 0.06. (c) Profile of dependence of haloalkane dehalogenase activity (▲) of the purified enzyme (redrawn from Dravis et al., Ref. [47]).

erythropolis could perform the hydrolysis of the 1-chlorobutane at 0.1 water activity (Figure 11.6).

The same behavior has already been observed with whole dehydrated cells of *Saccharomyces cerevisiae* [11] or immobilized ADH in the gas phase [45]. It seems also that the cellular matrix increases the need for hydration to perform catalysis.

In our case, raising water activity logically increases the rate of hydrolysis. This can simply be explained by the fact that water is one of the substrates, but it could also be partly due to an increase of molecular mobility with the increase of hydration.

For both bacteria, the increase in activity was concomitant with the decrease in the catalyst's stability. At high temperatures, this phenomenon can be explained by thermal denaturation. At low temperatures (lower than 40 °C for *R. erythropolis* and 30 °C for *X. autotrophicus*) the drop in activity could be attributed to the accumulation of acid, more rapidly produced at high water activity.

11.3.3
Effect of Buffer pH Before Dehydration

To prepare the catalyst, cells are suspended in a buffer and then lyophilized. The pH has to be controlled to maintain activity after drying. For *R. erythropolis*, using Tris/HCl as buffer, an optimal pH of 9.0 was found, in agreement with data reported for the isolated enzyme in the aqueous [52] or gas phase [47]. For dehydrated *X. autotrophicus* cells the optimal pH for activity in the gas phase was 8.5 (Tris/HCl buffer), whereas an optimal pH of 8.2 has been determined for the isolated enzyme by Keuning et al. [53] as shown in Figure 11.7.

This behaviour is in agreement with data obtained by Grizon [49], who observed a dependence of the ADH activity of whole dehydrated cells of *Saccharomyces cerevisiae* upon pH of the buffer. For practical application it seems also important to suspend cells at the optimal pH for the isolated enzyme.

11.3.4
Range of Substrates

Since the longer the carbon chain length of the solute, the lower is the water solubility and the more limited are the options for biological treatment. The efficiency of dehydrated cells in hydrolyzing a range of halogenated compounds was also examined [14].

The dehalogenase activity of dehydrated cells of *R. erythropolis*, for both chlorinated and brominated compounds, increases with the carbon chain length of the compound (up to 1-chlorohexane and 1-bromohexane). The opposite has been reported for the aqueous phase for both the isolated enzyme [54] and resting cells [14]. Eliminating solubilization requirements also clearly improves rates of

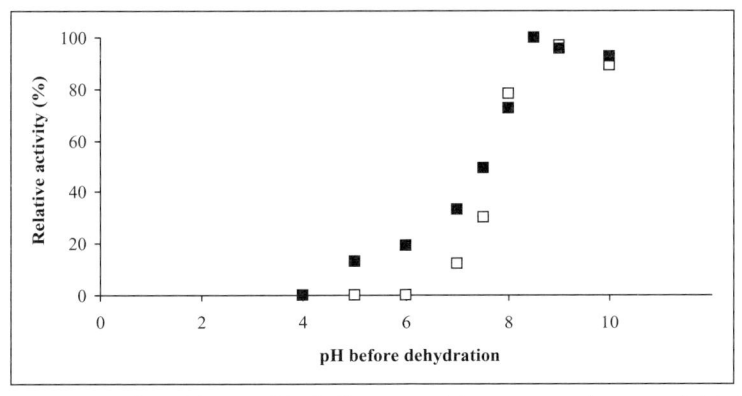

Figure 11.7 Effect of the pH of the buffer used to suspend cells before dehydration on the dehalogenase activity of (□) *R. erythropolis* cells maintained at 40 °C or (■) *X. autotrophicus* cells maintained at 30 °C in a solid/gas reactor. The maximal dehalogenase activity was taken as 100% reference in each case. The total flow through the reactor was 500 µmol min^{-1}, and 1-chlorobutane and water activity were respectively 0.06 and 0.7.

degradation of some poorly water-soluble compounds. It can also been expected that compounds with low water solublility could be much more easily degraded in the gas phase. Nevertheless, the compound has to be included in the selectivity range of the enzyme, since too large molecules such as 1-chlorodecane could not be hydrolyzed. This result was in agreement with the low or non-existent rate of degradation of this compound in the aqueous phase by the corresponding isolated enzyme.

Taken together, these results indicate that dehydrated cells could be an alternative catalyst for the conversion of poorly water-soluble compounds and that it might be possible to treat a mixture of pollutants by combining strains. Nevertheless, for future application the improvement of catalyst stability is still needed and is currently being studied.

11.3.4.1 Example 3: Scaling up the System; Application to Industrial Production

Natural esters are widely used by the aroma and fragrance industries because of their fruity or floral taste/odor. Many of them are alkyl esters of formic, acetic, propionic, and butyric acids. The development of an efficient biotechnological process, compatible with the "natural" label for the products but offering costs comparable to the costs of chemical processes, has been achieved, representing the first application of gas/solid technology on an industrial scale [51].

A process based on the use of an immobilized *C. antarctica* B lipase (Novozym 435) has been developed in order to perform direct esterification of natural alcohols with carboxylic acids in a solvent-free system. Based on the system represented in Figure 11.2, the process had to overcome the problem of recycling the nitrogen, once the reaction was over.

The process was then developed on a closed loop of nitrogen, circulating in three different zones, thus reducing the nitrogen consumption to zero during continuous operation.

As shown in Figure 11.8, a loop of nitrogen is realized between a low-pressure zone (flash evaporator EV01, heat exchanger EC01, and bioreactor R01), a medium-pressure zone (gas/liquid separator V04), and a high-pressure zone (final condenser V05).

Liquid substrates (acid and alcohol) and water are injected into the nitrogen loop in a flash evaporator, using a pressurized liquid loop with a mass controlled leak going to the flashing unit of each liquid substrate. Once the liquid/gas flash is realized in the flashing unit, the gas enters in a heat exchanger, and its outlet temperature is set to the working temperature of the bioreactor. Then, the thermodynamic activities in the gaseous stream are set at this stage by controlling each partial pressure as explained previously. A packed-bed type reactor follows the heat exchanger, and all this part of the system is maintained under regulated vacuum using the liquid ring vacuum pump (PV01) and the vacuum regulation valve. After the reaction step, the end of the process is solely dedicated to the removal of condensable molecules from the nitrogen fraction and the recovery of liquid products, and is performed by a two-stage condensation operation involving cooling of the gas stream coupled with an increase in the absolute pressure performed by the

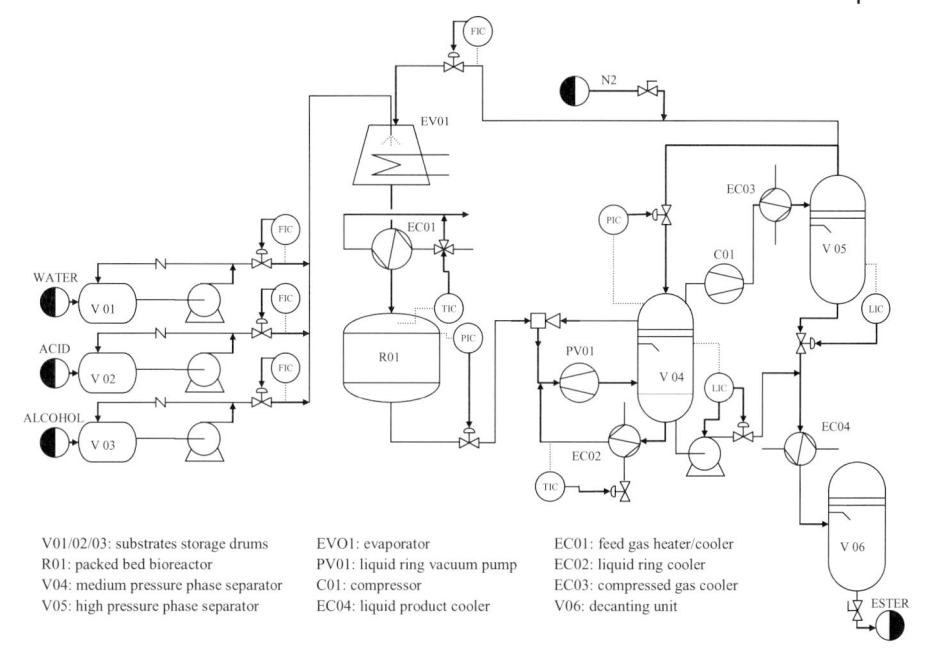

11.3 Influence of Operational Parameters on the Dehalogenase Activity of Whole Dehydrated Cells

Figure 11.8 Schematic diagram of an industrial continuous solid/gas bioreactor developed for the production of natural esters in a closed nitrogen loop.

V01/02/03: substrates storage drums
R01: packed bed bioreactor
V04: medium pressure phase separator
V05: high pressure phase separator
EV01: evaporator
PV01: liquid ring vacuum pump
C01: compressor
EC04: liquid product cooler
EC01: feed gas heater/cooler
EC02: liquid ring cooler
EC03: compressed gas cooler
V06: decanting unit

liquid ring vacuum pump itself and a compressor (C01) for finishing the condensation process. Then, clean nitrogen from the outlet of the pressurized heat exchanger can be depressurized and recycled to the flash evaporator.

Different syntheses coupling C1 to C4 carboxylic acids and C1 to C9 alcohols for the industrial production of natural aromas were optimized, and production tests were performed. Table 11.2 summarizes some examples of the production rates of the system that can be achieved with an acid conversion exceeding 94% and a maximal excess of alcohol of 2 (mole per mole) in the worst cases ($\Delta G° > -11\,kJ\,mol^{-1}$) for a residence time commonly observed as low as 0.5 to 0.15 s.

Results obtained showed that ester production rates of 3–5 kg h^{-1} were easily achievable, with on-line acid moiety conversion exceeding 94%.

Synthesis performed between 80 and 120 °C showed that good thermodynamic control of the system gave good stability of the catalyst, since the half-life of the reactor for the different syntheses ranged from 850 to 2000 h in continuous operation. More detailed analysis of the behavior of the catalytic bed showed that stability in this system is mainly governed by the acid thermodynamic activity. As a result, the packed-bed reactor is not homogeneous in terms of stability and the catalyst is more subject to denaturation in its first part (lower temperature, higher acid thermodynamic activity) than in the last part (higher temperature, lower acid thermodynamic activity).

Table 11.2 Examples of ester production rates and acid conversions achieved for different optimized syntheses of formic, acetic, propionic, butyric, and *iso*-butyric acid alkyl esters.

Acidic Moiety	Number of alcohols tested	Acid conversion [%]	Ester production rate [kg h^{-1}]	Catalyst charge [kg]
Formic	2	94.0–95.4	1.5–2.0	2.0–3.0
Acetic	5	95.0–97.4	3.8–4.0	2.0–3.0
Propionic	5	95.6–98.0	3.0–4.5	2.0–3.0
Butyric	5	97.5–98.1	2.6–3.2	2.0
iso-Butyric	5	94.0–97.8	2.2–3.5	2.0

Operation of this pilot plant demonstrated that solid/gas biocatalysis was able to compete with classical esterification bioprocesses (systems working with hexane as solvent), reducing any potential risk to a minimum by virtue of the absence of any solvent and greatly simplifying the downstream process by reducing the volumes that have to be treated by distillation to obtain a pure product.

11.4
Conclusion

The production of natural esters is the first example of a large-scale application of solid/gas biocatalysis, and many other systems are being studied today with a view to short or medium term development.

Lipases still offer the potential for an important range of applications since they are able to carry out the reactions of esterification, transesterification (acidolysis or alcoholysis), inter-esterification, or hydrolysis, often with high specificity or selectivity, suitable for the production of high-added-value molecules as shown in Example 1 above (stereospecific alkylation, acylation, or hydrolysis for the resolution of racemic mixtures of acids, alcohols or esters).

Dehydrogenases also represent a class of interesting enzymes since enantioselective reduction of ketones can lead to the production of enantiomerically pure secondary alcohols for the fine chemicals industry. Compared to liquid systems, in which the cofactor is often eliminated by the circulating phase in continuous systems, solid/gas catalysis can be highly suitable since it has been demonstrated that the cofactor is stable and its regeneration effective by addition of a second substrate. Also, stereoselective oxidation of secondary alcohols by these systems can help in the resolution of racemic mixtures.

More recently, enzymatic carboligation in a solid/gas bioreactor was demonstrated to be possible [55] in a model system based on the condensation of two propanal molecules to produce of propioin using thiamine diphosphate-dependent enzymes (benzaldehyde lyase from *Pseudomonas fluoreszens* and benzoylformate decarboxylase from *Pseudomonas putida*).

Solid/gas catalysis can also be of great help in obtaining interesting synthons, while complying with green chemistry requirements by avoiding the use of solvents, enabling high production rates to be achieved, and minimizing energy and processing costs by simplifying the downstream process.

Environmental application for bioremediation purposes must also get the benefit of the technology. Since haloalkane dehalogenase activity was tested successfully, as shown in Example 2 above for the dehalogenation of toxic VOCs, many others enzymatic activities should be of great interest, such as oxygenases, nitrilases, and organophosphorus hydrolases, for example.

The development of bioactive filters for air treatment is an attractive potential application of the technology.

Finally, as previously emphasized, solid/gas biocatalysis, because of its peculiarities, leads to a more accurate approach to the study of the effect of the microenvironment on enzymatic activity and stability. This allows access to intrinsic parameters of enzymes, thus providing a better molecular understanding of enzyme catalysis in general.

Solid/gas catalysis appears today as probably the most appropriate experimental tool for validating molecular modeling experiments, and this should help to answer the following questions:

- What are the energetics in solid/gas biocatalysis compared to liquid systems?
- How do the mechanisms of enzymes working on gaseous substrates differ from those in solution? How do they differ concerning the binding step and the catalytic process?
- What is the exact influence of water and organic molecules on the enzyme structure? Could its effects on properties such as selectivity, affinity, binding constants, and catalytic constants be predictable by controlling the hydration/solvation state?
- What is the partition behavior of water and organics inside more complicated systems such as whole cells, and how do the physico-chemical properties of the macro and micro environments affect this distribution?

The next challenge will probably be to unlock the bottlenecks still existing in the use of whole cells as micro reactors for effective multi-step transformations and to develop technological tricks to allow the number of compounds usable by the technology to increase.

As well as these major application fields, biosensors and analytical techniques should also benefit from the technology. Some examples have already been described. The detection of formaldehyde by a formaldehyde dehydrogenase coated onto a piezoelectric crystal has been performed at the ppm level. Detection of pesticides and organophosphorus compounds at the ppb level has been rendered possible by the same technique, involving either acetylcholinesterase or butyrylcholinesterase. And lastly, very specific biosensors have been developed along similar lines using antibodies directed against parathion or against benzoyl ecgonine for the specific detection of cocaine.

Abbreviations

R	Ideal gases constant (8.314 J/mol^{-1} K^{-1})
T	Temperature (K)
Pa	Absolute pressure in the system (atm)
$\dot{Q}v_{N_2}^n$ normalized	Normalized volumetric flow rate of carrier gas in line n (L h^{-1} at 273 K and 1 atm)
$\dot{Q}_{N_2}^n$	Molar flow rate of carrier gas in line n (mol h^{-1})
$\dot{Q}_X^n$	Molar flow rate of substrate X in line n (mol h^{-1})
$Ppsat_X^n$	Saturation pressure of compound X in line n (atm)
Pp_X^n	Partial pressure of compound X in the gas entering the bioreactor (atm)
a_x	Thermodynamic activity of compound X in the bioreactor (dimensionless)
$\dot{Q}v_{total}$	Total volumetric flow rate in the bioreactor (L h^{-1})
$\dot{Q}_{total}$	Total molar flow rate in the bioreactor (mol h^{-1})

References

1 T. Yagi, M. Tsuda, Y. Mori, H. Inokuchi, J. Am. Chem. Soc. 1969, **91**, 2801.
2 A. Kimura, H. Suzuki, T. Yagi, Biochim. Biophys. Acta 1979, **567**, 96–105.
3 S. Pulvin, M.-D. Legoy, R. Lortie, M. Pensa, D. Thomas, Biotechnol. Lett., 1986, **8**, 783–784.
4 S. Pulvin, F. Pavaresh, D. Thomas, M.-D. Legoy, Ann. NY Acad Sci. 1988, **545**, 434–439.
5 E. Barzana, A. Klibanov, M. Karel, Appl. Biochem. Biotechnol. 1987, **15**, 25–34.
6 E. Barzana, A. Klibanov, M. Karel, Biotechnol. Bioeng. 1989, **34**, 1178–1185.
7 F. Yang, A. J. Russell, Biotechnol. Bioeng. 1996a, **49**, 700–708.
8 S. Lamare, M.-D. Legoy, M. Graber, Green Chem. 2004, **6**, 445–458.
9 C. Hou, Appl. Microbiol. Biotechnol. 1984, **19**, 1–4.
10 M. Zilli, A. Converti, A. Lodi, M. Del Borghi, G. Ferraiolo, Biotechnol. Bioeng. 1992, **41**, 693–699.
11 T. Maugard, S. Lamare, M.-D. Legoy, Biotechnol. Bioeng. 2001, **73**, 164–168.
12 I. Goubet, T. Maugard, S. Lamare, M.-D. Legoy, Enzyme Microb. Technol. 2002, **31**, 425–430.
13 V. Grizon, M.-D. Legoy, S. Lamare, Biocat. Biotrans. 2004a, **22**, 177–182.
14 B. Erable, I. Goubet, S. Lamare, M.-D. Legoy, T. Maugard, Biotechnol. Bioeng. 2004, **86**, 47–54.
15 B. Erable, I. Goubet, S. Lamare, A. Seltana, M.-D. Legoy, T. Maugard, Biotechnol. Bioeng. 2005, **91**, 304–313.
16 B. Erable, I. Goubet, S. Lamare, M.-D. Legoy, T. Maugard, Chemosphere 2006, **65**, 1146–1152.
17 R. P. Bell, J. E. Crichtlow, M. I. Page, Chem. Soc. Perkin Trans. II, 1974, 66–70.

18 J. B. A. Van Tol, J. B. Odenthal, J. A. Jongejan, J. A. Duine, in: J. Tramper, M. H. Vermuë, H. H. Beeftink, and U. Von Stockar (eds.) *Biocatalysis in non-conventional media*, Elsevier, Amsterdam, 1992.

19 H. K. Hansen, P. Rasmussen, A. Fredenslund, M. Schiller, J. Gmehling, *Ind. Eng. Chem. Res.* 1991, **30**, 2355–2358.

20 A. E. M. Janssen, A. M. Vaidya, P. J. Halling, *Enzyme Microb. Technol.* 1996, **18**, 340–346.

21 A. E. M. Janssen, B. J. Sjursnes, A. V. Vakurov, P. J. Halling, *Enzyme Microb. Technol.* 1999, **24**, 463–470.

22 C. R. Johnson, G. W. Wells, *Curr. Opin. Chem. Biol.* 1998, **2**, 70–76.

23 J. Uppenberg, M. T. Hansen, S. Patkar, T. A. Jones, *Structure* 1994, **2**, 293–308.

24 V. Léonard, S. Lamare, M.-D. Legoy, M. Graber, *J. Mol. Catal. B: Enzym.* 2004, **32**, 53–59.

25 R. N. Patel, *Enzyme Microb. Technol.* 2002, **31**, 804–826.

26 J. L. L. Rakels, A. J. J. Straathof, J. J. Heijnen, *Enzyme Microb. Technol.* 1993, **15**, 1051–1056.

27 R. S. Phillips, *Enzyme Microb. Technol.* 1992, **14**, 417–419.

28 C. Orrenius, T. Norin, K. Hult, G. Carrea, *Tetrahedron-Asymmetry* 1995, **6**, 3023–3030.

29 S. Ueji, R. Fujino, N. Okubo, T. Miyazawa, S. Kurita, M. Kitadani, A. Muromatsu, *Biotechnol. Lett.* 1992, **14**, 163–168.

30 J. Ottosson, L. Fransson, J. W. King, K. Hult, *Biochim. Biophys. Acta – Protein Structure and Molecular Enzymology*, 2002, **1594**, 325–334.

31 J. Ottosson, K. Hult, *J. Mol. Catal. B: Enzyme*, 2001, **11**, 1025–1028.

32 V. Léonard, L. Fransson, S. Lamare, K. Hult, M. Graber, *ChemBiochem* 2007, accepted.

33 M.-P. Bousquet-Dubouch, M. Graber, N. Sousa, S. Lamare, M.-D. Legoy, *Biochim. Biophys. Acta* 2001, **1550**, 90–99.

34 R. Bouscaren in: P. Le Cloirec (ed.) Les composés organiques dans l'environnement, Lavoisier Publisher, France, 1998.

35 CITEPA (2006) Emissions dans l'air en France: substances impliquées dans les phénomènes d'acidification, d'eutrophization et de pollution photochimique. http://www.citepa.org/emissions/index.htm

36 P. Le Cloirec, *Les composés organiques dans l'environnement*, Le Cloirec P. Ed, Lavoisier Publisher, France, 1998.

37 H. Robert, S. Lamare, F. Pavaresh, M.-D. Legoy, *Prog. Biotechnol.* 1992, **8**, 23–29.

38 S. Lamare, M.-D. Legoy, *Trends Biotechnol.* 1993, **11**, 413–418.

39 S. Lamare, M.-D. Legoy, *Biotechnol. Bioeng.* 1995, **45**, 387–397.

40 M. Graber, M.-P. Bousquet-Dubouch, N. Sousa, S. Lamare, M.-D. Legoy, *Biochim. Biophys. Acta* 2003a, **1645**, 56–62.

41 M. Graber, M.-P. Bousquet-Dubouch, S. Lamare, M.-D. Legoy, *Biochim. Biophys. Acta* 2003b, **1648**, 24–32.

42 F. Létisse, S. Lamare, M.-D. Legoy, M. Graber, *Biochim. Biophys. Acta* 2003, **1652**, 27–34.

43 F. Yang, A. J. Russell, *Biotechnol. Bioeng.*, 1996b, **49**, 709–716.

44 C. Ferloni, M. Heinemann, W. Hummel, T. Daussmann, J. Büchs, *Biotechnol. Prog.* 2004, **20**, 975–978.

45 A. H. Trivedi, A. C Spiess, T. Daussmann, J. Büchs, *Appl. Microbiol. Biotechnol.* 2006a, **71**, 407–414.

46 A. H. Trivedi, A. C. Spiess, T. Daussmann, J. Büchs, *Biotechnol. Prog.* 2006b, **22**, 454–458.

47 B. C. Dravis, K. E. Lejeune, A. D. Hetro, A. J. Russel, *Biotechnol. Bioeng.* 2000, **69**, 235–241.

48 B. C. Dravis, P. E. Swanson, A. J. Russel, *Biotechnol. Bioeng.* 2001, **75**, 416–423.

49 V. Grizon, Etude de la réaction d'oxydoréduction catalysée par l'alcool déshydrogénase de la levure Saccharomyces cerevisiae en réacteur solide/gaz, Ph.D. Thesis. La Rochelle University, La Rochelle, France, 2004b.

50 D. B. Janssen, J. R. D. van der Ploeg, F. Pries, *Environ. Health Perspect.* 1995, **103**, 29–32.

51 S. Lamare, B. Caillaud, K. Roule, I. Goubet, M.-D. Legoy, *Biocat. Biotrans.* 2001, **19**, 361–377.

52 T. M. Stafford, *The microbiol degradation of chloroalkanes*, Ph.D. Thesis. The Queens University of Belfast, Belfast, United Kingdom, 1993.

53 S. Keuning, D. B. Janssen, B. Witholt, *J. Bacteriol.* 1985, **163**, 635–9.

54 J. F. Schindler, P. A. Naranjo, D. A. Honaberger, C. H. Chang, J. R. Brainard, L. A. Vanderberg, C. J. Unkefer, *Biochemistry* 1999, **38**, 5772–5778.

55 R. Mikolajek, A. C. Spies, M. Pohl, S. Lamare, *J. Büchs*, 2007, submitted.

56 S. O. Hwang, D. J. Trantolo, D. L. Wise, *Biotechnol. Bioeng.* 1993, **42**, 667–673.

57 P. A. Cameron, B. H. Davison, P. D. Frymier, J. W. Barton, *Biotechnol. Bioeng.* 2002, **78**, 251–256.

58 T. Debeche, C. Marmet, L. Kiwi-Minsker, A. Renken, M. A. Juillerat, *Enzyme Microb. Technol.* 2005, **36**, 911–916.

59 S. Lamare, M.-D. Legoy, International Patent WO 99/04894, 1999.

60 C. Kim, S.-K. Rhee, *Biotechnol. Lett.* 1992, **14**, 1059–1064.

61 F. Parvaresh, D. Thomas, M.-D. Legoy, *Biotechnol. Bioeng.* 1992, **39**, 467–473.

62 S. Lamare, M.-D. Legoy, *Biotechnol. Bioeng.* 1997, **57**, 1–8.

63 H. Uchiyama, Oguri K. Yagi O., E. Kokufata, *Biotechnol. Lett.* 1992, **14**, 619–622.

12
Biocatalysis with Undissolved Solid Substrates and Products

Alessandra Basso, Sara Cantone, Cynthia Ebert, Peter J. Halling and Lucia Gardossi

12.1
Introduction

Catalysis in reaction systems with undissolved substrates and products is not restricted to biocatalysis. High yields in solid-state synthesis, solid-to-solid reactions, and solvent-free systems have also been reported for aldol condensation, Baeyer-Villiger oxidation, oxidative coupling of naphthols, and condensation of amines and aldehydes [1, 2].

The study of enzymatic reactions in systems with substrates present at very high mass/volume ratios, often leading to suspensions rather than solutions, has also been pursued. The majority of the published work on this type of reaction was related to the synthesis of protected peptides [3–15], but the synthesis of beta-lactam antibiotics [16–19], glycosides [20, 21], glycamides [22], and esters [23–27] starting from suspended substrates has also been reported.

Enzymatic synthesis in reaction mixtures with mainly undissolved substrates and/or products is a synthetic strategy in which the compounds are present mostly as pure solids [28, 29]. It retains the main advantages of conventional enzymatic synthesis such as high regio- and stereoselectivity, absence of racemization, and reduced side-chain protection. When product precipitates, the reaction yields are improved, so that the necessity to use organic solvents to shift the thermodynamic equilibrium toward synthesis is reduced and synthesis is made favorable even in water.

The thermodynamics of these reaction systems have been investigated, resulting in methods to predict the direction of a typical reaction *a priori*. Furthermore, studies on kinetics, enzyme concentration, pH/temperature effects, mixing, and solvent selection have opened up new perspectives for the understanding, modeling, optimization, and possible large-scale application of such a strategy.

12.2
The Reaction System: Classification

Several different approaches, which involve suspended substrates or products with aqueous or organic solvent added (or with no separate solvent added), are reported in the literature of the last decade.

Although these reaction mixtures usually consist largely of solids, it has been recognized that a liquid phase is essential for enzymatic activity [6, 30]. In a reaction with two solid substrates, this usually means the addition of a solvent (sometimes referred to as "adjuvant") to the mixture [6, 31, 32].

One of the two substrates can be a liquid at the reaction temperature, so that it can then be used as the "solvent" to partially dissolve the other substrate [33–36].

In some cases a liquid phase can be formed from two solid substrates by eutectic melting, when the reaction temperature lies below the melting points of the pure substrates, but above their eutectic temperature [6, 7, 37]. A small liquid phase may also be formed from two solid substrates because of traces of associated water [10]. Table 12.1 summarizes the main features of different reaction systems with undissolved substrates and/or products described in the literature.

The physical appearance of such reaction mixtures can vary widely depending on the ratio of the different components and on the nature of the liquid phase used. Thus, there can be systems that are mainly solid, or dilute suspensions in a large liquid phase, e.g., in which a product can precipitate because its solubility in the solvent used is extremely low.

Although substrates are usually largely undissolved in such systems, very high conversion yields were observed in many of the reactions studied in the literature. In all these cases, the driving force for the very good yields obtained was the precipitation of the reaction products [30, 38]. Hence, the name *"precipitation-driven biocatalysis"* has been suggested to cover all those reaction systems (Figure 12.1) [39] that have been named as *"solid-to-solid"* [3], *"suspension-to-suspension synthesis"* [40], *"aqueous two-phase systems with precipitated substrates and/or products"* [41], or *"heterogeneous eutectic melting"* [37].

It is clear that several aspects of reaction systems with suspended substrates are significantly different from those in solution. The presence of solid substrates has important consequences for the reaction kinetics and thermodynamics, and it requires different strategies for reaction engineering.

12.3
Theory

The thermodynamic feasibility of a reaction with precipitating products can be assessed by comparing the mass action ratio to the equilibrium constant of a reaction. This requires estimation of solubilities, using melting points of reactants in combination with the reaction equilibrium constant [39].

Table 12.1 Classification of reaction systems used for biocatalysis with undissolved substrates and/or products. Y: Yes; N: No; L: Low; H: High; A: Aqueous media; O: Organic solvent.

	Presence of:			Characteristic of the system
Solid substrate?	Solid product?	Liquid fraction?	Organic solvent or aqueous media?	
Suspended substrates				One or both substrates are present mainly as solid suspended in the solvent. Products do not precipitate and yields tend to be low unless the equilibrium lies in the synthetic direction (e.g., in the presence of organic solvents).
Y	N	L[a]	A	Mainly solid substrate with a little liquid between particles.
Y	N	L[a]	O	Mainly solid substrate with a little liquid between particles.
Y	N	H	A	Excess solid substrate suspended in an aqueous phase.
Y	N	H	O	Excess solid substrate suspended in an organic phase.
Substrate suspended in a liquid substrate				One solid substrate is suspended in a substrate which is liquid at the reaction temperature and which acts also as liquid phase. Yields can vary greatly depending on several parameters.
Y	Y	H	O	One solid substrate is suspended in a liquid substrate that acts as solvent.
Y	N	H	O	One solid substrate is suspended in a liquid substrate that acts as solvent.
Suspended substrates and precipitating products				Both substrates and products are present mainly as solids suspended in a liquid phase. Very good yields are commonly observed.
Y	Y	L[a]	A/O	Mainly solid substrates giving precipitating products.
Y	Y	H	A/O	The so called "suspension-to-suspension", with excess solids.
Product precipitated from a reaction in solution				A homogeneous solution from which the product precipitates. These systems have been used for peptide synthesis since the 1930s.
N	Y	H[b]	O	Precipitation-driven synthesis of peptides in organic solvent.
N	Y	H[b]	A	Precipitation-driven synthesis of peptides in aqueous phase.
Eutectic melting				Two solid substrates form a liquid phase by eutectic melting.
N	N	H	N/A[c]	The two substrates melt and no additional solvent needs to be added.
N	Y	H	N/A[c]	The two substrates melt and no additional solvent needs to be added. When product precipitates the reaction mixture solidifies.
Y	Y	H	N/A[c]	Also possible, can have one solid substrate and eutectic melt of mixture.

a Working with suspended substrates, another component, normally termed an "adjuvant", has been added in a small amount (usually 5–30%), to cause or enhance formation of a liquid phase.
b Might be L at end of reaction.
c No solvent.

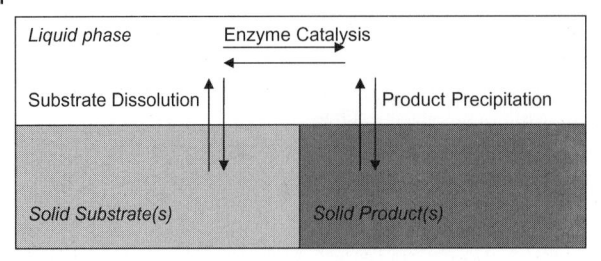

Figure 12.1 Precipitation-driven biocatalysis, where the dissolution of the substrates and the enzyme-catalyzed reaction happen with consequent precipitation of the products.

The yields of solid-to-solid reactions are governed by thermodynamics [3]. If in the case of product precipitation the reaction yields are nearly quantitative [8–10, 39, 42–46], much lower yields have been observed in biocatalysis with suspended substrates where the product does not precipitate [16–19].

For an acylation reaction with suspended substrates and precipitating product, the overall reaction can be described as follows (Equation 1):

$$AOH_{(s)} + BH_{(s)} \rightarrow AB_{(s)} + H_2O \tag{1}$$

where AOH is an acyl donor, BH is a nucleophile, and AB is the reaction product. It is desirable if the equilibrium lies to the side of solid product. The favored direction depends on the crystal energy difference between the product and substrates and the free energy change of the chemical reaction. It is useful to consider two thermodynamic constants. The thermodynamic equilibrium constant (K_{th}) of the chemical reaction is given by Equation 2:

$$K_{th} = \frac{a_{AB}^{eq} a_{w}^{eq}}{a_{AOH}^{eq} a_{BH}^{eq}} \tag{2}$$

where, at equilibrium, a indicates the thermodynamic activity of the reactants and a_w is the water activity. The thermodynamic mass action ratio (Z_{th}: the ratio of thermodynamic activities of the pure solid product over the solid substrates) is defined by Equation 3:

$$Z_{th} = \frac{a_{AB}^{s} a_w}{a_{AOH}^{s} a_{BH}^{s}} \tag{3}$$

Provided that the same reference standard state (for example the pure solid) is used in K_{th} and Z_{th} for each reactant, we can combine these to give the free energy of the solid-to-solid conversion according to Equation 4:

$$\Delta G_{solid-solid} = -2.3RT \log\left(\frac{K_{th}}{Z_{th}}\right) \tag{4}$$

The usual standard state chosen will be pure water, and the equation will be approximately correct whenever a dilute aqueous phase is present. If $\Delta G_{solid\text{-}solid}$ is negative, product precipitation can be expected, while, if it is positive, no product precipitation will be observed.*

A recent paper gives a method for the calculation of the Gibbs free energy changes and heats of reaction for the formation of amide bonds in the solid-to-solid approach [47].

If now we look at the Equation 4, but this time in terms of directly measurable parameters, provided that the reaction system is sufficiently dilute (ideal behavior), we can write Equation 5 and Equation 6 instead of Equation 2 and Equation 3.

$$K_{eq} = \frac{[AB]}{[AOH][BH]} \quad (M^{-1}) \tag{5}$$

$$Z_{sat} = \frac{S_{AB}}{S_{AOH}S_{BH}} \quad (M^{-1}) \tag{6}$$

where K_{eq} is the concentration based equilibrium constant and Z_{sat} is the ratio of product over substrate solubilities.

Only when $Z_{sat} < K_{eq}$ does the precipitation-driven reaction proceed toward near-quantitative yields. The equilibrium composition of the solid phase of the reaction mixture will then consist either of only solid substrates or of only solid products, never both. This explains the fact that very high yields are commonly observed.

Methods have been developed for the prediction of aqueous solubilities of organic compounds and therefore for the prediction of the mass action ratio Z_{sat} [48–50].

When the thermodynamics are unfavorable, one possible approach is to perform enzymatic reactions in a kinetically controlled strategy, and the reaction needs to be stopped at the maximum conversion (Figure 12.2) [16, 40].

In the kinetically controlled synthesis, an activated acyl donor (ester, amide, or anhydride) is used to form an acyl-enzyme intermediate.

Hydrolases that form an acyl-enzyme intermediate, such as some proteases and amidases, can be effectively used in this approach. On the other hand, this method is not applicable to metallo- and carboxyproteases.

Compared to thermodynamically controlled synthesis, the initial rates of kinetically controlled synthesis are usually faster and the conversion can be complete.

---

* When the three solid phases are present, as in Equation 1, their activities are fixed as those of the crystalline solids. However, water activity is also fixed, at least to an upper limit – that of an aqueous solution saturated with respect to all three reacting compounds. So if there is no other component present, the system is completely invariant thermodynamically, and reaction must proceed, to either synthesis or hydrolysis, until at least one solid phase disappears completely. The only exception can be if the solid phases are not pure crystalline (e.g. if solid solutions form). If another component is added, but the three solid phases remain the same, then water activity can only be reduced further, to the extent that the new component enters the solution phase.

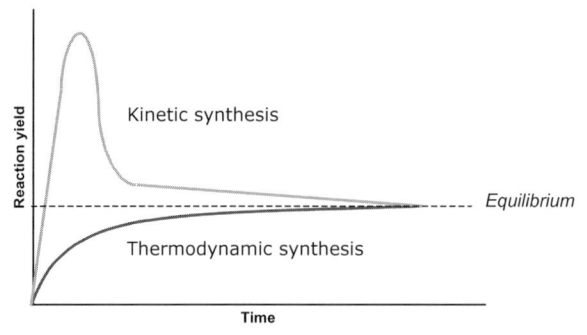

Figure 12.2 Time courses of kinetically controlled and thermodynamically controlled synthesis. The dotted line represents the position of the thermodynamic equilibrium.

However, once the maximum yield is reached, the enzyme begins to hydrolyze the product. Thus the yield will decrease to reach the equilibrium position.

12.4
Factors Affecting the Reactions in the Presence of Undissolved Substrates/Products

In reactions with undissolved substrates or products, the pH of the reaction mixture exerts an important influence on the rate of enzymatic reactions. In these systems, the concentration of the species is maximal and reaches the solubility value. Therefore, in the presence of ionized suspended substrates or products, the pH can be derived from a charge balance of the species present in the reaction mixture. The presence of any additional aqueous buffer will then not have a significant effect [9].

There was found to be a correlation between changes in the initial substrate ratio and the pH of the liquid phase [10], and it was demonstrated that even a slightly unequal substrate ratio can have a large effect on rate. For example, a two-fold higher initial rate could be achieved in a reaction mixture with 60% L-Leu-NH$_2$ (pK$_a$ around 7.8) and 40% Z-L-Gln (pK$_a$ around 3.6) as compared to a reaction with equimolar substrates [10]. However, an unequal substrates ratio means that at the equilibrium an excess of one unreacted substrate will be present, so other methods are generally preferred.

Following these considerations, the pH of these systems was actively controlled by the addition of inorganic salts (acids or bases) [51]. As a matter of fact, a 20-fold increase in the initial rate could be observed upon addition of KHCO$_3$ or K$_2$CO$_3$ in the synthesis of Z-L-Asp-L-Phe-OMe, but 1.5 equivalents of salt were optimal to obtain yields of more than 90%. The profile of initial rate as a function of K$_2$CO$_3$ concentration displayed a bell-shaped figure typically seen for pH effects on enzyme activity (Figure 12.3).

Similar effects could be observed during the thermolysin-catalyzed synthesis of Z-L-Gln-L-Leu-NH$_2$ when NaHCO$_3$ was used instead of KHCO$_3$. On the basis

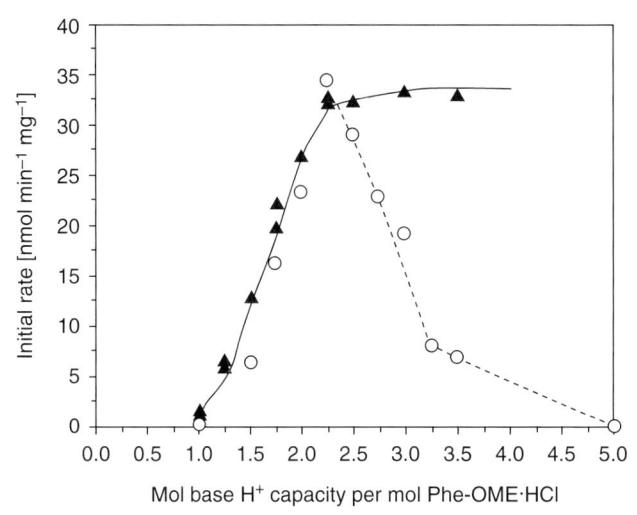

Figure 12.3 Effect of basic inorganic salt on initial rate of protease-catalyzed synthesis of Z-aspartame (0.1 mmol Z-Asp, 0.1 mmol Phe-OMe·HCl) with varied amounts of KHCO$_3$ (triangles) and K$_2$CO$_3$ (circles). The x-axis unit is equivalent to 1 mol of KHCO$_3$ or 0.5 mol K$_2$CO$_3$ per mole Phe-OMe·HCl. Reprinted with permission from Erbeldinger, M.; Ni, X.; Halling, P. J. Biotechnol. Bioeng., **2001**, 72, 69. Copyright (2001) American Chemical Society. [51].

of these results it was possible to create a model for the prediction of the initial pH and the change of pH during product formation of Z-L-Asp-L-Phe-OMe. This helped to explain both final yields and the enzyme kinetics in this system [51].

In such systems, salts can be used also when thermodynamics are unfavorable and precipitation is not achieved. The addition of specific counter-ions that form poorly soluble salts with an ionic form of the product of a reaction (and not with the substrates) lead to product salts precipitation from the reaction mixture.

There are a number of reported "solid-to-solid" reactions where products precipitated as salts rather than as neutral compounds. The thermolysin-catalyzed production of the potassium salt of Z-aspartame [51, 52], the commercialized process of aspartame synthesis, where a salt of cationic D-Phe-OMe and anionic Z-aspartame precipitates [53], and the enzymatic conversion of solid Ca-maleate to solid Ca-D-malate [44] are examples of such behavior.

In the thermolysin-catalyzed synthesis of the potassium salt of Z-aspartame (Scheme 12.1), the reaction rate was found to be strongly dependent on the amount of basic salt (KHCO$_3$, Figure 12.3) added to the system, as mentioned above [51].

For the conversion from solid Ca-maleate to solid Ca-D-malate (Scheme 12.2), kinetic models were developed to predict kinetics of dissolution of the substrates, the enzyme-catalyzed reaction, and the precipitation of the reaction product, all of which occur at the same time [44].

It has also been tested whether this approach could work for the synthesis of the beta-lactam antibiotic amoxicillin [46]. First, a number of organic and inorganic

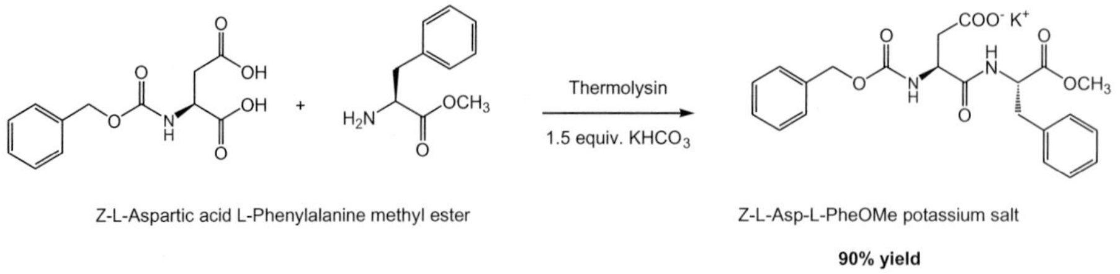

Scheme 12.1 Enzymatic solid-to-solid synthesis of Z-L-Asp-L-PheOMe potassium salt (Z-aspartame potassium salt) using inorganic salts [51].

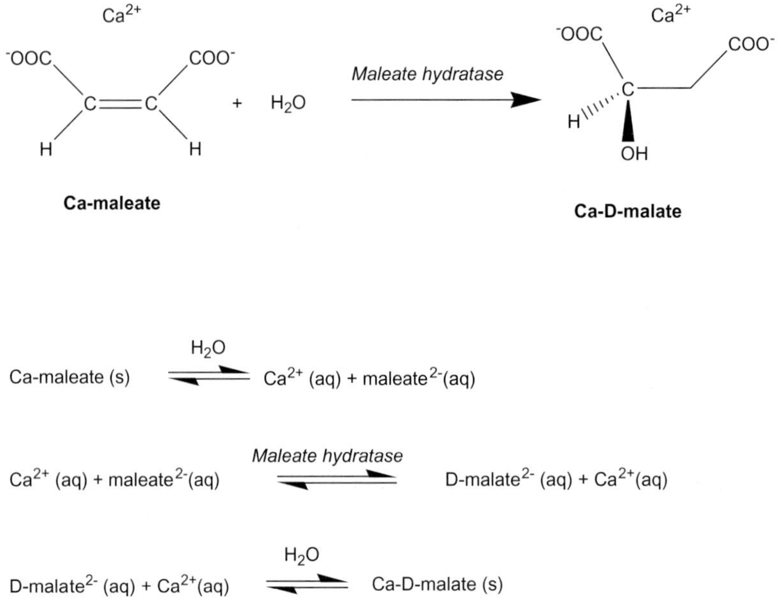

Scheme 12.2 Reaction of conversion of solid Ca-maleate to form solid Ca-D-malate by maleate hydratase and reaction steps [44].

counter-ions were tested to see whether insoluble salts were formed. From this screening, it was observed that zinc significantly decreased the concentration of amoxicillin in solution. Then, a 30-fold increase in the yield of amoxicillin was observed in an enzymatic synthesis reaction in the presence of 0.1 M ZnSO$_4$ [46]. Unfortunately, Zn^{2+} dramatically affected the stability of the beta-lactam nucleus to the extent that it could not be confirmed whether product precipitation eventually occurred during the reaction or not.

Substrate dissolution rates may be important in the overall kinetics of these reactions. Dissolution rates may be affected by interfacial limitation, but mass transfer in the boundary layer around the particles always plays a key role. Mass transfer rates are determined by equilibrium solubility, diffusion coefficient, and boundary layer thickness (which is agitation dependent). In many of these systems the concentration of substrates in solution is very high, so that the reaction mixture is often highly viscous [30]. In such viscous systems, the diffusion coefficient can be reduced and the boundary layer may be thicker, so that mass transfer limitation becomes more likely. However, both in the work of Wolff et al. [54] and Michielsen et al. [44], the reaction mixtures were relatively low-viscosity aqueous suspensions.

Kinetic descriptions of this type of biocatalyzed system in the presence of different amounts of water [9], substrates [10], or inorganic salts [51] have been reported. In some cases, solvents ("adjuvants") have been used to increase mass transfer [6, 7]. However, it was found that upon addition of organic co-solvents longer process times were actually required, even though the substrate solubility increased several times [54].

Furthermore, the type of enzyme formulation (free enzyme, immobilized enzyme, or whole cells) plays a key role in determining the progress of the overall reaction. For most applications, lyophilized enzyme powders have been used with good results; presumably they dissolve into the liquid phase. When poorly soluble products are formed, the enzyme can be recovered by washing with water [52]. For co-factor-dependent reactions permeabilized cells may be used [44]. When using immobilized enzymes, it has been demonstrated that the chemical nature and the pore size of the support are very important parameters to consider [8, 41].

Of course, the reactor design for reactions on a larger scale, with suspended substrates, plays a key role, since it largely depends on the consistency of the reaction mixture.

Eichhorn et al. reported the first successful reactions on a preparative scale [11]. Erbeldinger et al. studied the feasibility of large-scale reactions by discussing the reactor design and the possibility of enzyme recycling [52]. They suggested extracting the biocatalyst with water, taking advantage of the much lower solubility of reaction products. No loss in enzymatic activity could be observed in a 4-step extraction process after the synthesis of Z-L-Asp-L-Phe-OMe using an industrial enzyme preparation.

12.5
Precipitation-Driven Synthesis of Peptides

Precipitation-driven reactions show some very favorable advantages when compared to other low-water systems for enzymatic synthesis of peptides [55, 56].

Table 12.2 reports, as an example, a comparison of different methods for thermolysin-catalyzed synthesis of the aspartame precursor Z-L-Asp-L-Phe-OMe in terms of reaction yield and initial rate.

The initial rate of thermolysin-catalyzed Z-L-Asp-L-Phe-OMe synthesis using suspended substrates (Table 12.2, entry 3) was similar to that in aqueous solution (entry 1: around 60%).

Table 12.2 Comparison of different methods for thermolysin-catalyzed synthesis of Z-aspartame (Z-L-Asp-L-Phe-OMe).

Z-L-Asp + L-Phe-OMe $\xrightarrow{\text{Thermolysin}}$ Z-L-Asp-L-Phe-OMe + H_2O

Reaction systems	Z-L-Asp-L-Phe-OMe		Ref.
	$v_0^{[a]}$	Conv. (%)	
1 Dilute aqueous media	735	5	57
2 Aqueous solution with precipitated product	Not reported	88–90	58
3 Substrate suspension in aqueous media	429	>90	51, 52
4 Organic solvent	1–35	20–99[b]	58, 60
5 Substrate suspension in organic solvent	16	94	32

a nmol min^{-1} mg^{-1}.
b Depending on the solvent and on the enzymatic preparation.

However, yields were much higher working in substrate suspension (>90% product) [51, 52] as compared to aqueous solution (5%) [57]. The high yield observed is in line with those obtained in organic solvents (both in solution and in substrate suspension; entries 4 and 5) where, however, initial rates were at most 5% of those in dilute aqueous solution [32, 58–60].

The "solid-to-solid" approach clearly combines the good rates observed in aqueous solution, with the high yields typical of biocatalysis in organic solvents. Many successful examples of precipitation-driven reactions for the synthesis of peptides have been published in recent years (Table 12.3).

In all reported cases where neutral peptides were synthesized, product precipitation was observed in the (micro-) aqueous media. Z_{sat}^0 could be calculated on the basis of the melting points of a large number of neutral di- and tri-peptides and their amino acid substrates, and it was found that it was always between two and six orders of magnitude smaller in value than the reference equilibrium constant for amide synthesis [39]. Hence, conversion of solid substrates to solid product was predicted to be favorable.

For charged products the situation is significantly different. For many reactions involving zwitterions, the presence of undissolved substrates did not lead to precipitated products in enzymatic synthesis reactions [11]. These observations are related to differences in the solubility properties of zwitterions as compared to acids, bases, or uncharged compounds. Because of the low concentration of the

Table 12.3 Synthesis of peptides in reaction systems in the presence of undissolved substrates or precipitated products catalyzed by proteases.

Acyl donor	Nucleophile	Product		Maximum yield (%)	Ref.
		Neutral	Charged		
Thermodynamically controlled synthesis					
ACO-Tyr	Ile-Gly-OMe	ACO-Tyr-Ile-Gly-OMe		89	11
	Phe-NH$_2$	ACO-Tyr-Phe-NH$_2$		86	11
	Nvl-NH$_2$	ACO-Tyr-Nvl-NH$_2$		92	12
ACO-Leu	Phe-NH$_2$	ACO-Leu-Phe-NH$_2$		97	11
	Phe-OMe	ACO-Leu-Phe-OMe		81	11
Z-Phe-OH	Xaa-NH$_2$	Z-Phe-Xaa-NH$_2$		>80	11
Z-Phe-OH	Npg-NH$_2$	Z-Phe-Npg-NH$_2$		56	12
Z-Npg-OH	Npg-NH$_2$	Z-Npg-Npg-NH$_2$		55	12
Z-Xaa-OH	Leu-NH$_2$	Z-Xaa-Leu-NH$_2$	Z-His-Leu-NH$_2$	>80	3, 4, 11, 12, 56
Z-Arg-OH	Leu-NH$_2$		Z-Arg-Leu-NH$_2$	55	11
Z-His-OH	Leu-NH$_2$		Z-His-Leu-NH$_2$	95	11
Z-Lys-OH	Leu-NH$_2$		Z-Lys-Leu-NH$_2$	60	11
N-Ac-Tyr-OEt	Arg-NH$_2$		N-Ac-Tyr-Arg-NH$_2$	>90	41
Leu-Gly, Val-Leu			Oligomers	>30	62
Kinetically controlled synthesis					
N-Xaa-esters	Xaa-amides	Corresponding peptides		90	14
	Xaa-t-butylesters				
	peptides				
Eutectic melts					
Z-Asp(OEt)-OEt	D-Ala-NH$_2$		Z-Asp(OEt)-D-Ala-NH$_2$	53	64
Z-Asp-OAll	Glu-OEt		Z-Asp-Glu-Glu-OEt	>75	14
Z-Asp	Phe-OMe		Z-Asp-Phe-OMe	56	68
Z-Lys-OEt	Gly-OEt		Z-Lys-Gly-OEt	70	70
Z-Lys-Gly-OEt	Asp-OAll		Z-Lys-Gly-Asp-OAll	68	70
Z-Lys-Gly-Asp-OAll	Glu-OAll		Z-Lys-Gly-Asp-Glu-OAll	67	70
Ac-Phe-OEt	Ala-NH$_2$		Ac-Phe-Ala-NH$_2$	>97	65

uncharged form of zwitterions, product precipitation is more likely to occur when zwitterionic products are formed and less likely when substrates are zwitterions [39]. To what extent the ionization of zwitterions contributes to the value of Z_{sat}^0 depends on the difference in microscopic ionization constants of their acid and basic groups (ΔpK_a). With increasing ΔpK_a values, the concentration of the uncharged form decreases further [61].

This is particularly well illustrated by comparing the condensation reactions of three N-protected basic amino acids (Z-His, Z-Lys and Z-Arg) to Leu-NH$_2$ [11]. In all three cases, the acyl donor is a zwitterion, but the ΔpK_a values are very different

(2 to 3, 8, and 10, respectively). In this study the authors observed that product precipitation was only feasible for the peptide Z-L-His-L-Leu-NH$_2$. When the amino group in the side chain of Z-Lys was protected to create an acid acyl donor instead of a zwitterion, precipitation-driven synthesis of Z-L-Lys-(Z)-L-Leu-NH$_2$ became possible [11].

Stevenson et al. [62] studied the oligomerization of hydrophobic peptides such as Val-Leu, which are characterized by bitterness, with the aim of modifying peptide flavor. They found precipitation of oligomers in water/ethanol mixtures. By variation of the hydrophobicity of the model peptides, no significant product precipitation was observed in Gly-Gly, Ala-Gly, and Val-Gly. Product precipitation was observed for the more hydrophobic oligomers of Leu-Gly and Val-Leu.

Water is an excellent solvent for precipitation-driven biocatalysis for the synthesis of peptides, leading to nearly complete conversions in most cases. This is because of the specificity, restricted to hydrophobic amino acids, of proteases used in these studies. As a consequence, the synthesized peptides will exhibit poor solubilities in water, resulting in high yields of solid product. It has been shown that the yield does depend strongly on the solvent chosen. As a rule of thumb, Gill and Vulfson [37] and Cao et al. [26] suggested that rather hydrophilic solvents (low log P) give the best results for both peptides and sugar fatty acid esters. This behavior was related to the poor solubility of the synthesized product in these solvents [26].

It has been theoretically derived that the favorability of solid-to-solid conversion is solvent independent [30]. Practical evidence was presented that the mass action ratio and equilibrium constant varied in the same proportions as the solvent was changed (Figure 12.4) [45].

A comparison of the synthesis of Z-Phe-Leu-NH$_2$ in ten different solvents revealed that the highest overall yields could be expected in solvents where the substrate solubility is minimized. The highest yields in terms of solid product were found in solvents where both substrate and product solubility are minimized [45]. These simple rules may not hold when special factors apply, such as the formation of solid solvates. This may account for a few apparent exceptions, such as the product precipitation in dichloromethane of both a peptide and a sugar fatty acid ester [45, 63].

The type of solvent can affect the kinetics, and, as a consequence, the equilibrium is sometimes not reached within a reasonable time. For example, Kim and Shin studied the kinetically controlled synthesis of alitame precursor (Z-Asp(OEt)-D-Ala-NH$_2$), and better results were obtained in the presence of dimethyl sulfoxide and 2-methoxyethyl acetate as adjuvants. These solvents promoted product precipitation while maintaining the reaction mixture in a homogeneous state, thus improving the conversion [64].

In contrast to biocatalysis in solution, the worst solvents (i.e. lowest solubilities) will give rise to the best synthetic yields. Of course it is essential that the biocatalyst should show good activity/stability in the solvent of choice. In many cases, and in particular for reactions involving hydrophobic uncharged products, water is an

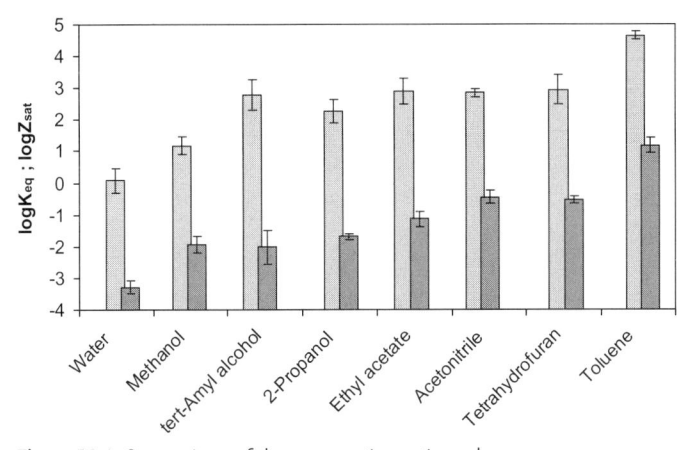

Figure 12.4 Comparison of the mass action ratio and equilibrium constants for peptide synthesis in different solvents at a_w 0.70, at 30 °C [45]. The light bars are equilibrium constants (K_{eq}) obtained from four individual experiments. The dark bars represent the measured saturated mass action ratio (Z_{sat}). Note that the difference between the two parameters is similar in all the solvents.

excellent choice. If the product to be formed is charged, then solvents where ionization is unfavorable will give rise to the best yields of solid product.

Some literature examples of peptide synthesis in the presence of organic (co-) solvents are summarized in Table 12.4. The uncharged peptide that was not reported to precipitate in the time scale of the reaction was Ac-Phe-Ala-NH$_2$ [65]. This was probably due to mass transfer limitations related to limited substrate solubility in the mixture of solvents used.

As reported in Table 12.1, as well as the solid-to-solid approach, peptide synthesis can be performed in systems where mixtures of substrates undergo eutectic melting.

Eutectic melting is a well-known phenomenon that lowers the melting point of a mixture below the melting point of each pure compound in the mixture [4].

Amino acid and peptide derivatives, commonly used as substrates in peptide synthesis, form low-melting-point mixtures when combined together in the absence of bulk solvents [66].

As an example, liquefied mixtures are easily formed by using some amino acids protected with acetyl, Boc-, Fmoc- and Z-groups. Some authors have called the resulting systems "eutectic mixtures", but this term should be reserved for the textbook definition as the composition of minimum melting point that freezes isothermally at the eutectic temperature. Hence we refer to them as "eutectic melts".

In many of the literature studies, a small amount (usually 5–30%) of another component, normally termed an adjuvant, has been added to cause or enhance formation of a liquid phase. There are several examples of application to efficient

Table 12.4 Precipitation-driven biocatalysis for the synthesis of peptides, catalyzed by proteases, in the presence of organic solvents.

Acyl donor	Nucleophile	Solvent	Yield (%)	Ref.
Z-Ala	Leu-NH$_2$	n-Hexane	92	13
Z-Pro			33	
Z-Lys(Z)	Leu-NH$_2$	H$_2$O/EtOH	61	7
		H$_2$O/EtOH/2-methoxyethyl acetate	80	
Z-Phe	Leu-NH$_2$	Toluene, n-hexane, THF, AcN, 1-hexanol, MeOH, i-PrOH	90–99	3, 5, 45
Z-Phe	D-Leu-NH$_2$	n-Hexane	3	13
	Ile-NH$_2$		41	
	Met-NH$_2$		45	
	Phe-NH$_2$		87	
Z-Ser	Leu-OEt	H$_2$O, glycerol	92	14
Z-Ala-Phe	Leu-NH$_2$	n-Hexane	78	13
Ac-Ala-Trp			84	
Ac-Gly-Trp			77	
Z-Phe	Tyr-OEt	Toluene	97–98	8
	Phe-OEt			
	Leu-NH$_2$			

enzymatic peptide synthesis [4, 6, 14, 37, 67]. In most cases the adjuvants used are themselves liquid at the reaction temperature, so formation of the liquid phase can no longer properly be attributed to eutectic melting. Essentially the process is indistinguishable from dissolution in a small quantity of a solvent (i.e. from the other types of reaction system discussed above).

Eutectic melting (and also similar systems with added adjuvants/solvents) has been used to prepare homogeneous substrate mixtures with extremely high concentration levels as media for enzymatic reactions [37, 68, 69].

An interesting example of scale-up of peptide synthesis in such low-melting point mixtures derived from eutectic melts has been described [70]. Neat combination of the pure substrates in the complete absence of water/solvent (adjuvant) provided simple heterogeneous systems consisting of the eutectic melts plus an excess of solid substrate (Figure 12.5).

12.6
Precipitation-Driven Synthesis of Esters and Surfactants

Synthesis in the presence of undissolved substrates has proven useful also in the lipase-catalyzed syntheses of surfactants such as glucose esterified with fatty acids or esters of glycosides [71], but also in more unusual examples such as ibuprofen resolution [72] and the syntheses of the cinnamic alcohol ester of glucuronic acid

Z-Lys(Z-Nε)-OEt

    | Gly-OEt
    | *Chymotrypsin*
    ↓

Z-Lys(Z-Nε)-Gly-OEt   **Yield 70%**

    | L-Asp(OAll)-OAll
    | *Subtilisin*
    ↓

Z-Lys(Z-Nε)-Gly-Asp(OAll)-OAll   **Yield 74%**

    | L-Glu(OAll)-OAll
    | *Chymotrypsin*
    ↓

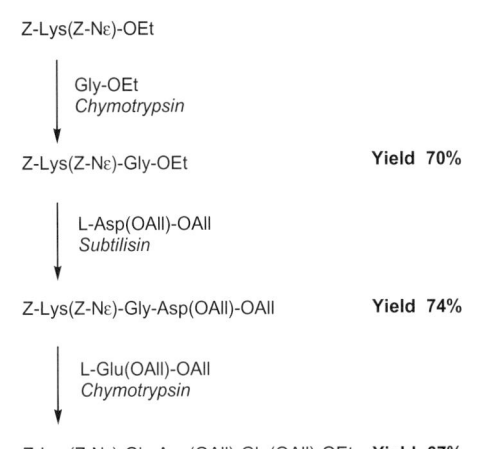

Z-Lys(Z-Nε)-Gly-Asp(OAll)-Glu(OAll)-OEt **Yield 67%**

Figure 12.5 Kilogram-scale production of a flavor-active peptide precursor conducted in heterogeneous, low-solvent mixtures [70].

[23], the phenylbutyric ester of vitamin C [25], or the more recent example of sorbitol-fatty acid esters [69].

In the synthesis of esters of sugars, high yields were obtained in many cases, as reported in Table 12.5. This can be attributed to the high temperatures used in these experiments (usually experiments are conducted at 40–80 °C), since the removal of water in open vials can shift the reaction toward the product [27]. The same purpose was successfully achieved by using azeotropic distillation [25]. An increase in ester yield with decreasing temperature was observed, and this behavior was related to the decreased product solubility.

The use of suspended substrates proved very useful in the production of arginine-based surfactants. In this two-step procedure, one amino group of a diaminoalkane was acylated with Z-Arg [55]. This step was done in a melted phase constituted by substrates, and quantitative yields were observed without a catalyst present. Subsequently, the second free amino group of the diaminoalkane reacted with an Z-Arg-alkyl ester (kinetic control). This step was performed in aqueous suspension or organic solvent using trypsin as the catalyst [73].

Similarly, glutaryl acylase was used as a biocatalyst in the synthesis of amides of glutaric acid derivatives using the acyl donor as the liquid phase in which the undissolved nucleophile was suspended [33]. In another example, immobilized lipase was used in a medium composed solely of substrate to resolve racemic ketoprofen esters via hydrolysis [36].

A hydrolytic reaction starting from solid substrates was used to improve the kinetic resolution of a solid di-ester to a solubilized monoester [74]. In suspension processes, the concentration of the substrate enantiomer to be converted is close to saturation, whereas in a conventional process the concentration gradually decreases. Calculations showed that the presence of solid substrates could lead to a 6-fold increase in productivity as compared to the reaction in solution.

Table 12.5 Synthesis of esters in the presence of undissolved substrates.

Product	Acyl donor	Nucleophile	Solvent	Yield (%)	Ref.
S-Ibuprofen	R,S-Ibuprofen	Fatty alcohols	None	Low ee	72
Aminoacid derivative	Z-Arginine	di-amino alkane	None	100	55
Amide	Glutaric acid ester	Tyrosynamide	None	48	33
Glucuronic acid esters	Glucuronic acid	Cinnamic alcohol / n-Butanol	tert-Butanol	15–22	23
Vitamin C ester	4-Phenylbutanoic acid	Vitamin C	tert-Butanol	91	23
Surfactant	Long-chain fatty acids	Sugar acetals	None	50–90	27
	Oleic acid	N-methyl-galactamine	2-Methyl-2 butanol	85	22
	Decanoic acid	N-methyl-glucamine	2-Methyl-2 butanol	83	22
	Ethyl butanoate	Lactose	tert-Butanol	1	71
		Sucrose	tert-Butanol	40	71
		Maltose	tert-Butanol	96	71
		Trehalose	tert-Butanol	84	71
	Caprylic acid	Glucose	2-Butanone	76	25
	Stearic acid	Glucose	2-Butanone	90	25
	Palmitic acid	Glucose	THF, DCM	63–66	63
		Glucose	2-Butanone	81	25
		Glucose	Dioxane, AcN, acetone, butyrolactone, tert-butanol, tert-amyl alcohol, 3-methyl-3-pentanol	30–88	26, 63

Shin and colleagues obtained high yields of sorbitol-fatty acid esters in media with extremely high (4–7 M) substrate concentration [69]. The media were entirely liquid, and contained small amounts of water and organic solvents, but are probably quite similar in composition to the liquid phases of related systems with excess solid substrates. The very high substrate concentrations positively influenced the initial rate of the reactions.

12.7
The Synthesis of β-Lactam Antibiotics in the Presence of Undissolved Substrates

The enzymatic synthesis of β-lactam antibiotics from β-lactam nuclei 6-aminopenicillanic acid (APA) and 7-aminodeacetoxycephalosporanic acid (ADCA) and appropriate side chain donors has been largely studied [75, 76].

An efficient synthesis of β-lactam with D-phenylglycine or its derivatives as a side chain can be accomplished only by using a kinetically controlled approach via acyl group transfer from an activated side chain donor, and D-phenylglycine can be used as ester, usually methyl (PGM) or ethyl (PGE), or as amide (PGA). As a

12.7 The Synthesis of β-Lactam Antibiotics in the Presence of Undissolved Substrates

matter of fact, thermodynamically controlled reactions with suspended substrates give low yields because of unfavorable thermodynamics and the consequent absence of product precipitation [16–19]. This has been related to the fact that both the substrates and the reaction product are zwitterions. Indeed, for the non-zwitterion analog of ampicillin (penicillin G), where only one of the substrates is a zwitterion (APA) with a moderate ΔpK_a of around 2, the thermodynamics are favorable, and indeed, precipitation-driven synthesis of penicillin G has been described [46].

Table 12.6 summarizes results obtained in the synthesis of antibiotics in the presence of undissolved substrates catalyzed by penicillin amidase, and

Table 12.6 Synthesis of antibiotics in the presence of undissolved substrates catalyzed by penicillin amidase.

Acyl donor	Nucleophile	Product	Maximum yield[a] (%)	Comment	Ref.
Kinetically controlled synthesis					
PGME	APA	Ampicillin	93	Enhanced solubility of APA in starting solution due to the presence of acyl donor.	77
PGME	APA	Ampicillin	87	Mathematical model for the process kinetics proposed.	78
PGME	APA	Ampicillin	97	Saturated concentration of APA maintained by repetitive addition of substrates.	79
PGME	APA	Ampicillin	70	Enzyme preparations stabilized by inorganic salt hydrates.	84
PGA	APA	Ampicillin	98	Supersaturated homogeneous solutions of substrates were used to improve kinetics.	82
HPGA	APA	Amoxicillin	91		
PGA	ADCA	Cephalexin	92		
PGME	ADCA	Cephalexin	93	Reaction in ethylene glycol; after 10 batches, biocatalyst maintains 87% of its activity.	83
PGA	APA	Ampicillin	>80	pH gradient used to optimize the process.	80
PGME	APA	Ampicillin	32	PGME used as solvent and reagent at the same time.	86, 87
	ADCA	Cephalexin	30		
PGME	ADCA	Cephalexin	100	Very high substrate concentrations (>750 mM) with PGME/ADCA = 3.	85
Thermodynamically controlled synthesis					
HPG	APA	Amoxicillin	Not reported	Using 50% (w/w) DMF as co-solvent the equilibrium constant is improved.	16
HPG	APA	Amoxicillin	<3	For amoxicillin synthesis 0.001–0.1 M $ZnSO_4$ used as counter-ion to favor product precipitation.	46
PhAc	APA	Penicillin G	80		

a Conversion of the nucleophile.

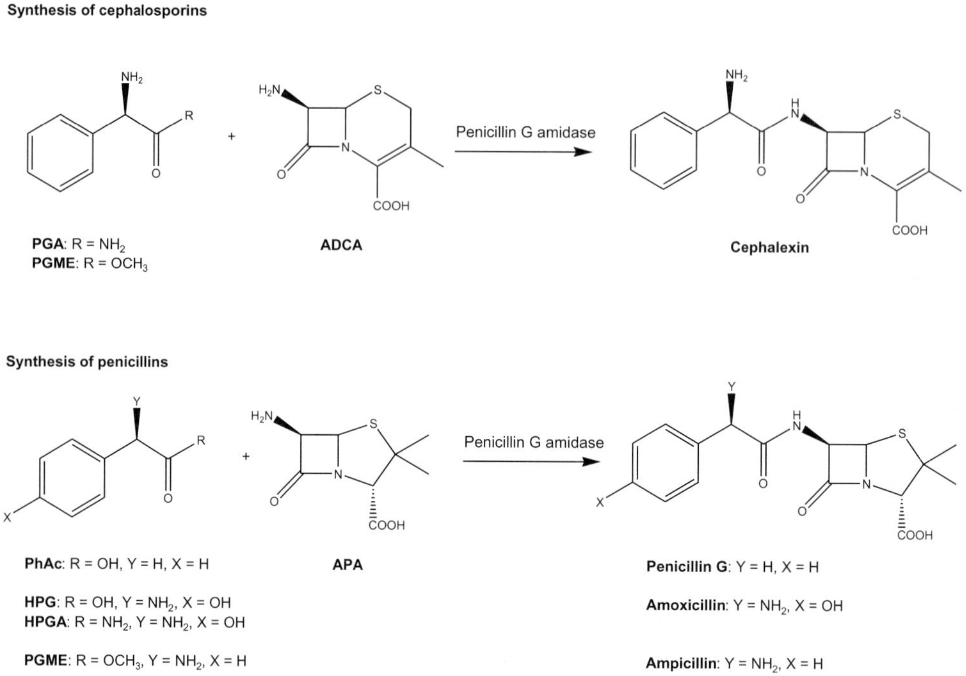

Synthesis of cephalosporins

PGA: R = NH₂
PGME: R = OCH₃

ADCA

Penicillin G amidase

Cephalexin

Synthesis of penicillins

PhAc: R = OH, Y = H, X = H
HPG: R = OH, Y = NH₂, X = OH
HPGA: R = NH₂, Y = NH₂, X = OH
PGME: R = OCH₃, Y = NH₂, X = H

APA

Penicillin G amidase

Penicillin G: Y = H, X = H
Amoxicillin: Y = NH₂, X = OH
Ampicillin: Y = NH₂, X = H

Scheme 12.3 Enzymatic production of β-lactam antibiotics (cephalosporins and penicillins) via kinetically or thermodynamically controlled synthesis.

Scheme 12.3 shows ways of preparing β-lactam antibiotics via kinetically or thermodynamically controlled synthesis.

The direction of a reaction can be assessed straightforwardly by comparing the equilibrium constant (K_{eq}) and the ratio of the product solubility to the substrate solubility (Z_{sat}) [39]. In the case of the zwitterionic product amoxicillin, the ratio of the equilibrium constant and the saturated mass action ratio for the formation of the antibiotic was evaluated [40]. It was found that, at every pH, Z_{sat} (the ratio of solubilities, called R_s in that paper) was about one order of magnitude greater in value than the experimental equilibrium constant ($Z_{sat} > K_{eq}$), and hence product precipitation was not expected and also not observed experimentally in a reaction with suspended substrates. The pH profile of all the compounds involved in the reaction (the activated acyl substrate, the free acid by-product, the antibiotic nucleus, and the product) could be predicted with reasonable accuracy, based only on charge and mass balance equations in combination with enzyme kinetic parameters [40].

Therefore, a "solid-to-solid" synthesis of amoxicillin in aqueous solution is not feasible.

Kinetically controlled strategies have been used to obtain high yields in the enzymatic synthesis of penicillins and cephalosporins. In the kinetically controlled approach, the main obstacle, however, is the competing hydrolysis of the side-chain donor and that of the product which necessitates the use of an excess of the side chain donor.

A breakthrough in the enzymatic approach was finally achieved when it was discovered that performing the reaction at high (over 0.3 M) concentrations or in so-called "supersaturated" solutions dramatically improved the yields.

Švedas and Sheldon have reported some interesting studies on the synthesis, catalyzed by native penicillin G amidase, of ampicillin and cephalexin in "aqueous solution-precipitate" systems and highly condensed aqueous systems, studying the effect of high substrate concentration and supersaturation [77]. They reported a positive effect of increasing the initial APA concentration to saturation, thus making it possible to improve the efficiency of the enzymatic acyl transfer. Working with a suspension of APA, an optimal acyl transfer efficiency can be guaranteed for a longer time, and the entire amount of a nucleophile present in a homogeneous supersaturated solution (over its thermodynamic saturation level) can take part in enzymatic synthesis [78–83].

In addition, at very low water contents, ampicillin accumulation curves do not exhibit a clear-cut maximum, inherent in the enzymatic acyl transfer reactions in aqueous medium (including quite concentrated heterogeneous aqueous solution-precipitate systems), because of the secondary hydrolysis of the target product by penicillin acylase (Figure 12.6) [84].

Recently, the kinetically controlled synthesis of cephalexin with penicillin amidase was studied at very high concentrations of substrates to the limit of

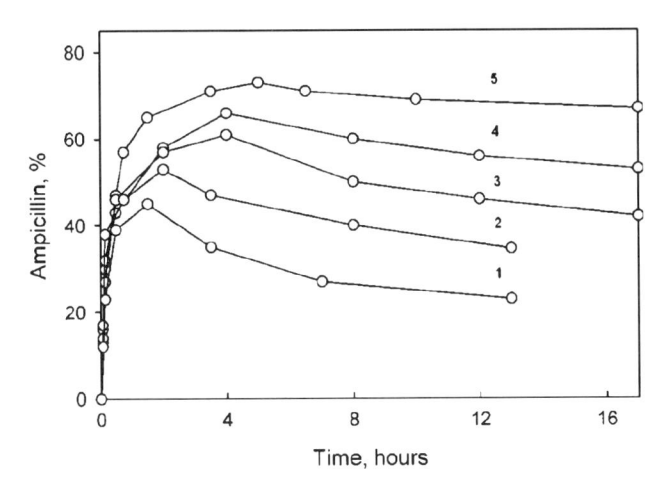

Figure 12.6 Curves for the accumulation of ampicillin in the PGA catalyzed-synthesis reaction at decreasing water content, from 35% w/w (curve 1) to 10% w/w (curve 5). Reprinted with permission from Youshko M. I., Svedas V. K., *Adv. Synth. Catal.*, **2002**, *344*, 894. Copyright (2002), American Chemical Society [84].

nucleophile solubility, and an increase of 30 times in volumetric productivity was obtained, keeping conversion yield close to 100% [85].

This approach can be extended by working in highly condensed systems formed with mainly undissolved substrates for the enzymatic synthesis of ampicillin and cephalexin, where the reaction mixture had no aqueous phase for dispersion of the reagents, and no organic solvents were used. The absence of an apparent aqueous phase in the reaction mixture reduces the incidence of the hydrolytic reaction [86–87].

12.8
Conclusion

A large number of successful biocatalytic syntheses are now being reported in systems with undissolved solid substrates and/or precipitating products. These reactions are applicable to a wide range of compounds, are easily carried out with standard laboratory equipment, and are environmentally acceptable and cost-efficient [88]. The systems vary quite widely in factors such as the size of the liquid phase and the concentration of solutes within it. The systems have been named in many different ways, but the terms used mostly describe overlapping groups. With the increased understanding of the thermodynamics, kinetics, scale-up, pH control, and solvent effects on these systems, it is now increasingly possible to develop or optimize these reactions rationally. The composition of the liquid phase, which may contain water or an organic solvent, has a key effect on kinetics. Successful syntheses have been described for many different peptides, as well as esters, surfactants, and antibiotics, and new synthetic targets will be identified using predictive methods [39]. Particularly promising are enzyme-catalyzed isomerization reactions or enantioselective resolutions [89].

Abbreviations

Ac	Acetyl- (*N-protecting group*)
ACO	Aminocarbonyl- (*N-protecting group*)
ADCA	7-Amino-3-desacetoxycephalosporanic acid
APA	6 Aminopenicillanic acid
HPG	D-*p*-Hydroxyphenylglycine
HPGA	D-*p*-Hydroxyphenylglycine amide
K_{eq}	Concentration based equilibrium constant
K_{th}	Thermodynamic equilibrium constant
Npg	Neopentylglycine
Nvl	Norvaline
OAll	Allyl group (*COOH-protecting group*)
PGA	D-Phenylglycine amide
PGME	D-Phenylglycine methyl ester
PhAc	Phenylacetic acid

Xaa   General aminoacids
Z   Carbobenzyloxy (*N-protecting group*)
Z-Npg Carbobenzyloxy-neopentylglycine
Z_{sat} Ratio of solubilities of product and substrate
Z_{th} Ratio of thermodynamic activities of pure solid product and solid substrate

All aminoacids have L-configuration except where indicated.

References

1 Rothenberg, G.; Downie, A. P.; Raston, C. L.; Scott, J. L. *J. Am. Chem. Soc.*, 2001, **123**, 8701.
2 Cave, G. W. V.; Raston, C. L.; Scott, J. L. *Chem. Commun.*, 2001, 2159.
3 Kuhl, P.; Eichhorn, U.; Jakubke, H. D. *Biotechnol. Bioeng.*, 1995, **45**, 276.
4 Gill, I.; Vulfson, E. N. *J. Am. Chem. Soc.*, 1993, **115**, 3348.
5 Halling, P. J.; Eichhorn, U. E.; Kuhl, P.; Jakubke, H. D. *Enzyme Microb. Technol.*, 1995, **17**, 601.
6 Lopez-Fandino, R.; Gill, I.; Vulfson, E. N. *Biotechnol. Bioeng.*, 1994, **43**, 1016.
7 Lopez-Fandino, R.; Gill, I.; Vulfson, E. N. *Biotechnol. Bioeng.*, 1994, **43**, 1024.
8 Basso, A.; De Martin, L.; Ebert, C.; Gardossi, L.; Linda, P. *Chem. Commun.*, 2000, 467.
9 Erbeldinger, M.; Ni, X.; Halling, P. J. *Biotechnol. Bioeng.*, 1998, **59**, 68.
10 Erbeldinger, M.; Ni, X.; Halling, P. J. *Biotechnol. Bioeng.*, 1999, **63**, 316.
11 Eichhorn, U.; Bommarius, A. S.; Drauz, K.; Jakubke, H. D. *J. Pept. Sci.*, 1997, **3**, 245.
12 Krix, G.; Eichhorn, U.; Jakubke, H. D.; Kula, M. R. *Enzyme Microb. Technol.*, 1997, **21**, 252.
13 Kuhl, P.; Eichhorn, U.; Jakubke, H. D. In *Biocatalysis in Non-conventional media*, Tramper, J.; Vermue, M. H.; Beeftink, H. H.; von Stockar, U., Eds.; Elsevier Science Publisher: Amsterdam, 1992, pp. 513–518.
14 Jorba, X.; Gill, I.; Vulfson, E. N. *J. Agric. Food. Chem.*, 1995, **43**, 2536.
15 Khul, P.; Halling, P. J.; Jakubke, H. D. *Tetrahedron Lett.*, 1990, **31**, 5213.
16 Diender, M. B.; Straathof, A. J. J.; Van der Wielen, L. A. M.; Ras, C.; Heijnen, J. J. *J. Mol. Catal. B: Enzymatic*, 1998, **5**, 249.
17 Svedas, V. K.; Margolin, A. L.; Berezin, I. V. *Enzyme Microb. Technol.*, 1980, **2**, 138.
18 Nierstrasz, V. A.; Schroen, C. G. P. H.; Bosma, R.; Kroon, P. J.; Beeftink, H. H.; Janssen, A. E. M.; Tramper, J. *Biocatal. Biotransform.*, 1999, **17**, 209.
19 Schroen, C. G. P. H.; Nierstrasz, V. A.; Kroon, P. J.; Bosma, R.; Janssen, A. E. M.; Beeftink, H. H.; Tramper, J. *Enzyme Microb. Technol.*, 1999, **24**, 498.
20 De Roode, B. M.; Franssen, M. C. R.; Van der Padt, A.; De Groot, A. E. *Biocatal. Biotransform.*, 1999, **17**, 225.
21 Ducret, A.; Trani, M.; Lortie, R. *Biotechnol. Bioeng.*, 2002, **77**, 752.
22 Maugard, T.; Remaud-Simeon, M.; Petre, D.; Monsan, P. *Tetrahedron*, 1997, **53**, 5185.
23 Otto, R. T.; Bornscheuer, U. T.; Scheib, H.; Pleiss, J.; Syldatk, C.; Schmid, R. D. *Biotechnol. Lett.*, 1998, **20**, 1091.
24 Yan, Y.; Bornscheuer, U. T.; Schmid, R. D. *Biotechnol. Lett.*, 1999, **21**, 1051.
25 Yan, Y.; Bornscheuer, U. T.; Cao, L.; Schmid, R. D. *Enzyme Microb. Technol.*, 1999, **25**, 725.
26 Cao, L.; Bornscheuer, U. T.; Schmidt, R. D. *J. Mol. Catal. B: Enzym.*, 1999, **6**, 279.
27 Fregapane, G.; Sarney, D. B.; Vulfson, E. N. *Enzyme Microb. Technol.*, 1991, **13**, 796.
28 MacManus, D. A.; Millqvist-Fureby, A.; Vulfson, E. N. In *Methods in Biotechnology, vol 15 – Enzymes In Nonaqueous Solvents*, Vulfson, E. N.; Halling, P. J.; Holland,

H. L.; Eds.; Humana Press: Totowa, New Jersey, 2001, pp. 545–552.
29 Kasche, V.; Spieß, A. In *Methods in Biotechnology, vol. 15 – Enzymes in Nonaqueous Solvents*; Vulfson, E. N.; Halling, P. J.; Holland, H. L., Eds.; Humana Press: Totowa, New Jersey, 2001, p. 553–564.
30 Erbeldinger, M.; Ni, X.; Halling, P. J. *Enzyme Microb. Technol.*, 1998, **23**, 141.
31 Kuhl, P.; Halling, P. J.; Jakubke, H. D. *Tetrahedron Lett.*, 1990, **31**, 5213.
32 De Martin, L.; Ebert, C.; Gardossi, L.; Linda, P. *Tetrahedron Lett.*, 2001, **42**, 3395.
33 Biffi, S.; De Martin, L.; Ebert, C.; Gardossi, L.; Linda, P. *J. Mol. Catal. B: Enzym.*, 2002, **19–20**, 135.
34 Van der Heijden, A. M.; Zwijderuijn, F. J.; Van Rantwijk, F. *J. Mol. Catal. A: Chem.*, 1998, **134**, 259.
35 Litjens, M. J. J.; Straathof, A. J. J.; Jongejan, J. A.; Heijnen, J. J. *Tetrahedron*, 1999, **55**, 12411.
36 Jin, J. N.; Lee, S. H.; Lee, S. B. *J. Mol. Catal. B: Enzym.*, 2003, **26**, 209.
37 Gill, I.; Vulfson, E. *TIBTECH*, 1994, **12**, 118.
38 Straathof, A. J. J.; Litjens, M. J. J.; Heijnen, J. J. In *Methods in Biotechnology, vol. 15 – Enzymes in Nonaqueous Solvents*; Vulfson, E. N.; Halling, P. J.; Holland, H. L., Eds.; Humana Press: Totowa, New Jersey, 2001, p. 603–610.
39 Ulijn, R. V.; Janssen, A. E. M.; Moore, B. D.; Halling, P. J. *Chem. Eur. J.*, 2001, **7**, 2089.
40 Diender, M. B.; Straathof, A. J. J.; van der Does, T.; Zomerdijk, M.; Heijnen, J. J. *Enzyme Microb. Technol.*, 2000, **27**, 576.
41 Kasche, V.; Galunsky, B. *Biotechnol. Bioeng.*, 1995, **45**, 261.
42 Michielsen, M. J. F.; Reijenga, K. A.; Wijffels, R. H.; Tramper, J.; Beeftink, H. H. *J. Chem. Technol. Biotechnol.*, 1998, **73**, 13.
43 Michielsen, M. J. F.; Meijer, E. A.; Wijffels, R. H.; Tramper, J.; Beeftink, H. H. *Enzyme Microb. Technol.*, 1998, **22**, 621.
44 Michielsen, M. J. F.; Frielink, C.; Wijffels, R. H.; Tramper, J.; Beeftink, H. H. *Biotechnol. Bioeng.*, 2000, **69**, 597.
45 Ulijn, R. V.; De Martin, L.; Gardossi, L.; Janssen, A. E. M.; Moore, B. D.; Halling, P. J. *Biotechnol. Bioeng.*, 2002, **80**, 509.
46 Ulijn, R.V.; De Martin, L.; Halling, P. J.; Janssen, A. E. M.; Moore, B. D. *J. Biotechnol.*, 2002, **99**, 215.
47 Ulijn, R. V.; Halling, P. J. *Green Chem.*, 2004, **6**, 488.
48 Yalkowsky, S. H. *Solubility and Solubilisation of Aqueous Media*, ACS/Oxford, 2000.
49 Dannenfelser, R.; Yalkowsky, S. H. *Ind. Eng. Chem. Res.*, 1996, **35**, 1483.
50 Grant, D. J. W.; Higuchi, T. *Techniques in Chemistry: Solubility of Organic Compounds*, Wiley: New York, 1990.
51 Erbeldinger, M.; Ni, X.; Halling, P. J. *Biotechnol. Bioeng.*, 2001, **72**, 69.
52 Erbeldinger, M.; Halling, P. J.; Ni, X. *AIChE J.*, 2001, **47**, 500.
53 Schoemaker, H. E.; Boesten, W. H. J.; Broxterman, Q. B.; Roos, E. C.; Van Den Tweel, W. J. J.; Kamphuis, J.; Meijer, E. M.; Rutjes, F. P. J. T. *Chimia*, 1997, **51**, 308.
54 Wolff, A.; Zhu, L.; Wong, Y. W.; Straathof, A. J. J.; Jongejan, J. A.; Heijnen, J. J. *Biotechnol. Bioeng.*, 1999, **62**, 125.
55 Clapes, P.; Moran, C.; Infante, M. R. *Biotechnol. Bioeng.*, 1999, **63**, 333.
56 Ulijn, R. V.; Erbeldinger, M.; Halling, P. J. *Biotechnol. Bioeng.*, 2000, **69**, 633.
57 Oyama, K.; Irino, S.; Harada, T.; Hagi, N. In *Enzyme Engineering VII*; Laskin, A. I.; Tsao, G. T.; Wingard jr., L. B., Eds.: The New York Academy of Science: New York, 1984, p. 95.
58 Isowa, Y.; Ohmori, M.; Ichikawa, T.; Mori, K. *Tetrahedron Lett.*, 1979, **28**, 2611.
59 Nagayasu, T.; Miyanaga, M.; Tanaka, T.; Sakiyama, T.; Nakanishi, K. *Biotechnol. Bioeng.*, 1994, **43**, 1118.
60 Oyama, K.; Irino, S.; Hagi, N. In *Methods in Enzymology, vol. 136*; Colowick, S. P.; Kaplan, N. O., Eds.; Academic Press: San Diego, 1987, pp. 503–518.
61 Ulijn, R. V.; Moore, B. D.; Janssen, A. E. M.; Halling, P. J. *J. Chem. Soc., Perkin Trans II*, 2002, **5**, 1024.

62 Stevenson, D. E.; Ofman, D. J.; Fenton, G. A. *J. Mol. Catal. B: Enzym.*, 1998, **5**, 39.
63 Cao, L.; Fischer, A.; Bornscheuer, U. T.; Schmid, R. D. *Biocatal. Biotransform.*, 1997, **14**, 269.
64 Kim, C.; Shin, C.-S. *Enzyme Microb. Technol.*, 2001, **28**, 611.
65 Björup, P.; Adlercreutz, P.; Clapes, P. *Biocatal. Biotransform.*, 1999, **17**, 319.
66 Kim, H. J.; Kim, J. H.; Youn, S. H.; Shin, C. S. *Thermochim. Acta*, 2006, **441**, 168.
67 López-Fandino, R.; Gill, I.; Vulfson, E. N. *Biotechnol. Bioeng.*, 1994, **43**, 1030.
68 Shin, G. H.; Kim, C.; Kim, H. J.; Shin, C. S. *J. Mol. Catal. B: Enzym.*, 2003, **26**, 201.
69 Kim, H. J.; Youn, S. H.; Shin, C. S. *J. Biotechnol.*, 2006, **123**, 174.
70 Gill, I.; Valivety, R. *Org. Process Res. Develop.*, 2002, **6**, 684.
71 Woudenberg-van Oosterom, M.; Van Rantwijk, F.; Sheldon, R. A. *Biotechnol. Bioeng.*, 1996, **49**, 328.
72 Pepin, P.; Lortie, R. *Biotechnol. Bioeng.*, 1999, **63**, 502.
73 Piera, E.; Infante, M. R.; Clapes, P. *Biotechnol. Bioeng.*, 2000, **70**, 323.
74 Wolff, A. V.; Straathof, A. J. J.; Heijnen, J. J. *Biotechnol. Prog.*, 1999, **15**, 216.
75 Bruggink, A.; Roos, E. C.; De Vroom, E. *Org. Process Res. Dev.*, 1998, **2**, 128.
76 Kallenberg, A. I.; van Rantwijk, F.; Sheldon, R. A. *Adv. Synth. Catal.*, 2005, **347**, 905.
77 Youshko, M. I.; Van Langen, L. M.; De Vroom, E.; Moody, H. M.; Van Rantwijk, F.; Sheldon, R. A.; Svedas, V. K. *J. Mol. Catal. B: Enzym.*, 2000, **10**, 509.
78 Youshko, M. I.; Švedas, V. K. *Biochemistry (Moscow)*, 2000, **65**, 1367.
79 Youshko, M. I.; van Langen, L. M.; de Vroom, E.; van Rantwijk, F.; Sheldon, R. A.; Švedas, V. K. *Biotechnol. Bioeng.*, 2001, **73**, 426.
80 Youshko, M. I.; van Langen, L. M.; de Vroom, E.; van Rantwijk, F.; Sheldon, R. A.; Švedas, V. K. *Biotechnol. Bioeng.*, 2002, **78**, 589.
81 Youshko, M. I.; Chilov, G. G.; Shcherbakova, T. A.; Švedas, V. K. *Biochim. Biophys. Acta: Proteins Proteom.*, 2002, **1599**, 134.
82 Youshko, M. I.; Moody, H. M.; Bukhanov, A. L.; Boosten, W. H. J.; Švedas, V. K. *Biotechnol. Bioeng.*, 2004, **85**, 323.
83 Illanes, A.; Altamirano, C.; Fuentes, M.; Zamorano, F.; Aguirre, C. *J. Mol. Catal. B: Enzym.*, 2005, **35**, 45.
84 Youshko, M. I.; Švedas, V. K. *Adv. Synth. Catal.*, 2002, **344**, 894.
85 Illanes, A.; Wilson, L.; Altamirano, C.; Cabrera, Z.; Alvarez, L.; Aguirre, C. *Enzyme Microb. Technol.*, 2007, **40**, 195.
86 Linda, P.; Gardossi, L.; Toniutti, M. Università degli Studi di Trieste, Trieste, Italy Patent PCT/EP04/00182.
87 Basso, A.; Spizzo, P.; Toniutti, M.; Ebert, C.; Linda, P.; Gardossi, L. *J. Mol. Catal. B: Enzym.*, 2006, **39**, 105.
88 Erbeldinger, M.; Eichhorn, U.; Kuhl, P.; Halling, P. J. In *Methods in Biotechnology, vol. 15 – Enzymes in Nonaqueous Solvents*; Vulfson, E. N.; Halling, P. J.; Holland, H. L., Eds.; Humana Press: Totowa, New Jersey, 2001, pp. 471–477.
89 Straathof, A. J. J.; Wolff, A.; Heijnen, J. J. *J. Mol. Catal. B: Enzym.*, 1998, **5**, 55.

Index

a
Accurel EP-100   104
acetate
    – allylic   127
(S)-[1-(acetoxyl)-4-(3-phenyl)butyl]phosphonic acid diethyl ester   172
6-O-acetyl-N-acetyl-D-glucosamine   151
N-acetyl-L-phenylalanine ethyl ester (APEE)   37, 58ff.
acid   104f.
    – derivative   104ff.
    – hydroxy   137
activity
    – ratio of thermodynamic activity   282
    – thermodynamic   258ff.
N-acylamino acid ester   243
acylation   78ff., 94, 145ff.
    – chemoselectivity   146f.
    – enantioselective   242
    – N   104, 147
    – O   124, 147
    – penicillin G acylase   232
    – polyol   148
    – regioselectivity   148ff., 180
acyloin   119
ADCA, see 7-amino-3-desacetoxycephalosporanic acid
adenosine   145f., 153
adjuvant   280ff.
alanine ester   181
Alcaligenes sp. lipase (ASL)   91, 231
alcohol
    – allylic   127
    – chiral   241ff.
    – primary   82ff.
    – racemic   170
    – secondary   88ff., 128
    – tertiary   100
alcohol dehydrogenase   11, 208, 256f.
alcohol oxidase   256f.
alcoholyse   158
alitame   290
alkaloid glycoside   158
alkylamine   101
π-allyl(palladium) intermediate   127
aluminium complex   131
Amano   121
Amano AK   84
    – lipase   86ff., 174
Amano PS   83
amination
    – intramolecular reductive   138
amine   101ff., 132
    – chemoenzymatic cyclic deracemization   138
    – chiral   242
    – racemic   174
    – secondary   138
amino acid   59
    – acyclic   137
    – cyclic   136
    – D   124
    – deracemization   136
    – derivative   104
    – sugar   150
L-amino acid amide   124
D-amino acid oxidase   136, 196
amino alcohol   178
cis-(1S,2R)-1-amino-2-indanol   37
8-amino-5,6,7,8-tetrahydroquinoline   134
γ-aminobutyric acid (GABA) analog   40
7-amino-3-desacetoxycephalosporanic acid (ADCA)   294ff.
aminoester   104
aminolysis
    – enantioselective   241
6-aminopenicillanic acid (APA)   294ff.
amphetamine   101

ampicillin 295ff.
amygdalin 158
t-amylalcohol (TAME) 104
antibiotics 294ff.
antileukaemic agent 180
antiviral agent 180
APEE, see *N*-acetyl-L-phenylalanine ethyl ester
aqueous solution-precipitate system 297
aqueous two-phase system with precipitated substrates and/or products 280
1-arylalkanol 89
arylalkylamine 101
2-arylpropanoic acid (profen) 106
Z-aspartame 242, 285ff.
azlactone
 – racemic 181
azobisisobutyronitrile (AIBN) 125
azole derivative
 – antifungal 179

b

bacteria
 – dehydrated 267f.
Baeyer-Villiger monooxygenase (BVMO) 37
(S)-benzyl-L-*tert*-leucine butyl ester 181
meso-*N*-benzylpiperidinediol 99
bio-surfactant 151
biocatalysis 3ff.
 – biphasic system 191ff., 211ff.
 – combinatorial 149
 – precipitation driven 280ff.
 – solid/gas 255ff.
 – undissolved solid substrate and product 279ff.
biocatalyst
 – activity 290
 – immobilization 204
 – stability 290
bioreactor 259ff., 272
 – industrial 273
 – solid/gas 265
biotransformation 237
 – ionic liquid 236f.
 – whole-cell 236
biphasic system 211ff.
 – buffer/organic solvent 212ff.
 – liquid/liquid 212ff.
 – liquid/solid 212ff.
 – organic-aqueous 191ff.
(R,R)-bis(amidoester) 102
bis(hydroxymethyl)phenylphosphine borane 87

Brij 56 89
ω-bromoaldehyde cyanohydrin 217
buffer
 – biological 22
 – ionic liquid 215
 – organic solvent 212
 – pH 271
 – solid state 58
Burkholderia cepacia lipase (BCL) 87
S-N(*tert*-butoxycarbonyl)-3-hydroxymethylpiperidine 172
1-*n*-butyl-2,3-dimethylimidazolium [BMMIM]$^+$ 242
 – triflate [TfO]$^-$ 242
1-butyl-3-methylimidazolium [BMIM]$^+$ 228ff.
 – bis(trifluoromethylsulfonyl)amide [NTf$_2$]$^-$ 228ff., 246
 – hexafluorophosphate [PF$_6$]$^-$ 228ff., 243ff.
 – tetrafluoroborate [BF$_4$]$^-$ 228ff.
rac-2-butylamine 175f.
tert-butylmethyl ether (*t*BME) 83ff., 100ff., 213ff., 242
BVMO, see Baeyer-Villiger monooxygenase

c

Candida antarctica lipase (CAL) 86, 151
Candida antarctica lipase A (CALA) 95ff., 121ff., 161
 – ionic liquid 231
Candida antarctica lipase B (CALB) 31ff., 82ff., 95ff., 119ff., 131, 147ff., 159, 170
 – catalysis 98, 263
 – enantiopreference 263
 – enantioselectivity 265
 – ionic liquid 230ff.
 – stereospecificity pocket 266
Candida bombicola 158
Candida cylindracea lipase 125, 149, 264
Candida rugosa lipase (CRL) 95ff., 133, 149ff., 232
ε-caprolactone 152
carbohydrate 150ff.
celite® 104, 215ff.
cellulose 152
 – acetate 152
cephalexin 297f.
chemoselectivity 145f.
 – hydrolase 145f.
Chirazyme L2 93ff.
chitobiose derivative 151

(R)-2-chloromandelic acid 183f.
chloroperoxidase (CPO) 208, 246
choline [HOEtMe3N]
 – [citrate] 246
Chromobacterium viscosum lipase 149
α-chymotrypsin 52ff., 232
clopidogrel 183
cocaine 276
colchicoside 158
conduritol 160
conformational effect 42
COSMO-RS method 21
[CPMA][MeSO$_4$] 237
cross-linked enzyme aggregate
 (CLEA®) 31, 185f., 219ff.
cross-linked enzyme crystal (CLEC) 31, 219
cross-linking 219
 – oxynitrilase 219
crown ether 60
cyanohydrin 121, 183ff., 212ff.
 – formation in biphasic system 221
cycloamylose 61
cyclodextrin 60f.
trans-cyclopentanediamine 102
meso-cyclopentenediol 99
cytochrome *c* 235

d

cis-decalin 99
dehalogenase 268f.
deracemization 113
 – biocatalyzed 139
 – chemoenzymatic 113, 136f.
 – cyclic 116, 135ff.
desoxyspergualin 173
desymmetrization 78, 179
dextran 153
Diels-Alder reaction 131
digitonin 155
dihydroxybiaryl 100
diisopropyl ether (DIPE) 84, 102, 213, 231
dimethyl sulfoxide (DMSO) 153
dimethylformamide (DMF) 153
dimethylimidazolium [MMIM]$^+$ 244, 246
 – dimethylphosphate 246
 – methylsulfate 244
displacement
 – S$_N$2 125
domino reaction
 – hydrolase-catalyzed 97
doxurobicin 147
dynamic kinetic asymmetric transformation
 (DYKAT) 130

dynamic kinetic resolution (DKR) 89ff., 114ff., 126ff.
 – chiral alcohol 241ff.
 – deracemization 140
 – enzyme-catalyzed 140

e

E-value 26ff., 75ff., 101
Einstein-Smoluchovski relation 35
emtricitabine (emtriva) 178
enantiomeric excess value (*ee*) 26, 80
enantiomeric ratio 26ff.
enantiomeric resolution
 – hydrolase catalyzed 102
enantioselectivity 10, 26ff., 37ff., 75ff.
 – determination 81f.
 – solvent engineering 264
encapsulation
 – sol-gel matrix 222
engineering 28
 – genetic 82
 – solvent 264
enzymatic catalysis
 – water 48ff.
enzyme 47ff.
 – activating 47ff.
 – activation 53ff.
 – chemoselectivity 145f.
 – crude preparation 216
 – enantioselectivity 264
 – immobilization 217
 – inactivation 193
 – ionic liquid 230ff.
 – non-aqueous solvent 169ff.
 – organic solvent 47ff.
 – protection 195f.
 – regioselectivity 148ff.
 – salt-activated 61ff.
 – selectivity 25ff.
 – stability 235
 – structure 51
 – transformation 227
enzyme activity 13, 48, 266
 – non-aqueous media 53
 – solvent effect 13
enzyme specificity 14
epoxidation 39
epoxide hydrolase 208
equilibrium 18
 – thermodynamic 284
 – thermodynamic constant 282
 – solvent effect 20
 – water effect 19

erythritol 152
ester
— activated 118
— natural 272
— resolution 178
— synthesis 170, 292ff.
esterase 232
esterification 170, 247
ethionamide monooxygenase (EtaA) 37
ethyl butyrate 102
1-ethyl-3-methylimidazolium [EMIM]$^+$ 235ff.
— bis(trifluoromethylsulfonyl)amide [NTF]$^-$ 235f.
— tetrafluoroborate [BF$_4$]$^-$ 242ff.
— triflate [TfO]$^-$ 242
eutectic melting 280, 291f.
excipient 59
Eyring-type relationship 33

f

farnesyl protein transfer inhibitor
— non-sulfhydryl 177
fatty acid ester 293
flavonoid ester synthesis 155
flavonoid glycoside 155f.
(S)-γ-fluoroleucine ethyl ester 182
flux
— product 203
— substrate 201f.
formaldehyde dehydrogenase 275
free energy 282
FTC butyrate 178f.
(R)-2-(2-furyl)-2-hydroxyacetonitrile 215

g

GABA analog 40
β-galactosidase 208
Gibbs free energy 283
— difference 33
ginsenoside 154
D-glucitol 152
glucopyranoside 152ff.
glucose
— ester 292
— oxidase 196
glutaryl acylase 293
glycidol 31ff.
L-glycopyranoside 161
glycosidase 11, 151, 208
— ionic liquid 244

glycoside 151
— acylation 153f.
— *endo-trans* 153
— ester 150, 292
— flavonoid 155f., 238
— natural 150ff.
— terpene 154
(—)-goniothalamin 98
Grubbs' catalyst 98

h

haloalkane dehalogenase 268
halophile 49
Hatta number 204f.
hemiaminal 133
hemithioacetal 122
heterogeneous eutectic melting 280
Hildebrand solubility parameter 29
HIV protease inhibitor 37
[HOEtMe3N][citrate] 246
hydantoin 119
hydrocyanation 185
hydrogel
— inulin-containing 152
— polymeric 152
hydrogenase 256f.
hydrolase 75ff., 117, 158, 208, 283
— chemoselectivity 146f.
— regioselectivity 148ff.
R-(+) hydromethyl glutaryl coenzyme A (HMG CoA) reductase inhibitor 171
hydroxy acid 137
5-hydroxy-2-(5H)-furanone 133
4-hydroxyacetophenone monooxygenase (HAPMO) 37
(S)-hydroxycarboxylic acid 184
hydroxynitrile lyase (HNL) 183ff., 208, 211ff., 222f.
— *Hevea brasiliensis* (HbHNL) 211ff., 222f.
— *Linum usitatissimum* (LuHNL) 211ff.
— *Manihot esculenta* (MeHNL) 211ff., 223
— *Prunus amygdalus* (PaHNL) 211ff., 222f.
— *Sorghum bicolor* (SbHNL) 211ff.
1-(3-hydroxypropyl)-3-methyl-imidazolium [HOPMIM][glycolate] 16, 234
hysteresis effect 8

i

ibuprofen 247, 292
— ester 240

immobilization 55, 218ff.
 – encapsulation in sol-gel matrix 222
industrial production 272
inert carrier 218
interfacial area 198
inulin 152
ion
 – chaotropic 64f.
 – kosmotropic 64f.
ionic liquid 16ff., 215, 227ff.
 – enzymatic transformation 227ff.
 – solvent property 228
 – whole-cell biotransformation 236
isoquercitrin 156

j
Jasmal 94
Jones-Dole B coefficient (JDB) 64f.

k
ketone cyanohydrin 217
ketoprofen
 – ester 293
 – (S) 106
kinetic resolution (KR) 114f., 133f.
 – dynamic, see dynamic kinetic resolution
 – limitation 78

l
laccase 208
β-lactam antibiotics 294
lactate 137
lactoside 151
linear solvation energy relationship (LSER) 29
lipase 9ff., 30ff., 80ff., 117, 147ff., 208, 237
 – Amano 121
 – Amano PS 96
 – LIP 97
 – MY 118
 – PS, see Pseudomonas cepacia lipase
 – QL 134
 – selectivity 126
 – Toyobo LIP-300 177
Lipozyme® 41, 119
Lipozyme IM 89
Lipozyme IM20 106
lotrafiban 239f.
lyophilization 8, 51ff.
 – excipient 59
lyoprotectant 59

m
macro- and microheterogeneous system 193f.
maleate hydratase 286
mandelate racemase 124
mandelic acid 124
mass action ratio 282
mass transfer coefficient 198, 200ff.
medium engineering 28
Meerwein-Ponndorf-Verley-Oppenhauer (MPVO) 131
membrane reactor 205
methoxypolyethyleneglycol (MPEG) 85
methyl-β-cyclodextrin (MβCD) 38
Michaelis-Menten kinetics 33ff.
micro-aqueous organic reaction system 216f.
microscopic ionization constant 289
molecular dynamics (MD) 66
monoamine oxidase 138
morphine dehydrogenase 234
MTBE 181ff.
Mucor javanicus lipase 147
Mucor miehei lipase (Mml) 159, 170, 182

n
NADH 193
NADP cofactor 234
NADPH 193
naproxen 240
naringin 156, 238
nitrilase 121
non-polar saturated area (NPSA) 197
non-polar unsaturated area (NPUA) 197
non-steroidal anti-inflammatory drug (NSAID) 106, 239
Novozym 435 89ff., 147ff., 182, 247
Novozym SP525 235
nucleophile 11, 161
nucleoside 156f.

o
octyl-β-D-glycopyranoside 89
1-n-octyl-3-methylimidazolium [OMIM]$^+$
 – tetrafluoroborate [BF$_4$]$^-$ 229
oligomerization
 – peptide 290
oligosaccharide 150f.
 – milk 151
organic solvent 3ff.
 – activating enzyme 47ff.
 – biocatalysis 3ff.
 – enantioselectivity 75ff.

–enzyme selectivity 25ff.
–pH 21
organogel
  –self-assembled 151
oxabicyclohexene 97
oxazolone 118
oxidation
  –stereoselective 274
oxidoreductase 245
oxycodone 234
oxynitrilase 211ff.

p
palladium complex 127ff.
pancreatin, see porcine pancreatic lipase
penicillin
  –amidase 295
  –G 295
  –G acylase 208, 232
  –G amidase 297
2,4-pentanediol 130
peptide
  –precipitation driven synthesis 287
percent of equilibrium (POE) 116
pH
  –buffer 271
phenoxybenzaldehyde cyanohydrin 184
(S)-m-phenoxymandelonitrile 215
phenylacetone monooxygenase (PAMO) 37
D-phenylglycine 294
2R4S phenylsulfonate 179
phosphatidylcholine 19f.
pig liver esterase (PLE) 85ff.
polar surface are (PSA) 197
polydispersity index (PDI) 152
polyester 152
  –sweet 152
polyethylene glycol (PEG) 235
polymer 59f.
  –sugar-based 152
polymerization
  –ring-opening 152
polyol 59, 148
  –sugar 152
polypropylene 104
polysaccharide 152
  –natural 152
porcine pancreatic lipase (PPL) 39, 81ff.
  –acylation 81ff.
process modelling 200
processing
  –downstream 206, 247
product mass transfer coefficient 201

(S)-profen 106
proleather 152
proline 136
protease 117, 151, 230ff.
Pseudomonas aeruginosa lipase (LIP) 102
Pseudomonas cepacia (Amano) lipase (lipase PS, PCL, PS-C) 86ff., 121, 149ff.
  –ionic liquid 231
Pseudomonas fluorescens lipase (PFL) 83ff., 128
Pseudomonas sp. lipase (PSL) 84, 100, 233ff.
purity
  –enantiomeric 81
pyrethroid insecticide 184
3-pyridinecarboxyaldehyde 185f.
pyruvate decarboxylase 208

q
quercitrin 156

r
racemization 115ff., 133ff.
  –acid-catalyzed 119
  –aldehyde 122
  –base-catalyzed 117
  –enzyme-catalyzed 124
  –metal-catalyzed 126
  –radical 125
  –S_N2 displacement 125
reaction
  –irreversible 76f.
  –reversible 76f.
reaction system
  –classification 280f.
reactor 205
  –bio 259ff.
  –solid/gas 263ff.
regioselectivity 148ff.
  –hydrolase 148
resolution
  –alcohol 170, 241
  –branched alkanoic acid 105
  –ester 178
  –racemic mixture 274
reversibility problem 79
Rhizomucor miehei lipase (RmL) 106, 119, 231
rhodium complex 128
Rhodococcus erythropolis NCIMB 13064 268ff.
ribavirin 180
riboflavin 153
ring-closing metathesis (RCM) 98

ring opening
  – asymmetric   181
ring-opening polymerization   152
ruthenium
  – complex   129ff.
  – *p*-cymene binuclear complex   243
rutin   156, 238

s

Saccharomyces cerevisiae   270
salicin   153
salt
  – hydrate   5
  – saturated salt solution   4f.
salt-activation   61ff.
  – mechanism   63
  – molecular dynamics   66
  – structural dynamics   66
Sauter mean droplet diameter   198f.
Schiff-base intermediate   123
selectivity   10f.
sensor   5
silica   233
sol-gel matrix   222f.
solid/gas reactor   263
  – enantioselective reaction   263
solid/gas system   258ff., 275
solid-to-solid system   280ff.
solubilization   53
solution
  – supersaturated   297
solvent   280
  – biocompatibility   197
  – content   82
  – engineering   264
  – ionic liquid   228ff.
  – non-aqueous   169ff.
  – selection   196
solvent effect   13
  – *E*-value   30
  – enzyme activity   13
  – non-hydrolytic enzyme   36
  – stability   16
solvent-accessible polar surface are (PSA)   197
sophorolipid   158
sorbitol-fatty acid ester   293
specificity constant   26
  – kinetic analysis   33
  – thermodynamic analysis   33
stability   11ff.
stereoinversion   139

steviolbioside   155
styrene oxide   171
substrate   271
  – formation-cleavage   120
subtilisin   38, 50f., 88, 161
  – Carlsberg   38, 52ff., 130, 147
sucrose   104
sugar   59, 150ff., 293
  – derivative   150f.
  – polyol   152
sulcatol   30
supersaturation   297
surfactant   292f.
  – bio   151
suspension to suspension synthesis   280
synthesis
  – asymmetric   113

t

TAME, see *t*-amylalcohol
*t*BME, see *tert*-butylmethyl ether
temperature   269
temsirolimus   148
terpene glycoside   154
tetrahydrofuran (THF)   53ff.
tetramethylpiperidine 1-oxyl free radical (TEMPO)   130
thermodynamic activity   258ff.
thermolysin   152, 208, 284ff.
Thermomyces lanuginosus lipase (TlL)   231
thermostability   12
thiocolchicoside   158
thioester   118
toxicity   16
  – molecular   16
  – phase   16
ω-transaminase (ω-TA)   37
transesterification   30ff., 59ff., 126, 179, 232f.
tri-*iso*-octylamine   21
tri-*n*-propylorthoformate   107
triacylglycerol hydrolase   39
N-trifluoroacetyl doxorubicin-14-valerate   147
trypsin   193, 196

u

UNIFAC model   15ff.
uridine   154

v

valrubicin   147
3-vinylcyclohex-2-en-1-ol   98

volatile organic compound (VOC) 267
volume phase ratio 198
volumetric flow rate 260

w
water 3ff.
– content 82
– distribution 5f.
– effect 8ff.
– low-water system 3f.
– mobility 67
water activity 4ff., 269f.
– control 4f.
water stripping model 66

x
Xanthobacter autotrophicus GJ10 268ff.
xylitol 152
xyloglucan 153